Fulleranes

CARBON MATERIALS: CHEMISTRY AND PHYSICS

A comprehensive book series which encompasses the complete coverage of carbon materials and carbon-rich molecules from elemental carbon dust in the interstellar medium, to the most specialized industrial applications of elemental carbon and its derivatives. A great emphasis is placed on the most advanced and promising applications ranging from electronics to medicinal chemistry. The aim is to offer the reader a book series which not only consists of self-sufficient reference works, but one which stimulates further research and enthusiasm.

Series Editors

Dr. Professor Franco Cataldo
Director of Actinium Chemical Research Institute
Via Casilina 1626/A
00133 Rome
Italy

Professor Paolo Milani
University of Milan
Department of Physics
Via Celoria, 26
20133, Milan, Italy

VOLUME 2:
FULLERANES: THE HYDROGENATED FULLERENES

Volume Editors

Dr. Professor Franco Cataldo
Istituto Nazionale di Astrofisica
Osservatorio Astrofisico di Catania
Via S. Sofia 78
95123 Catania, Italy

Dr. Susana Iglesias-Groth
Inst. de Astrofisica Canarias
Dept. Fisica Fundamental II
Via Lactea S/N
38200 La Laguna, Tenerife
Canary Islands, Spain

For other titles published in this series, go to www.springer.com/series/7825

Franco Cataldo · Susana Iglesias-Groth
Editors

Fulleranes

The Hydrogenated Fullerenes

With a foreword by Sir Harry Kroto, Nobel Laureate

Springer

Editors
Dr. Franco Cataldo
Actinium Chemical Research Institute
Rome
Italy
franco.cataldo@fastwebnet.it

Dr. Susana Iglesias-Groth
Instituto de Astrofisica de Canarias
Tenerife
Spain
sigroth@iac.es

ISSN 1875-0745 eISSN 1875-0737
ISBN 978-1-4020-9886-4 e-ISBN 978-1-4020-9887-1
DOI 10.1007/978-1-4020-9887-1
Springer Dordrecht Heidelberg London New York

Library of Congress Control Number: 2010925498

Printed on acid-free paper

Springer is part of Springer Science+Business Media (www.springer.com)

Foreword

The discovery of the Fullerenes (Kroto et al. 1985) is an iconic example of the unpredictable way fundamental science often leads to the ground-breaking advances that open up totally unexpected, new and disparate fields of scientific study. For instance no-one predicted that a simple suggestion, to simulate the reactions in the atmospheres old red giant stars by applying the brilliant technology devised by Rick Smalley and co-workers to study refractory clusters (Dietz et al. 1981), would uncover a totally new and novel field of organic chemistry. Furthermore, the discovery of the Fullerenes led to the discovery (Ijima 1991) or actually the re-discovery (Oberlin et al. 1976), of their elongated cousins, the carbon nanotubes. Together they promise revolutionary advances in electrical, magnetic and tensile strength materials behaviour. These new materials with their unique nanoscale infrastructures have initiated a tidal wave of fundamental research in Nanoscience and its strategic twin, Nanotechnology (N&N). In the next few decades this wave promises to lead to completely new and revolutionary technologies with socio-economic impact which should rival or surpass the breakthrough technologies of the last half of the twentieth century which has seen phenomenal advances in materials applications, laser applications, computers etc., and of course the Internet. Perhaps the most incredible aspect of the whole Fullerene story is the fact that this third (well-defined) molecular allotrope of carbon has actually been under our noses all the time – in fact created whenever hydrocarbons are burned under soot-forming conditions. This is not a rare event yet C_{60} evaded discovery until almost the end of the twentieth century.

Curiously cage carbon molecule concept did occur to the fertile minds of a few scientists and major credit must go to Eiji Osawa who published the idea in 1970 (Osawa 1970) and the next year described it again in a book with Yoshida et al. (1971). Even more fascinating is the fact that David Jones published the idea of large carbon cage structures a few years earlier in 1966 in a delightful article in the New Scientist when he ruminated imaginatively over ways to introduce pentagonal disclinations into extended graphene sheets – so enabling them to close into what I called Giant Fullerenes (Kroto 1990). It was from this fascinating article that I learned that the introduction of exactly 12 pentagons will close a sheet of hexagons of any number (except one as I learned later) into a cage. I guess this is something that most mathematicians know as an interesting consequence of Euler's Law – but I must admit that I did not. As soon as I found this out – a few weeks after the discovery of C_{60} – the molecule's apparent stability became almost obvious as it is clear that to close the cage without abutting pentagons at least 60 atoms are needed. Molecules with abutting pentagons are well known to be unstable structural configurations in organic chemistry – which mathematicians might not be expected to know but most chemists should – though not all apparently do (Stone et al. 1986) – but then theoreticians often seem able to get the answers they are looking for. These considerations thus led me to formulate the Pentagon Isolation Rule (PIR) concept (Kroto 1986) as a criterion of Fullerene stability. Furthermore the PIR led, almost immediately, to important and highly satisfying circumstantial support of the basic Fullerene cage concept; I subsequently concluded, on the basis of simple trial-and-error model-building, that no closed fullerene structures with isolated pentagons could be constructed between C_{60} and C_{70} (Kroto 1986). This very welcome support came at a time when some half-a-dozen papers were published casting significant doubt on our structural contention. This conclusion was discovered simultaneously, and proven correct, by Tom Schmalz, Doug Klein and co-workers in Galveston (1988).

From the point of view of this book I think it might be of interest to recount the fact that during the period between the discovery of C_{60} in 1985 and the brilliant extraction breakthrough of Wolfgang Kraetschmer, Lowell Lamb, Konstantinos Fostiropoulos and Don Huffman in 1990, I probed some semi-quantitative theoretical ideas about how hydrogenation might stabilize some of the smaller fullerenes – C_{28} in particular (Kroto 1986). I also found it quite intriguing to stand back and consider, from the Fullerene cage perspective, the fact that dodecahedrane, the elegant molecule that Louis Paquette's group had synthesised in 1982 (Ternansky et al. 1982), is basically C_{20}, the smallest and surely the least stable Fullerene, stabilized by complete hydrogenation. David Walton and I, somewhat later, published a paper which used related fairly simple, semi-quantitative, arguments to predict small fullerene structures (with less than 60 atoms) that might stabilized by hydrogenation. These small fullerenes, such as C_{24}, C_{28}, C_{32}, C_{50}, etc., all of course must possess abutting pentagons. A paper published some ten years later, on $C_{50}Cl_{10}$ (Xie et al. 2004), gave some nice support to this stabilization idea. During the 1985–1990 period a parallel experimental programme at Sussex, aimed at creating macroscopic amounts of C_{60}, resulted in its isolation and mass spectrometric identification (Taylor et al. 1990) essentially simultaneously with the breakthrough of Kratschmer and colleagues (Kraetschmer et al. 1990). This work culminated in the highly satisfying NMR characterization of C_{60} by detection of the single line which proved that all 60 carbon atoms were indeed equivalent (Taylor et al. 1990). The study also showed that C_{70} did indeed possess the originally conjectured a rugby ball structure (Heath et al. 1985) and pioneered the chromatographic separation of fullerenes.

So in 1990 the Fullerenes became available in viable quantities and initiated a new and exciting field of chemistry. My colleagues Roger Taylor, David Walton and I set up a Fullerene and Nanoscience research programme at Sussex to explore the chemistry of the Fullerenes as well as the properties of carbon nanotubes.

This led to several nice contributions to the burgeoning new field of Fullerene and Nanotube Science. David Walton suggested that C_{60} should add halogen atoms and this resulted in a favourite synthetic breakthrough from this period by Paul Birkett who discovered that C_{60} formed a specific $C_{60}Cl_6$ structural analogue in which five Cl atoms added to the C atoms surrounding (at one C–C bond distance) a particular pentagon to which sixth Cl atom attached (Birkett et al. 1993).

Thus it is very satisfying to be asked to write a short piece as an introduction to this book as it is most gratifying realize how much beautiful science has grown out of ideas and experiments carried out by us and one-or-two other groups during the earliest period of Fullerene story. Franco Cataldo and Susana Iglesias-Groth are to be congratulated in assembling this compendium of elegant contributions to this monograph on "Fulleranes: The Hydrogenated Fullerenes" which is a most exciting new field of Synthetic Organic Chemistry. The range of contributions is quite fascinating ranging, as it does, from such down-to-Earth studies as: synthesis, spectroscopy (UV, IR, NMR, HREELS), low temperature spectroscopy, thermodynamics and reaction equilibria, dissociation and ion chemistry of C_{60}, high pressure hydrogenation reactions, deuterium isotope effects, topological modeling as well as the possible existence of Fulleranes in space.

Harry Kroto
Chemistry and Biochemistry Department
Florida State University
Tallahassee
Florida

References

Birkett PR, Avent AG, Darwish AD, Kroto HW, Taylor R Walton DRM (1993) J Chem Soc Chem Commun 1230
Dietz TG, Duncan MA, Powers DE, Smalley RE (1981) J Chem Phys 74: 6511–6512
Heath JR, O'Brien SC, Zhang Q, Liu Y, Curl RF, Kroto HW, Tittle FK, Smalley RE (1985) J Am Chem Soc 107:7779–7780
Iijima S (1991) Nature 354: 56–58
Jones DEH (1966) New Scientist 32:245
Kraetschmer W, Lamb LD, Fostropoulos K, Huffman DR (1990) Nature 347:354–358
Kroto HW (1986) Nature 322:766
Kroto HW (1990) Chemistry Britain 26: 40–45
Kroto HW, Heath JR, O'Brien SC, Curl RF, Smalley RE (1985) Nature 318:162–163
Oberlin A, Endo M, Koyama T (1976) J Cryst Growth 32:335–349
Osawa E, Kagaku (1970) 25:854–863 in Japanese [also Chem Abs 74 75698v (1971)]
Schmalz TG, Seitz WA, Klein DJ, Hite GE (1988) J Am Chem Soc 110:1113–1127
Stone AJ, Wales DJ (1986) Chem Phys Lett 128:501–503
Taylor R, Hare JP, Abdul-Sada AK, Kroto HW (1990) J Chem Soc Chem Commun 1423–1425
Ternansky RJ, Balogh DW, Paquette LA (1982) J Am Chem Soc 104:4503–4504
Xie S-Y, Gao F, Lu, Huang R-B, Wang C-R, Zhang X, Liu M-L, Deng S-L, Zheng L (2004) Science 304:699
Yoshida Z, Osawa E (1971) Aromaticity. Kagakudojin 174–178 in Japanese

Preface

Fullerenes show a remarkable stability both toward γ photons and cosmic rays suggesting that once formed they can survive in the harsh interstellar and circumstellar environment (Cataldo et al. 2009).

The main sources of carbon dust and molecules in the interstellar medium are the late-type carbon-rich stars, but there is a class of stars which lies in the transition between the asymptotic giant branch (AGB) to the planetary nebula stage which is very promising as sources of fullerenes. The stars in such class are rather rare, are helium rich and extremely depleted in hydrogen content in their gaseous shell so that the carbon vapour ejected from the star could be cooled in appropriate and ideal environment for the fullerene formation. The prototype of such stars where fullerene may be formed is the *R Coronae Borealis* class of stars (Kroto 2006). In the interstellar medium C_{60} fullerene should be present as neutral molecule or should undergo ionization to C_{60}^{+}. Such cation may be responsible for some spectral features in the diffuse interstellar bands (DIBs). Fullerene cations undergo multiple addition of atomic hydrogen forming hydrogenated derivatives (Petrie et al. 1995) but also neutral fullerenes add easily atomic hydrogen at very low temperatures (Howard 1993), it is reasonable to think that fulleranes, the hydrogenated fullerenes derivatives should be present in the interstellar medium (Petrie and Bohme 2000).

The first chapter from Iglesias-Groth and Cataldo reviews the potential contribution of single fullerenes and buckyonions to interstellar light extinction. Fulleranes prepared by atomic hydrogen treatment of fullerenes are the object of the study of Carraro, Maboudian and Stoldt. In particular laboratory spectra of fullerane show remarkable agreement with important features detected in the observational data, among others the so-called canonical interstellar spacing of infrared spectra and the single bump in the ultraviolet absorption spectra. The problem of molecular hydrogen formation in space starting from atomic hydrogen and interstellar carbon dust surfaces is addressed in Chapter 3. In particular the role played by fullerene-like structures and fullerenes present in the carbon dust in the molecular hydrogen formation is underlined. In Chapter 4, G. Kabo and colleagues present a detailed work on the thermodynamic properties of fullerene hydrides and their equilibria reaction. In particular, Kabo and colleague show that $C_{60}H_{36}$ can act as hydrogen accumulator. This point lead us from the astrophysical/astrochemical context through the physical chemistry into the potential practical applications of fullerene hydrides. The possibility to link chemically on a single C_{60} fullerene cage up to 30 hydrogen molecules and to reversibly get back such hydrogen is highly seductive and paves the way to use fullerenes or better fullerenic carbon as a substrate for the reversible accumulation and release of hydrogen used as energy vector in thermal engines. The advantage of such systems of hydrogen storage based on fullerenic carbon is the lightness and the high density of hydrogen molecules stored per unit volume in contrast to the classic metal hydrides systems which are extremely heavy. A. Talyzin in Chapter 5 presents a beautiful work dealing with the hydrogenation of fullerenes under high pressure and temperature. Indeed the systems analyzed by Talyzin show an exceptional tendency to accumulate hydrogen but the reverse reaction was not found fully reversible since the hydrogenated fullerene derivatives do not release back only hydrogen but also other decomposition products. In the following chapter, J.B. Briggs and G.P. Miller show a very surprising synthesis of fulleranes using polyamine. For example, the combination of microwave and polyamine leads to produce $C_{60}H_{18}$ in very high yields and in very short reaction times. Chapters 7 and 8 are dedicated respectively to the synthesis of perdeuterated derivatives of C_{60} and C_{70} fulleranes and to the study of their thermal and UV light stability. In the photolysis of fulleranes a remarkable isotope effect has been measured for the first time and these studies address again to the topic of the stability of the hydrogenated and deuterated fullerenes in space as well as the possible deuterium enrichment of the fulleranes in space (Cataldo and Iglesias-Groth 2009). Chapter 9 is a very detailed and interesting work on the NMR characterization of the chemical structures of fulleranes and their symmetries. Chapter 10 is dedicated to the low temperature infrared spectra of C_{60} and C_{70} fullerenes as well as $C_{60}H_{18}$ fullerane. The spectra were taken at temperature as low as 80 K and compared with high temperature spectra taken at 523 K and extrapolated to >1,000 K. The knowledge of the infrared spectra of fullerenes and fulleranes paves the way for their detection in the space. The chapter of Bazhenov, Bashkin and Meletov deals with the high temperature and high pressure hydrogenation of carbon nanostructures in a review that covers both potential materials science applications and the astrophysical implications. Finally, the last chapter, Chapter 12, discusses the application of topology in the calculation of the stability and symmetry of hydrogenated fullerenes. It is a purely theoretical chapter but whose conclusions are in agreement with the current knowledge about the symmetry and stability of certain fulleranes like $C_{60}H_{36}$. Thus, Chapter 12 shows how powerful is the modern topological approach in the calculation of the structures and stability of fulleranes and other fullerenes derivatives.

In conclusion, the book is a bridge among different disciplines ranging from chemistry to materials science to physics, astrophysics and topology with a unique final subject: showing the state of the art of the research on fulleranes, the hydrogenated derivatives of fullerenes.

November 2009

Susana Iglesias-Groth
Franco Cataldo

References

Cataldo F, Iglesias-Groth S (2009) Mont Not Roy Astronom Soc 400:291
Cataldo F, Strazzulla G, Iglesias-Groth S (2009) Mont Not Roy Astronom Soc 394:615
Howard JA (1993) Chem Phys Lett 203:540
Kroto HW (2006) In: Rietmeijer FJM (ed) Natural fullerenes and related structures of elemental carbon, Chapter 1. Springer, Dordrecht
Petrie S, Bohme DK (2000) Astrophys J 540:869
Petrie S, Becker H, Baranov VI, Bohme DK (1995). Int J Mass Spectrom 145:79

Contents

Chapter 1
Fulleranes and Carbon Nanostructures in the Interstellar Medium

Susana Iglesias-Groth[1] and Franco Cataldo[2]

Abstract We review the potential contribution of single fullerenes and buckyonions to interstellar extinction. Photoabsorption spectra of these molecules are compared with some of the most relevant features of interstellar extinction, the UV bump, the far UV rise and the diffuse interstellar bands. According to semiempirical models, photoabsorption by fullerenes (single and multishell) could explain the shape, width and peak energy of the most prominent feature of the interstellar absorption, the UV bump at 2,175 Å. Other weaker transitions are predicted in the optical and near-infrared providing a potential explanation for diffuse interstellar bands. In particular, several fullerenes could contribute to the well known strong DIB at 4,430 Å comparing cross sections and available data for this DIB and the UV bump we estimate a density of fullerenes in the diffuse interstellar medium of 0.1–0.2 ppm. These molecules could then be a major reservoir for interstellar carbon.

We give an estimation of the carbon fraction locked in these molecules. We discuss the rotation rates and electric dipole emission of hydrogenated icosahedral fullerenes in various phases of the interstellar medium. These molecules could be the carriers of the anomalous microwave emission detected by Watson et al. (Astrophys. J. 624:L89, 2005) in the Perseus molecular complex and Cassasus et al. (2006) in the dark cloud LDN 1622. Hydrogenated forms of fullerenes may account for the dust-correlated microwave emission detected in our Galaxy by Cosmic Microwave Background experiments.

[1]Instituto de Astrofísica de Canarias, 38200 La Laguna, Tenerife, Spain
e-mail: sigroth@iac.es

[2]Istituto Nazionale di Astrofisica di Catania, 95123 Catania, Italy
Actinium Chemical Research, 00133 Rome, Italy
e-mail: franco.cataldo@fastwebnet.it

F. Cataldo and S. Iglesias-Groth (eds.), *Fulleranes: The Hydrogenated Fullerenes*,
Carbon Materials: Chemistry and Physics 2, DOI 10.1007/978-1-4020-9887-1_1,

1.1 Introduction

Research on samples of vaporized graphite using laser beams, originally aimed to reproduce the chemistry of the atmospheres of carbon enriched giant stars, led to the unexpected discovery of the C_{60} (the 60 carbon atoms molecule known as fullerene) (Kroto et al. 1985). Subsequent experiments produced carbon aggregates with a larger number of atoms (C_{70}, C_{84}, C_{240}) and established that for an even number of atoms larger than 32, these aggregates were stable. While other molecules have serious difficulties to survive in the interstellar medium, the robustness of C_{60} and of fullerenes in general, strongly support a long survival in the harsh conditions of interstellar space. It has been claimed the detection of fullerenes with 60–240 carbon atoms in the Allende and Murchison meteorites (Becker et al. 2000). This is an indication that fullerenes were present in the presolar nebula and strongly support (Kroto 1988) their existence in the interstellar medium. The UV bump at 2,175 Å is the most prominent feature in the interstellar medium extinction curve (Fitzpatrick et al. 1986; Fitzpatrick 1999). The carrier of this band has been frequently associated to some form of carbonaceous material (graphitic spheres, polycyclic aromatic hydrocarbons (PAHs), hydrogenated amorphous carbon, etc.) but there is no definitive conclusion on the ultimate responsible for the band. The increase in the extinction at higher energies and the existence of many much weaker diffuse interstellar bands (DIBs) in the optical (of unknown origin are also some of the intriguing observational properties of the interstellar material. These absorption bands are present in the optical and near infrared spectra of reddened stars. The widths of these bands range from 0.5 to 30 Å (Herbig 1995). Hundreds have been detected so far but association with any known atomic or molecular species has not been demonstrated yet.

Several laboratory experiments (Chhowalla et al. 2003) and theoretical studies (Henrard et al. 1997; Iglesias-Groth 2004, 2006) of the photoabsorption by fullerenes and buckyonions (multishell fullerenes) suggest that these molecules could be responsible of the UV bump located at 2,175 Å. A significant fraction of interstellar carbon (10–30%) could reside in fullerene related molecules (Iglesias-Groth 2004). It has also been suggested (Watson et al. 2005; Webster 1991, 1992) that fullerenes could be carriers of some DIBs. According to theoretical spectra obtained using semiempirical models (Iglesias-Groth et al. 2001, 2002; Ruiz et al. 2001), icosahedric fullerenes and buckyonions (from C_{60} to C_{6000}) present numerous low intensity bands in the optical and near-infrared, several with wavelengths very similar to well known DIBs. Fullerenes deserve further study as potential carriers of DIBs. So far only two DIBs may have been identified as caused by the cation of C_{60} (Foing et al. 1994).

Finally, hydrogenated fullerenes have been proposed as carriers of the anomalous microwave emission recently detected by several experiments on the Cosmic Microwave Background (Iglesias-Groth 2005, 2006). In the interstellar medium these molecules should spin with rates of several to tens of gigaHertz, if as expected they have a small dipole moment, then they would emit electric dipole radiation in a frequency range very similar to that observed for the anomalous microwave emission.

In this chapter, we review the possible contribution of single fullerenes, fulleranes and buckyonions to the interstellar extinction and the anomalous microwave emission. We compare their photoabsorption spectra with some of the observational properties of interstellar extinction, in particular with the characteristics of the UV bump and with the distribution of diffuse interstellar bands (DIBs) in the optical and near infrared and discuss the potential of fulleranes as carriers of anomalous microwave emission.

1.2 General Properties of Fullerenes

1.2.1 Icosahedric Fullerenes

The arquetype of the icosahedric fullerenes is the C_{60} a closed cage formed by 60 carbon atoms distributed on the surface of a sphere of radius ~3.55 Å following the symmetry of a truncated icosahedrum. In 1970, Osawa et al. suggested that a molecule with this geometry would be chemically stable. In this molecule, each carbon atom with four valence electrons is bounded to the three nearest carbon atoms. Two of the bounds are single and link an hexagon with a pentagon, the third is double and links two hexagons. The molecule is formed then, by 30 double and 60 single bounds. The distances between atoms in single and double bounds are 1.45 and 1.37 Å, respectively. In the study of the electronic properties of the C_{60} it is frequent to consider only the four valence electrons 2s, $2p_x$, $2p_y$, $2p_z$ of each carbon atom. Each of the 60 atomic orbitals $2p_z$ is aligned in the radial direction, leading to the p molecular states. The three atomic orbitals 2s, $2p_x$, $2p_y$ are distributed in the tangential plane to the molecular surface at the carbon atom position producing σ molecular states in the direction of the molecular bounds. The lineal combination of these three orbitals at each atomic site produces three hybrid orbitals sp^2, one is in the direction of the double bound and the other two are aligned with the single bounds. The 60 π orbitals are relevant for the conductivity properties of the molecule, as in the case of graphite, and the 180 σ orbitals are mainly responsible for the elastic properties (see Fig. 1.1).

1.2.1.1 Photoabsorption Spectra

The most prominent features of the empirical photoabsorption spectrum of C_{60} are two bands located at about 6 and 23 eV (Fig. 1.2, 50–200 nm). These bands have been interpreted as collective excitations similar to the π and σ plasmon. In the fullerene C_{60} these type Mie plasmons are due to the strong delocalization of the valence electrons. The σ-type is much more intense than the π-type plasmon. It has been proposed (Braga et al. 1991) that the π transition could be related to the UV bump in the interstellar extinction.

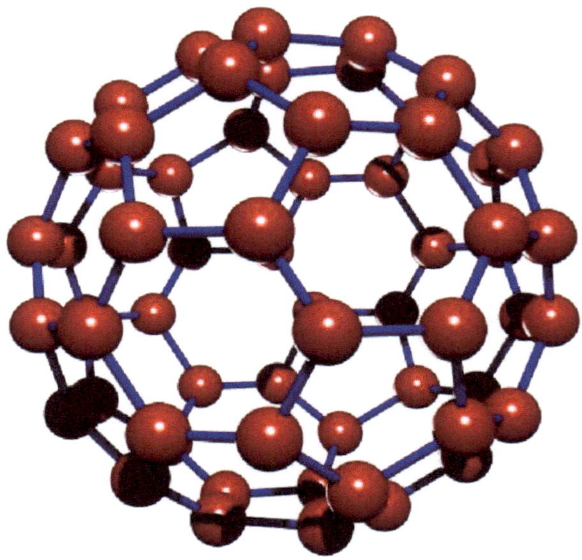

Fig. 1.1 Fullerene C_{60} (Credit: A. Pérez-Garrido)

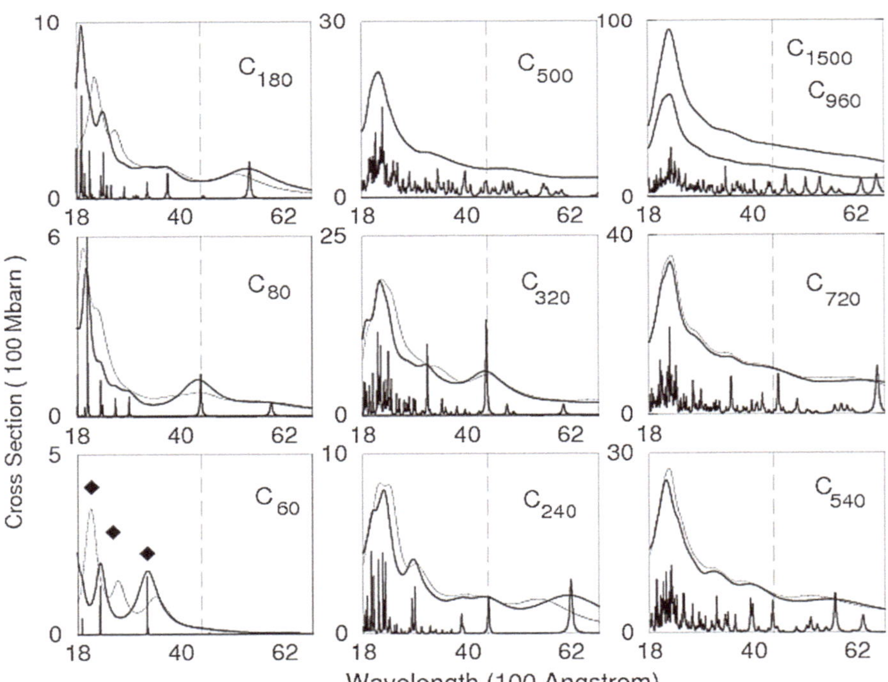

Fig. 1.2 Photoabsorption spectra of the icosahedral individual fullerenes of the Goldberg family as a function of wavelength considering two models, a simple Hückel (solid thick line) and a PPP (thin solid line)

Under some simplifications associated with the symmetry of fullerenes, it has been possible to perform calculations of type Hartree–Fock in which the interelectronic correlation has been included up to second order Møller–Plesset (Moller et al. 1934; Purcell 1979; Cioslowski 1995), and calculations based on the density functional (Pople et al. 1976). However, given the difficulties faced by ab initio computations when all the electrons of these large molecules are taken into account, other semiempirical methods of the Hückel type or tight-binding (Haddon 1992) models have been developed to determine the electronic structure of C_{60} (Cioslowski 1995; Lin and Nori 1996) and associated properties like polarizabilities (Bonin and Kresin 1997; Rubio et al. 1993) hyperpolarizabilities (Fanti et al. 1995) plasmon excitations (Bertsch et al. 1991) etc. These semiempirical models reproduce the order of monoelectronic levels close to the Fermi level. Other more sophisticated semiempirical models, like the PPP (Pariser–Parr–Pople) (Pariser and Parr 1953; Pople 1953) obtain better quantitative results when compared with photoemission experiments (Savage 1975).

Semiempirical Hückel-type and Pariser–Parr–Pople molecular models and the random phase approximation for valence electrons were used to predict the electronic photoabsorption spectra of the icosahedral fullerenes C_{60}, C_{240}, C_{540}, C_{960}, C_{1500} which belong to the $60n^2$ Goldberg family of polyhedral and for C_{180} and C_{720} which belong to the family $20(n + 1)^2$ (Webster 1997). The model parameters were first derived by fitting the available experimental photoabsorption spectrum of C_{60}, and then suitably modified to describe larger fullerenes (Iglesias-Groth et al. 2002; Iglesias-Groth 2003). We show in Fig. 1.2 photoabsorption spectra for sevaral icosahedral fullerenes.

1.2.2 Fullerenes with Multiple Spheric Layers: Buckyonions

In 1992 Ugarte discovered that fullerenes can adopt multilayered configurations where one is encapsulated inside each other (see Fig. 1.3a), like onion layers (Tomanek et al. 1991). He used electron bombarding techniques on carbon dust with current densities between 100 and 200 A/cm² and observed the transformation of polymeric particles of carbon in others with multiple spheric layers. Diverse laboratory experiments have proven that these carbon groups, commonly known as buckyonions can be formed by tens of layers. Electronic microscopy has determined that the separation between layers is of the order of 3.4–3.5 Å, that is approximately the separation between sheets of graphite. The buckyonions have also been synthesized exposing carbon dust to thermal treatments. Several theoretical results seem to indicate that the multilayered spheric fullerens are the most stable form of carbon groups. Very little is known about the electronic structure of these molecules.

The microscopic electronic structure of the buckyonion can be derived using an effective one-electron model where the screening effects are treated within the random phase approximation (RPA). The particular spherical geometry of these

a

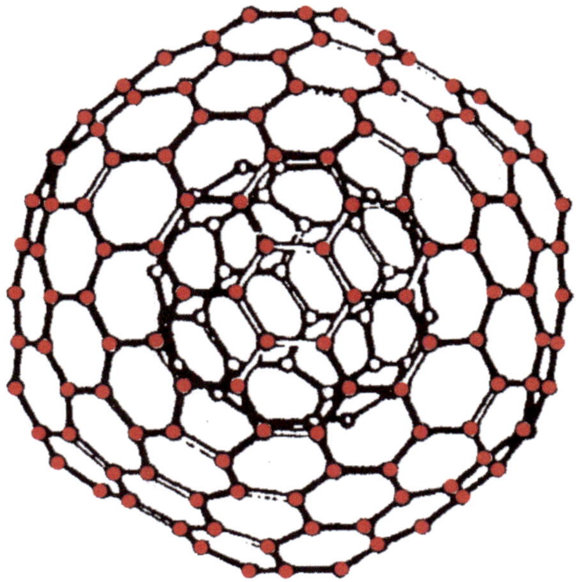

b

Fig. 1.3 (**a**) Buckyonion C_{60} @C_{240}. (**b**) Allende metorite picture (6, 7)

multishell fullerenes makes possible the use of electrostatic arguments to derive a simple expression for the RPA which gives the polarizability of the buckyonion and the dipole moment induced on each shell in terms of either the screened or unscreened polarizabilities of the isolated shells. A systematic analysis as a function of the buckyonion size has been performed (Iglesias-Groth et al. 2003).

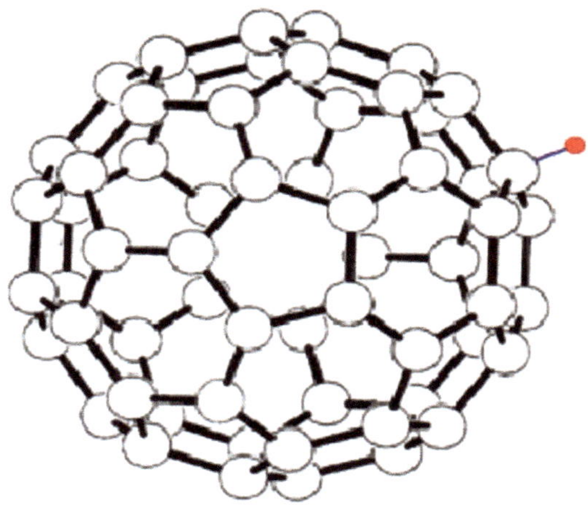

Fig. 1.4 Fullerane $C_{60}H$

1.2.3 Fulleranes: Hydrogenated Fullerenes

A significant fraction of the fullerenes and buckyonions in the interstellar medium could be hydrogenated as discussed by Webster in 1992 (see Fig. 1.4). These molecules, named generically as fulleranes have deserved attention as potential carriers of diffuse intestellar bands and other interstellar and circumstellar features (Webster 1991, 1992, 1993a). Both, fullerenes and fulleranes have been detected in samples of the Allende meteorite (Becker et al. 1994), see Fig. 1.3b.

1.3 Formation of Fullerenes and Fulleranes in Astrophysical Environments

The C_{60} molecule is difficult to destroy by ultraviolet radiation or by collisions with other particles. While other molecules have serious difficulties to survive in the interstellar medium, the robustness of C_{60} and of the other fullerenes allows their long survival. The bonds between carbon atoms make them at least as robust against dissociation in the interstellar medium as polycyclic aromatic hydrocarbons (PAHs) can be.

1.3.1 Meteorites

It is estimated that the C_{60} content of the Allende meteorite is of 0.1 ppm (Becker and Bunch 1997). The isotopic ratios of noble gases found encapsulated in these meteoritic fullerenes are very different to the terrestrial values. It is postulated that these molecules were not generated during the impact of the meteorite with Earth but instead were present in the original matter from which the meteorites formed. PAHs that have been proposed as precursor molecules for formation of fullerenes in the gas phase, have also been detected in meteorites. The hydrogenated form of the C_{60}, that is the C_{60} fullerane, may be present in meteorites. Although little is known about which could be the properties of hydrogenated fullerenes in the physical circumstances and conditions of the presolar cloud.

1.3.2 Carbon Stars and Planetary Nebulae

Fullerenes and graphite may originate abundantly in stellar atmospheres rich in carbon like those of some giant stars and some progenitors of planetary nebulae (Fig. 1.5). These objects display for important mass loss rates and are therefore able to greatly enrich the interstellar medium.

Carbon rich giant stars are highly evolved stars that have a greater concentration of carbon than oxygen in their atmospheres. This superficial enrichment is the consequence of nuclear reactions that take place in their interiors and of the efficient convection that transports the nucleosynthesis products to the most external layers. Carbon stars have effective temperatures in the range 2,000–3,000 K and are some 10,000 times brighter than the Sun. They often present important mass loss

Fig. 1.5 Egg Planetary Nebulae image was taken with NASA's Hubble Space Telescope

rates higher, in some cases, than 10^{-5} M_0/year. The brightest stars can remain in this phase for more than 10^5 year. Thus, they are an important enriching agent of carbon for the interstellar medium. In the extended envelopes surrounding these stars, a very active chemistry takes place, in particular, solid dust grains can be nucleated. These particles could grow to sizes of 1 μm although the most typical are likely of size 0.005 μm or smaller. The evidence of solid grains in the mass expulsed by carbon stars comes from the continuum infrared emission much greater than expected from a photosphere. This excess of radiation is thought to be due to the emission of the circumstellar dust grains.

Many carbon rich stars also present an important emission at 11.3 μm associated with solid carbon and some of them present nebulosity of reflection as a consequence of the scattering of the circumstellar grains. There are indications that in the material ejected by these stars, carbon must exist, apart from CO molecules and solid grains, in some other form or species until now unknown, fullerenes are a possibility. Unfortunately, there is very little information about the presence of molecules of intermediate size (between 10 and 10^6 atoms) in circumstellar regions. There are bands in carbon rich planetary nebulae, for example those of 3.3, 6.2, 7.7, 8.6 and 11.3 μm which have not been detected in carbon stars but are observable in transition objects evolving between the giant red phase and the planetary nebula as for example, the Egg Nebula (Fig. 1.5) and the Red Rectangle. These infrared bands are normally associated with the vibration modes of materials based on carbon, possibly PAHs. But until now it has not been possible to make a conclusive identification of the carrier.

Evidence of a relation between carbon particles and DIBs can be found in the analysis of the Red Rectangle spectrum. This object is a losing mass carbon star probably evolving to a planetary nebula phase. Diverse spectroscopic studies have revealed the good agreement between the emission lines found at 5,797, 5,850, 6,379, and 6,614 Å and some of the most intense diffuse bands of the interstellar medium. It is likely that the carrier of some of these interstellar bands is also present in the material ejected by this object.

1.3.3 Formation of Fullerenes and Buckyonions

The mechanism of fullerene formation is only partially known (Curl 1993). These molecules may be produced at relatively high temperatures by "annealing" of large plane monocyclic carbon molecules. Buckyonions may be produced from carbon soot and from nanodiamonds (Ugarte 1992). It is known that nanodiamonds of size (30–60 Å) can be transformed in particles very similar to buckyonions when heated to temperatures of 1,200–1,800 K (Kuznetsov et al. 1994) which are typical of carbon rich giant stars. The process may work as follows (Ugarte 1992) the (111) surfaces of nanodiamonds are the first to graphitize when treated at high temperature, a first graphitic layer is generated on the surface and later new concentric layers are added toward the interior in an analogous way to the process of buckyonion formation through energetic electron irradiation. Since diamond is more dense than graphite, the final particle has a larger size. Since the transformation takes place from the surface to the interior, the final particle is of spheroidal form and very compact with the inner layer of less than 10 Å of diameter. Nanodiamonds have been detected in meteorites (Blake et al. 1988), where they are much more abundant than graphite, and possibly in the circumstellar envelope of post-AGB stars.

1.4 Fullerenes in the Interstellar Medium

Soon after the discovery of the fullerenes it was suggested that these or particles of similar nature, could be related to one of the most intriguing problems of astrophysics: the diffuse interstellar bands discovered more than eight decades ago, but not yet explained, and with the ultraviolet band centered in 2,175 Å, which is the most intense band in the interstellar medium discovered more than 30 years ago. The origin of the UV bump is attributed to carbon particles of small size whose characteristics are not yet conclusively established.

Although the high level of symmetry of C_{60} indicates that this particular fullerene is unlikely as a carrier of the complex spectrum of the diffuse interstellar bands, diverse studies have investigated possible mechanisms for which this molecule can acquire a complex spectrum of absorption in the optical. Ballester et al. already presented a model to describe the lines of resonance of several atoms (O, N, Si, Mg, Al, Na and S) trapped in the molecule of C_{60} and argued that these systems, quite stable according to laboratory tests, could be responsible for some of the DIBs. Leger et al. (1988), suggested that the spectrum of ionized C_{60} is much more complex than the neutral molecule and could produce absorption bands in the optical and in the infrared. Foing and Ehenfreund (1994) found two diffuse bands at 9,577 and 9,632 Å coinciding within 0.1% with the laboratory measurements of the bands of C_{60} observed in a Neon mould. This is considered evidence of the existence of the C_{60}^+ in the interstellar medium. There are also several proposals associating DIBs with the hydrides of the C_{60} ($C_{60}H_n$). There are also alternative suggestions that the carrier of these bands could be related to PAHs and hydrogenated amorphous carbon (HACs) compounds.

1.4.1 Interstellar Extintion and the UV Bump

When radiation propagates a given distance z through a medium with extinction coefficient $\tau_{ext}(\lambda)$, it is progressively attenuated according to the law

$$I_0(z) = I0 \exp(-\tau \text{ ext } z) \tag{1.1}$$

where I_0 is the intensity at the origin (z = 0). The extinction at a certain wavelength is frequently measured in magnitudes and denoted as $A(\lambda_i)$. The colour index of an interstellar cloud is defined as

$$E_{i-j} = A(\lambda_i) - A(\lambda_j) \tag{1.2}$$

The reddening factor is defined as the absorption in the V-band relative to the colour index B–V

$$R_V = A(\lambda_V) / E_{B-V} \tag{1.3}$$

This value in the diffuse interstallar medium is $R_V \sim 3.1$ (Fitzpatrick et al. 1986).

The normalized absorption at a certain wavelength λ also known as reddening function or extinction law can be written as

$$k(\lambda)=E_{\lambda-V} / E_{B-V} = (A(\lambda)\text{-}A(V))/E_{B-V} \approx 1.086/E_{B\text{-}V}\,[\tau_{ext}(\lambda)\text{-}\ \tau_{ext}(V)] \tag{1.4}$$

It is rather usual to express the extinction law as a function of the reddening factor. Following the parameterization proposed by Fitzpatrick and Massa (1986) and Fitzpatrick (1999) it is possible to reproduce the extinction curves in different lines of sight in the range 3–8 eV with the function:

$$k(\lambda) = E_{\lambda-V} / E_{B-V} = a_1 + a_2x + a_3D(x,x_0,\gamma) + a_4F(x) \tag{1.5}$$

where x is energy or λ^{-1}. The linear component is associated to extinction by silicates and the Drude function with center in x_0 and width γ is defined as

$$D(x,x_0,g) = x^2 / [(x^2\text{-}x_0^2)^2 + \gamma^2 x^2] \tag{1.6}$$

Finally, the function F(x) takes into account the UV rise at energies higher than 7 eV. This extinction is described as

$$F(x) = 0.5392\,(x-x_0)^2 + 0.0564\,(x\text{-}x_0)^3 \tag{1.7}$$

The shape, wavelength and intensity of the 2,175 Å band has been measured in a large number (more than 50) lines of sight (Fitzpatrick et al. 1986). It shows a profile similar to a Lorentzian or to a Drude curve (Ruiz et al. 2001). The central wavelength of the bump is very stable, with changes of less than 1% from one direction to another. This small variation is, however, beyond the uncertainty of the measurements, typically smaller than 5 Å. The width of the band, γ, changes significantly (more than 30%) with line of sight. The average value is ~1.23 eV (0.99 μ^{-1}) and the observed range is $0.96 \leq \gamma \leq 1.5$ eVμ^{-1} (Fitzpatrick et al. 1986). The changes in width and peak position are apparently not correlated (Foing et al. 1994). At energies higher than 7 eV the extinction curve is similar for the various lines of sight and its intensity is not correlated with the lineal component associated to silicates, but it is weakly correlated with the height and width of the UV bump (Jenniskens and Greenberg 1993).

1.4.2 Theoretical Spectra and the 2,175 Å Band

In general, the photoabsorption cross section of individual and multishell fullerenes reproduce the behaviour of the interstellar extinction curve in the near UV. The theoretical spectra show a prominent absorption band around 5.7 eV which fits well the position and width of the 2,175 Å bump. It is also predicted an increased extinction towards higher energies (down to 1,000 Å) with a shape very similar to that measured in interstellar extinction curves. The fullerenes that better reproduce the UV bump have radii in the range 7–13 Å. Buckyonions with a complete number of shells can also reproduce the observations.

In Fig. 1.6a the cross sections for a fullerenes with two layers represented. It suitably reproduce the peak energy and width of the interstellar band. The similarity between the theoretical cross sections and the observed extinction in the low energy range is also remarkable.

1.4.3 Carbon Fraction in Fullerenes and Buckyonions

The comparison of the computed cross sections of fullerenes and buckyonions with observations of the UV bump for $R_V = 3.1$ allow an estimate of the number of these molecules in the diffuse interstellar medium. Let us describe the extinction curve as $a_1 + a_2x + a_3T(x)$ where $T(x)$ is the theoretical cross section computed for each fullerene or buckyonion. Here we assume that indeed the extinction at the energy of the bump is the result of the fullerene plus silicate contributions. We obtain via a least squared fit the relative contribution of the two components (see Fig. 1.6b). The coefficients of this lineal component do not depend significantly on the particular fullerene under consideration taking typical values of $a_1 \approx 1.6$ and $a_2 \approx 0.07$ with a relative error of 20%.

It is known experimentally that the hydrogen column density is related with the excess colour index (Braga et al. 1991) through

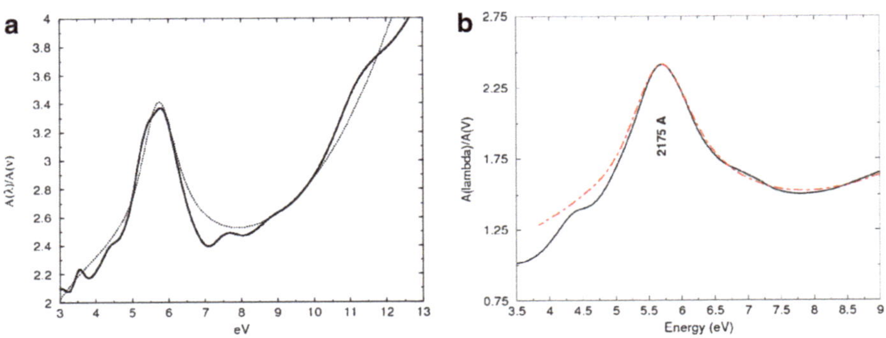

Fig. 1.6 (**a**) Comparison between photoabsorption cross section for the buckyonion C_{180} @C_{720} (continuous line) with the interstellar extinction curve for the reddening factor R_V α 3.1 (*dashed line*) (**b**) We show a redicted absorption curve for fullerene and buckyonion mixtures in comparison with observations of the UV 2,175 Å bump (5.7 eV). The solid line is the best-fit models to the UV bump obtained for a power-law size distribution n(R) α R^{-m} of buckyonions and individual fullerenes ranging from C_{60} to C_{2160}, with index $m = 3.5$ (*solid line*). Extinction curve of the diffuse ISM (*dashed red line*). Theoretical models are from reference (Iglesias-Groth 2003, 2004)

$$N(H)/E_{B-V} = 5.9\times10^{-21}\ cm^{-2} \tag{1.8}$$

Assuming A(V) ≈ 3.1 E_{B-V} for the diffuse interstellar medium, we obtain

$$A(V) \approx 5.3\times 10^{-22}\ \text{mag cm}^2 N(H) \tag{1.9}$$

and the relative number density of any particular fullerene can be then estimated from the a_3 coefficient

$$N(full)/N(H) \approx 5 \times 10^{-22}\ a_3 T\,(x_0) \tag{1.10}$$

The typical values for a_3 are of order 0.3 $\times 10^{15}$ (when cross sections are expressed in barns). This coefficient changes according to the type of fullerene considered but less than 50%, corresponding to the lowest values of the largest buckyonions. We find then that fullerenes and buckyonions may have densities in the range 0.2–0.08 particles per million (ppm) hydrogen atoms. Remarkably similar to the values found in meteorites. In the unlikely case that all types of fullerenes considered in this study had the same density in interstellar space, it would require that of order 100–200 atoms of carbon per 10^6 hydrogen atoms be locked in these molecules. This is a large fraction of the total carbon expected in the interstellar medium which we can assume equal to the solar atmosphere abundance of 355 ± 50 (Jenniskens and Greenberg 1993). It is also known from observations with the Goddard High Resolution Spectrograph of the Hubble telescope that in interstellar space there are 140 ± 20 carbon atoms per million hydrogen atoms in gaseous form (Cardelli et al. 1996), therefore the maximum number of carbon atoms available for fullerenes cannot be larger than 200 per million hydrogen atoms.

Indeed, the actual carbon fraction in fullerenes depend of the proper mixture of these molecules in the interstellar medium. It is likely that the number density of fullerenes and buckyonions will decrease with increasing radius (R). A distribution of the type $N(full)\ \alpha\ R^{-m}$ has been frequently considered in the literature on interstellar grain populations (Mathis et al. 1977). A mixture of fullerenes and buckyonions following such size distribution may reproduce the observed UV bump. The best fits to the shape, peak energy and width of the bump are obtained for m values in the range 2.5–4.5 (Fig. 1.6b).

1.4.4 *Diffuse Interstellar Bands*

Since the discovery of diffuse interstellar bands (DIBs) by Heger (1921) more than 300 interstellar features have been identified in this category (Galazutdinov et al. 2000; Herbig 2000; Jenniskens and Greenberg 1993). The origin of the DIBs is still a mistery. The list of potential carriers goes from negative hydrogen and linear carbon chains to polycyclic aromatic hydrocarbons (PAHs), fullerenes, fulleranes and dust grains. In most DIBs some substructure can be seen, see e.g. (Ehrenfreund et al. 1996), suggesting that gas-phase molecules are likely carriers, but so far no satisfying explanation for the origin of these intriguing interstellar bands could be given.

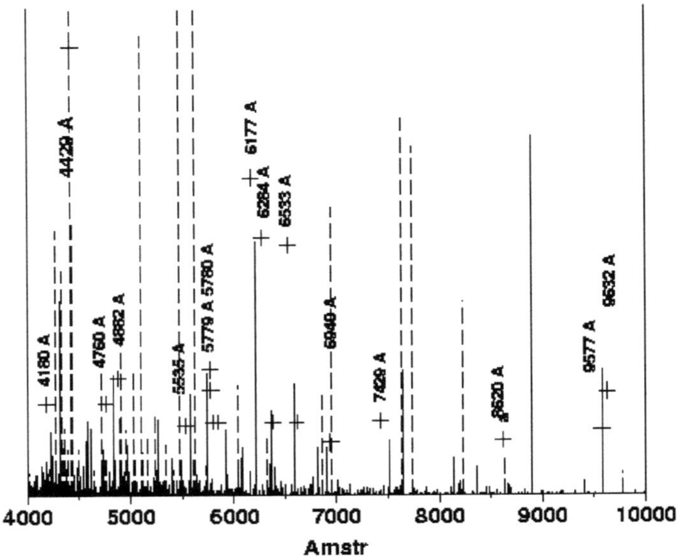

Fig. 1.7 Comparison between predicted transitions and the positions of the stronger DIBs known, see Table 1.1 from (42)

The same models mentioned above predict for both fullerenes and buckyonions a large number of weak bands in the optical and near infrared. These bands have strength consistent with those of the strongest DIBs and the number per wavelength interval appears to decrease towards longer wavelengths as it is the case for the DIBs. In Fig. 1.7 the predicted wavelengths for these transitions are compared with the positions of the 16 stronger DIBs known (Iglesias-Groth 2003). At least 30% of these DIBs coincide (within the precision of the model ~10 Å) with a theoretical transition of a fullerene or buckyonion, leading to a tentative association of these molecules with the DIBs carriers. Several fullerenes (C_{80}, C_{240}, C_{320}, C_{540}) and buckyonions, present a relatively strong band at energies close to the DIB at 4,430 Å.

Among optical DIBs, the 4,430 Å band is the strongest. This band is remarkably broad with a width (FWHM) of order a few tens of Å. Krelowski and Walker (1987) assign the two broad DIBs 4,430 and 6,177 Å to the same family and Krelowski et al. (1989) and McIntosh and Webster (1993) note that the carrier of this family appears to prefer denser interstellar gas than other carriers. There is also evidence of a positive correlation between the 4,430 Å band and the strongest feature in the interstellar extinction curve, the UV bump at 2,175 Å (Webster 1992; Nandy and Thompson 1975). It is therefore plausible that these two bands are produced by the same type of molecule.

Using the theoretical cross sections and available empirical data on the 4,430 Å band it is derived a fullerene abundance of 0.1 molecules per million hydrogen atoms in regions of the interstellar medium with excess color index of E(B – V) = 1.0.

Table 1.1 The most intense diffuse bands in the interstellar medium, where W is the equivalent width (39)

Wavelength (Å)	W (Å)	Wavelength (Å)	W (Å)
4,180	0.70	6,284	1.95
4,429	3.40	6,533	1.89
4,760	0.70	6,940	0.40
4,882	0.89	7,429	0.56
5,535	0.53	8,620	0.42
5,779	0.95	9,577	0.50
5,780	0.80	9,632	0.78
6,177	2.40	13,170	0.4

The same interstellar abundance is obtained from observations of the UV bump and the 4,430 Å band. The abundances inferred for fullerenes are consistent with estimates for the carbon budget in the ISM.

Hydrides of the fullerene C_{60} have also been investigated as potential DIB carriers (Webster 1992, 1993b). No specific identification has been suggested but it should be noted that the conjugated systems of π-electrons are predicted to have transitions in the visible range. The predicted optical and near-infrared transitions of fullerene based molecules may offer a potential explanation for the long-standing problem of the diffuse interstellar bands and other interstellar and circumstellar features.

In Fig. 1.2 we plot the resulting transitions up to 8 eV for icosahedral fullerenes with total number of carbon atoms N < 1,500. In each subpanel we present the theoretical cross-sections for the bands convolved with a Lorentzian function of width Γ_ω (the parameter in Eq. 1.2 of (Iglesias-Groth et al. 2002)). Results are plotted for two values of the Γ_ω parameter: 0.01 eV (set of narrow transitions) and 0.3 eV (broad features).

Iglesias-Groth et al. (2003b) carried out computations for the most stable buckyonions with an external shell of N = 240 and 540 carbon atoms, i.e. the C_{60} @C_{240} and C_{60} @C_{240} @C_{540}. The top panels of Fig. 1.8 show the results. Essentially, the addition of internal shells do not change significantly the number, strength and wavelength of the bands predicted for the external fullerene.

As shown in Fig. 1.8a other transitions are also predicted for fullerenes at optical wavelengths. The next larger cross-section is found for the C_{240} molecule at a wavelength remarkably close to the second strongest optical DIB at 6,177 Å (position marked with a vertical dotted line in Fig. 1.8a). The measurements available for this DIB (see e.g. Table A1 of Herbig 1995) give an equivalent width approximately 60% of the 4,430 Å band and FWHM of order 20 Å. A similar relative strength is found in our computations. The two broad DIBs 4,430 and 6,177 may belong to the same family according to Krelowski and Walker (1987).

The third strongest DIB in the optical is found at 6,284 Å very close to one of the transitions predicted for the fullerene C_{540} for which it is predicted a transition close to the location of another two well known strong DIBs at 5,778 and 5,780 Å. Predictions for these and other weaker bands in the optical are however, more uncertain and deserve a careful analysis. We remark that the relatively simple models employed to study the

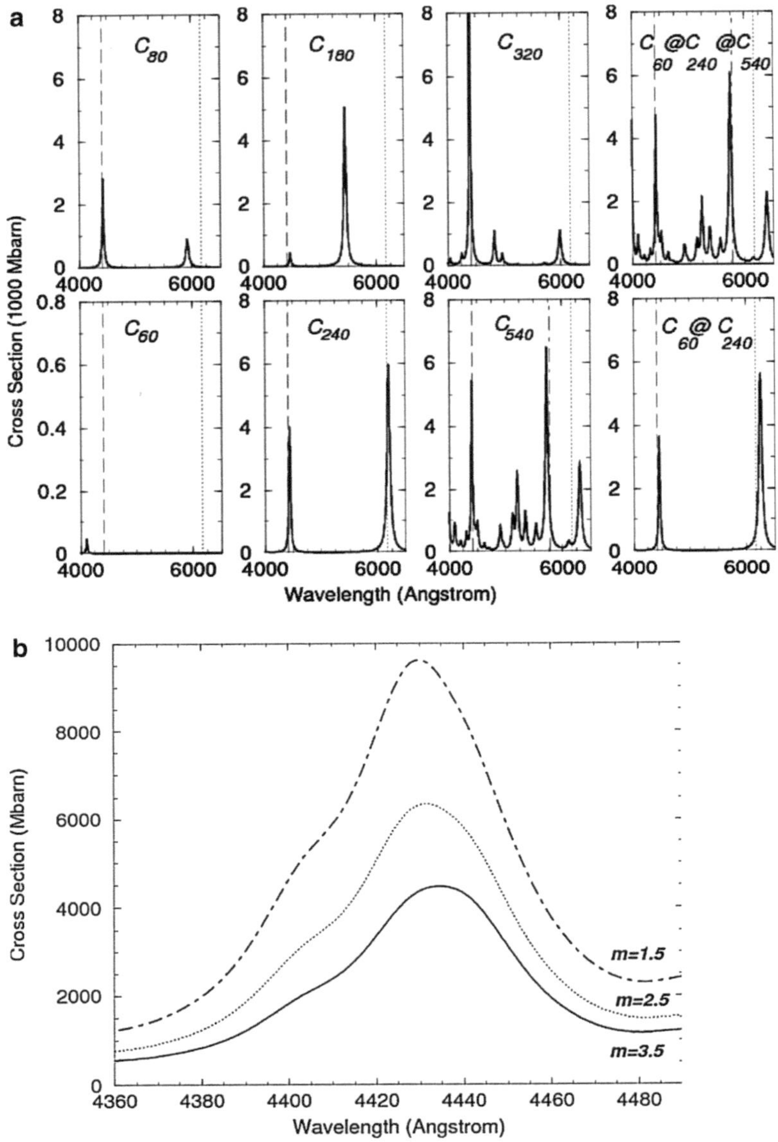

Fig. 1.8 (**a**) Photoabsorption cross sections for fullerenes as a function of wavelength. The thin dashed vertical lines indicate the position of the λλ 4,430 interstellar banda. The dotted line marks the position of the λλ 6,177 interstellar band. The dotdashed line marks the position of the λλ 5,780 band (from (Iglesias-Groth 2005)). (**b**) Photoabsorption cross sections as a function of wavelength for a mixture of single-shell fullerenes in the spectral region of the DIB at 4,430 Å. Curves are plotted for various indexes m of the size distribution power-law. The spectra have been convolved with a Lorentzian function of line-width parameter $\Gamma = 0.01$ eV (from (Iglesias-Groth 2005))

photoabsorption of fullerenes in the optical predict their strongest optical transitions in the vicinity of the 4,430 Å band. Fullerenes of various sizes up to R ~ 10.5 Å may therefore play a role as carriers of this and several other strong DIBs.

1.4.5 The Hydrogenation of Fullerenes in the Interstellar Medium

Petrie and Bohme (2000) and Millar (1992) have studied the interstellar fullerene chemistry focusing on ion/molecule chemistry in various astrophysical environments. Fullerene based molecules could play a relevant role in the chemistry and physics of the interstellar medium and circumstellar environments (see also Watson et al. 2005; Webster 1991; Cataldo 2003).

Adopting for the C_{60} a fractional abundance of 4×10^{-7} (15% of the total elemental carbon in gas phase), Millar's model for a dark interstellar cloud at 10 K gives an appreciable abundance of protonated ion, $C_{60}H^+$, and the double charged cation, C^{++} and also that the fractional abundance of $C_{60}H^+$ is larger than C_{60}^+. Petrie et al. (1993) reported the tendency of fullerene ions (e.g. C_{60}^+ and C_{60}^{++}) to add atomic hydrogen repeatedly in the gas phase at 294 K. The repeated addition of hydrogen to neutral C_{60} in the gas phase at high temperatures (673 K) and pressures (65 atm) has also been reported by Attalla et al. (1993).

The process of hydrogenation of fullerenes involves formation of C–H bonds as a result of breakage of C=C double bonds of fullerenes and H–H bonds of molecular hydrogen to form hydrogen atoms. Although the hydrogenation reaction is exothermic (the heat is released as a result of reaction), additional energy is required to break these bonds, the activation energy for the hydrogenation has been estimated to be around 100 kJ/mol (1.0 eV/H_2).

Petrie and Bohme (1994) proposed that the most likely sites of hydrogenation upon a fullerene under interstellar conditions are at adjacent "defects" which are present upon the carbon surfaces of some fullerenes. These authors argue that the C–H bond will be stronger for an adjacent-pentagon C atom than for a C atom not involved in the junction of two pentagons. The dissociation of an adjacent pentagon C–H bond requires much more energy than dissociation of an isolated pentagon C–H bond. The migration of hydrogen atoms upon the fullerene surface could then lead to a preferential hydrogenation at those sites having the highest C–H bond strength. Fullerenes containing adjacent pentagons will tend to undergo preferential hydrogenation at these adjacent-pentagon sites and will tend to maintain a higher level of hydrogenation than fullerenes lacking adjacent pentagons.

The interstellar light extiction curve at 2,175 Å can be modeled using fulleranes with different degree of hydrogenation (Webster 1995; Cataldo 2003; Cataldo et al. 2009). The fraction of interstellar carbon abundance needed to produce fulleranes is high, but not prohibitively and carbon stars and planetary nebulae could provide these molecules in sufficient quantity. The experimental determination of optical and infrared spectra for laboratory-isolated fulleranes would be very valuable in assessing their role as carriers of the DIBs and of the unidentified infrared emission features.

1.5 Anomalous Microwave Emission and Hydrogenated Fullerenes

Recent experiments dedicated to the study of the anisotropy of the cosmic microwave background have found evidence for a new continuum microwave emission in the range 10–100 GHz correlated at high galactic latitudes with thermal emission (DIRBE 100 μm map) from interstellar dust (Kogut et al. 1996; de Oliveira-Costa et al. 1999, 2004; Hildebrandt et al. 2007). Unambiguous evidence for this new emission mechanism is provided by observations in the Perseus molecular complex (Watson et al 2005) and in the dark cloud LDN 1622 (Casassus et al. 2006). An explanation for this dust correlated microwave emission based on electric dipole radiation from small, fast rotating carbon based molecules has been proposed (Draine et al. 1998). These models appear to reproduce the major features of the so-called anomalous microwave emission (Finkbeiner et al. 2004) but do not identify the actual carrier of the emission. Fullerenes and buckyonions fit the basic properties of the carriers proposed in some of the Draine and Lazarian (1998) models, particularly the spherical shape and a number of atoms in the range 30–1,000 (required to dominate emission in the 10–100 GHz range). If fullerenes and buckyonions are broadly distributed in the Galaxy and conform a reservoir for interstellar carbon as we have discussed above, they may also play a role as carriers of microwave emission. The reason is that these molecules will be hydrogenated in the interstellar medium. The inhomogeneous distribution of hydrogen atoms in the surface of fullerenes will then cause a net dipole moment because of the polar nature of the C–H bond (0.3 Debyes) and therefore, these molecules are potential carriers of electric dipole emission. As shown by Iglesias-Groth (2006) the same mixture of fullerenes able to explain the UV bump is expected to produce significant microwave emission via electric dipole radiation.

1.5.1 Electric Dipole and Inertia Moment of Fulleranes

It is expected that a fraction of fullerenes in the interstellar medium will be hydrogenated, would have a much higher dipole moment because of the polar nature of the C–H bond. Experimental results on the electric dipole moment of isolated alkali-C_{60} molecules show an increase from 12.4 D (Debye) for C_{60}Li to 21.5 D for C_{60}Rb and indicate a strong electron transfer from the alkali atom to the C_{60} cage that is almost complete for the largest alkalis (Antoine et al. 2000). Given this behavior it is reasonable to expect that the dipole moment of fulleranes is higher than that of the C–H bond (0.3 D) and therefore, these molecules are potential carriers of electric dipole emission.

We will consider here the various families of fulleranes $C_N H_p$ that can be formed from icosahedral fullerenes with N = 60, 180, 240, 540, 960, 1,500, and 2,160 carbon atoms. Buckyonions can also be hydrogenated. We will consider here only those buckyonions with a complete number of shells and with hydrogen atoms bonded only in the outermost shell. At most, one hydrogen atom may be bound to each of the carbon atoms of a fullerene, thus, the number, P, of hydrogen atoms of any fullerane or buckyonion will lie in the range P = 1 to N. The ratio between the number of hydrogen and carbon atoms will be, s = P/N. We will deal with the following types of chemical species: $C_N H_P$, $C_N^- H_P$, $C_N^+ H_P$, $C_N H^+$, and the corresponding hydrogenated buckyonion analogs (C_{60} @C_{240} @ … $C_N H_P$, etc.). The hydrogen atoms of the fulleranes are expected to be located radially outward from the carbon atoms. The length of the C–H bond will be taken as in benzene 1.07 Å, see for example (Braga et al 1991). We assume the moment of inertia, I_C, of a fullerene as that of a spherical cage with radius R: $I_C = 2/3Nm_cR^2$, where m_c is the mass of the carbon atom. For the ensemble of fulleranes with P hydrogen atoms we adopt as moment of inertia the sum of the moment of inertia of the relevant fullerenes, I_C, with the moment of inertia of an hypothetical spherical cage of mass P times the mass of the hydrogen atom, and radius R + 1.07 Å. The moment of inertia of the hydrogenated buckyonion will be taken as the sum of the moments of inertia of all the individual fullerenes conforming the molecule plus that of the layer of hydrogen atoms.

It can be assumed in what follows that the dipole moment of a fullerane (or hydrogenated buckyonion) is proportional to the dipole moment of the C–H bond ($\kappa \sim 0.3$ D) and that the hydrogen atoms are randomly distributed on the surface of the cage in such a way that the intrinsic dipole moment of the fullerane with P hydrogen atoms can be approximated by $\mu \sim \kappa 0.3P^{1/2}$. We will adopt a value $\kappa = 1$ in the calculations. Owing to collisions with ions and electrons and also because of photoelectric emission, fullerenes in interstellar space may be charged. An additional electric dipole component could be expected as a consequence of any displacement between the centroid of the molecule charge and the center of mass (Webster 1991). Values for the polarizabilities of the fullerenes and buckyonions have been adopted from the semiempirical model of Iglesias-Groth (2005, 2006, 2007). In particular, this formalism gives 5.9 D for the dipole moment of $C_{60} H^+$. Since this formalism may overestimate the dipole moment of these molecules it can be conservatively assumed in these cases that the value of the dipole moment is that of the $C–H^+$ bond 1.68 D (Follmeg et al. 1987).

1.5.2 Rotation and Electric Dipole Emisivity from Fulleranes

The rotational damping and excitation mechanisms affecting fulleranes and hydrogenated buckyonions in the ISM can be studied using the formalism developed by (Draine et al. 1998). In thermal equilibrium, fulleranes share the same temperature as the gas and estimation of the rms rotation frequency is straightforward. When the temperature of fulleranes and gas are different, some processes will produce rotational damping and others will lead to rotational excitation. Draine and Lazarian (1998) discuss both types of processes comprehensively. Following their treatment, it can be computed the rotational damping produced by *collisional drag*, associated

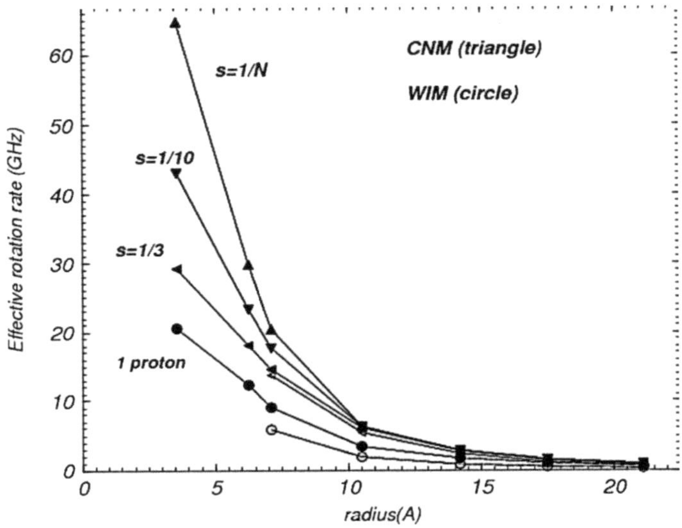

Fig. 1.9 Effective rotation rates, $\omega_{rad}/2\pi$, as a function of radius for neutral fulleranes with various levels of hydrogenation (*filled triangles*), hydrogenated buckyonions (*open triangles*) in CNM conditions. The case of a fullerenes and buckyonions with one proton is shown as filled circles or open circles respectively

with H, H_2, and He species (neutral and ions) temporarily stuck and subsequently desorbed from the surface of the fullerenes, *plasma drag* caused by the interaction of the electric dipole moment of the fullerene and the electric dipole of the passing ions, and the i*nfrared emission*, due to the thermal emission of photons that previously heated the molecule. Rotation excitation rates associated with the impact of neutral particles and ions, plasma drag, and infrared emission can be computed. We have assumed the temperature of the fulleranes is 20 K in the CNM and WIM phases of the ISM. In Fig. 1.9 we show results for the effective rotation rates,

$$\omega_{rad} = (5/3)^{0.25} < \omega^2 >^{0.5} \tag{1.11}$$

of fulleranes with different degree of hydrogenation. We illustrate the case for the CNM phase of the ISM. The rotation rate decreases as the radius of the molecule increases. Buckyonions show slower rotation rates due to the higher moment of inertia. Increasing the level of hydrogenation for molecules of a given radius (increasing the dipole moment) leads to lower rotation rates. We display in Fig. 1.9 results for three levels of hydrogenation s = 1/N, 1/10, and 1/3, where N is the number of carbon atoms in the fullerane or in the most external shell of the buckyonion. Assuming $\kappa = 1$, in the case of C_{60} these hydrogenation ratios correspond to dipole moments of 0.3, 0.74, and 1.34 D, respectively, and 0.3, 4.4, and 8.1 D for C_{2160}, the largest fullerane in this study. Hydrogenation ratios as high as s = 1/3 are not rare among other carbon based molecules, for example PAHs, and, as discussed by Webster (1991) may be rather common in fulleranes. For comparison, we also display the rotation rates of singly protonated fullerenes in WIM conditions. We find lower rotation rates than in the CNM case.

1.5.3 Electric Dipole Emissivity from Fulleranes

In Fig. 1.10a we plot the empirical results obtained by Watson et al. (2005) and emissivity curves for mixtures of fulleranes and buckyonions computed by Iglesias-Groth. The measurements are corrected following Watson et al. (2005) for the contribution of vibrational dust (T_{dust} = 19 K and emissivity index 1.55) and for the contribution of free–free in the Perseus region. We consider all the measurements available in the frequency range where anomalous emission appears to make a major contribution, i.e. from 10 to 90 GHz and overplot emissivity curves computed for several minimal cases either using cold neutral medium (CNM) conditions or a combination of CNM and warm interstellar medium (WIM). A few examples show how the predicted emissivities are indeed very close to the observed ones. Fitting the observations require some fine tuning between the number of molecules under CNM and WIM conditions and the level of hydrogenation s (ratio of hydrogen to carbon atoms in a given molecule). First, it is apparent that a significant level of hydrogenation is required if fulleranes are to explain the observed emission. Increasing the level of hydrogenation leads to a higher dipole moment and shifts the bulk of emission to lower frequencies. However, the tail emission at high frequencies (40 GHz) also requires a contribution of fulleranes with a small dipole moment. Molecules under WIM conditions alone would not be able to explain this tail. The dashed–dotted curve in Fig. 1.10a corresponds to the emissivity of a mixture of fulleranes and buckyonions in CNM, where any possible degree of hydrogenation has equal weight. The curve has been scaled to match approximately the measurement at 22 GHz, but the shape is inconsistent with the observations. The dotted line present the results for a mixture assuming a range of hydrogenation peaking at $s = 1/10$, such that 40% of the molecules of a given radius have s = 1/10, 30% have a factor 2 higher dipole moment and the remaining 30% a factor 2 lower. The emissivity curves obtained for other hydrogenation ranges do not give significantly better fits. The experimental data are best reproduced (solid line) for a combination (with equal weight) of the previous curve and the emissivity curve of singly protonated fullerenes in WIM conditions, again scaled to match the measurement at 22 GHz. This scale factor is close to 3 and results from a possible underestimation of the dipole moments and/or number density of fulleranes in the region. The dashed line illustrates the same combination between CNM and WIM but modifying the hydrogenation range of the molecules in CNM, which are now assumed to be uniformly distributed among $s = 1/10$, 1/20 and 1/30.

In Fig. 1.10b we plot the total flux emission curve for a mixture of fulleranes (with a number of carbon atoms N = 60–2,160) following a size distribution law $n(R) \propto R^{-m}$, with $m = 3.5 \pm 1.5$ and compare it with the *Cosmic Background Imager*

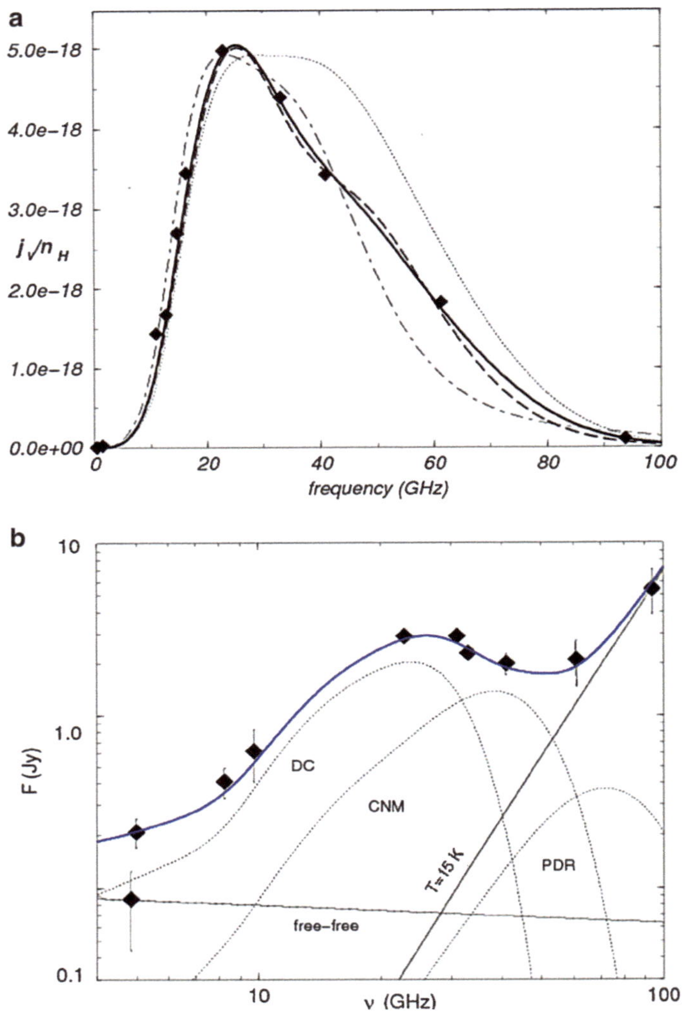

Fig. 1.10 (**a**) Observations of Perseus anomalous microwave emission by Watson et al. 2005 (*filled diamonds*, (76)) and predicted rotational emissivity per H (Jy $cm^2sr^{-1}H^{-1}$) of a mixture of fulleranes and hydrogenated buckyonions in CNM conditions (*dotted and dotted-dashed line*) and a combination of CNM and WIM conditions (*dashed line and solid line*). See text for details and Iglesias-Groth (43). (**b**) Flux emission curves for fulleranes for a degree of hydrogenation s = 1/3 under a combination of plausible physical conditions in LDN 1622 (After Iglesias-Groth 2006). Dotted curves: Predicted emission curves for fulleranes in the dark cloud (assuming ionization fraction f = 3% and a emission in a region equivalent to 17% the area of the cloud in CNM and also for the PDR the 5% the area, we predicted emission curves for fulleranes in the dark cloud). Total contribution of DC, CNM and PDR emission is the solid line. Thick solid line: DC, CNM and PDR including free–free (1.7 emissivity index) and thermal dust emission (15 K) Dyamonds: Observations of LDN 1622 anomalous microwave emission by Casassus et al. (2006)

observations of LDN 1622 listed in Table 3 of Casassus et al. (2006), taking into account the free–free and thermal dust component reported by these authors.

In summary, it is possible to understand the Perseus molecular complex and the dark cloud LDN 1622 microwave anomalous emission in terms of electric dipole emission of fulleranes if these molecules follow a size distribution similar to that proposed in the study of the UV extinction bump. The dominant microwave emission would be associated in both cases to the smaller fulleranes. These molecules could also be responsible for the diffuse microwave dust correlated emission at high Galactic latitude detected by the COSMOSOMAS experiment and WMAP.

1.6 Conclusions

We have summarized the evidence supporting the existence of fullerene-based molecules in the interstellar medium.

The cross section obtained for single fullerenes and buckyonions reproduce the behaviour of the interstellar medium UV extinction curve. A power-law size distribution $n(R) \sim R^{-m}$ with $m = 3.5 \pm 1.0$ for these molecules can explain the position and widths observed for the 2,175 Å bump and, partly, the rise in the extinction curve at higher energies. We infer ISM densities of 0.2 and 0.1 ppm for small fullerenes and buckyonions (very similar to the densities measured in meteorites). If as expected the cosmic carbon abundance is close to the solar atmosphere value, individual fullerenes may lock up 20–25% of the total carbon in the diffuse interstellar space.

Fullerenes, fulleranes and buckyonions are expected to present weaker transitions in the optical and near infrared with their number decreasing towards longer wavelengths. These transitions may be responsible for some of the known but unexplained diffuse interstellar bands. It would be very important to obtain high sensitivity, high resolution laboratory spectra of these molecules in the optical and near infrared for a more precise comparison with the very detailed observations of DIBs.

Hydrogenated fullerenes and buckyonions may produce rotationally based electric dipole microwave radiation under the conditions of the diffuse interstellar medium. These molecules are potential carriers for the anomalous Galactic microwave emission recently detected by several cosmic microwave experiments. Their precise contribution to this emission should be fully investigated.

References

Antoine R, Rayane D, Benichou E, Dogourd Ph, Broyer M (2000) Eur Phys J D 12:147
Attalla MJ, Vassallo AM, Tattam BN, Hanna JV (1993) J Phys Chem 97:6329
Ballester JL, Antoniewicz PR, Smoluchowski R (1990) Astrophys J 356:507
Becker L, Bunch TE (1997) Meteoritics Planet Sci 32:479
Becker L, Poreda RJ, Bunch TE (2000) Ann Lunar Planet Sci Conf 31:1803
Bertsch GF, Bulgac A, Tomanek D, Wang Y (1991) Phys Rev Lett 67:2690

Blake DF et al (1988) Nature 332:611
Bonin KD, Kresin VV (1997) Electric dipole polarizabilities of atom molecules and clusters. World Scientific, Singapore
Braga M, Larsson S, Rosén A, Volosov A (1991) Astron Astrophys 245:232
Cardelli JA, Meyer JM, Savage D (1996) Astrophys J 467:334
Casassus S, Cabrera G, Forster F, Pearson TJ, Readhead ACS, Dickinson C (2006) Astrophys J 639:951
Cataldo F (2003) Fullerenes Nanotubes Carbon Nanostruct 11:295
Cataldo F, Iglesias-Groth S, Manchado A (2009) MNRAS 400:291
Chhowalla M, Wang H, Sano N, Teo KBK, Lee SB, Amaratunga GA (2003) J Phys Rev Lett 90:155504
Cioslowski J (1995) Electronic structure calculations on fullerenes and their derivatives. Oxford University Press, New York
Curl RF (1993) Nature 363:14
de Oliveira-Costa A et al (1999) Astrophys J 527:L9
de Oliveira-Costa A et al (2004) Astrophys J 606:L89
Désert FX, Jenniskens P, Dennefeld M (1995) Astronom Astrophys 303:223
Draine BT, Lazarian A (1998) Astrophys J Lett 19:494
Fanti M, Orlandi G, Zerbetto F (1995) J Am Chem Soc 117:6101
Finkbeiner DP, Langston GI, Minter AH (2004) Astrophys J 617:350
Fitzpatrick EL (1999) PASP 111:63
Fitzpatrick EL, Massa D (1986) Astrophys J 307:286
Foing BH, Ehrenfreund P (1994) Nature 369:296
Foing BH, Ehrenfreund P (1997) Astronom Astrophys 317:L59
Follmeg B, Rosmus P, Werner H-J (1987) Chem Phys Lett 136:562
Galazutdinov GA, Musaev FA, Krelowski J, Walker GAH (2000) PASP 112:648
Haddon RC (1992) Acc Chem Res 25:127
Heger ML (1921) Lick Obs Bull 10:141
Henrard L, Lambin Ph, Lucas AA (1997) Astrophys J 487:719
Herbig GH (1995) Ann Rev Astron Astrophys 33:19
Herbig GH (2000) Astrophys J 542:334
Hildebrandt SR, Rebolo R, Rubíño-Martín JA, Watson RA, Gutiérrez CM, Hoyland RJ, Battistelli ES (2007) MNRAS 382:594
Iglesias-Groth S (2003) Doctoral thesis, Universidad de La Laguna, Spain
Iglesias-Groth S (2004) Astrophys J 608:L37
Iglesias-Groth S (2005) Astrophys J 632:L25
Iglesias-Groth S (2006) MNRAS 368:1925
Iglesias-Groth S (2007) Astrophys J 661:L167
Iglesias-Groth S, Ruiz A, Bretón J, Gómez Llorente JM (2002) J Chem Phys 116:1648
Iglesias-Groth S, Ruiz A, Bretón J, Gómez Llorente JM (2003) J Chem Phys 118:7103
Jenniskens P, Greenberg JM (1993) Astron Astrophys 274:439
Kogut A et al (1996) Astrophys J 460:1
Krelowski J, Walker GAH (1987) Astrophys J 312:860
Krelowski J, Bonetti A, Greenberg JM, Aiello S (1989) Evolution of interstellar dust and related topics. Elsevier, North-Holland, Amsterdam, p 285
Kroto HW (1988) Science 243:1139–1143
Kroto HW, Heath JR, Obrien SC, Curl RF, Smalley RE (1985) Nature 318:162
Kuznetsov VL, Chuvilin AL, Butenko YV, Mal'kov IY, Titov VM (1994) Chem Phys Lett 202:343
Léger A, d'Hendecourt L, Verstrate L, Schmidt W (1988) Astron Astrophys 203:145
Lin YL, Nori F (1996) Phys Rev B 53:1641
Mathis JS, Rumpl W, Nordsieck KH (1977) Astrophys J 217:425
McIntosh AD, Webster AS (1993) MNRAS 261:l13

Millar TJ (1992) MNRAS 259:35
Moller C, Plesset MS (1934) Phys Rev 46:618
Nandy K, Thompson GI (1975) MNRAS 173:23
Osawa E (1970). Kagaku 25:854 (in Japanese [also (1970) Chem Abs 74:75698])
Pariser R, Parr RGJ (1953) Chem Phys 21:466
Petrie S, Bohme DK (1994) MNRAS 268:938
Petrie S, Bohme D (2000) Astrophys J 540:869
Petrie S, Javahery G, Bohme DK (1993) Astron Astrophys 271:662
Pople JA (1953) Trans Faraday Soc 49:1375
Pople JA, Binkley JS, Seeger R (1976) Int J Quant Chem Symp 10:1
Purcell EM (1979) Astrophys J 231:404
Rubio A, Alonso JA, López JM, Stott MJ (1993) Physica B 183:247
Ruiz A, Hernández-Rojas J, Bretón J, Gómez Llorente JM (1998) J Chem Phys 109:3573
Ruiz A, Bretón J, Gómez Llorente JM (2001) J Chem Phys 114:1272
Savage BD (1975) Astrophys J 199:92
Tomanek D, Guo T, Jiu C, Haufer RE, Chibante LPF, Fure J, Wang L, Alford JM, Smalley RE (1991) J Phys Chem 95:7564
Ugarte D (1992) Nature 359:707
Ugarte D (1995) Astrophys J 443:L85
Watson RA et al (2005) Astrophys J 624:L89
Webster AS (1991) Nature 352:412
Webster AS (1992) Astronom Astrophys 257:750
Webster AS (1993a) MNRAS 263:385
Webster AS (1993b) MNRAS 265:421
Webster AS (1995) Astrophys Space Sci Lib 202:349
Webster AS (1997) MNRAS 288:221

Chapter 2
Infrared and Ultraviolet Spectra of Fulleranes: HREELS Studies and Implications for the Interstellar Medium

Carlo Carraro[1], Roya Maboudian[1], and Conrad R. Stoldt[2]

Abstract The reaction of atomic hydrogen with few-layer buckminsterfullerene films deposited on silicon in ultrahigh vacuum produces fullerane mixtures, which are studied in situ by high resolution electron energy loss spectroscopy. The vibrational and electronic excitation spectra are measured and discussed. Their main features are compared to observational data, including the infrared emission from interstellar and circumstellar clouds and the interstellar extinction curve. Laboratory spectra show remarkable agreement with important features detected in the observational data, among others the so-called canonical interstellar spacing of infrared spectra and the single bump in the ultraviolet absorption spectra. These studies suggest that members of the buckminsterfullerane family may be the primary carriers of the unidentified spectral features seen in interstellar dust.

2.1 Introduction

Various forms of molecular carbon, from ions to radicals, have been detected in the diffuse interstellar medium (ISM) using electronic, rotational, and vibrational spectroscopies (Henning and Salama 1998; Snow and Witt 1995). Discrete absorption and emission bands seen toward diffuse interstellar clouds indicate the presence of numerous two-atom molecules such as CO, CN and C_2. In addition to these interstellar features, a large family of spectral bands observed from the far-UV to the far-IR still defies explanation. Currently, it is the general consensus that many of the unidentified spectral features are formed by a complex, carbonaceous species that show rich chemistry in interstellar dust clouds (Ehrenfreund

[1]Department of Chemical Engineering, University of California, 201 Gilman Hall, Berkeley, CA 94720-1462, USA
e-mail: carraro@berkeley.edu; maboudia@berkeley.edu

[2]Department of Mechanical Engineering, University of Colorado, Boulder, CO 80309-0427, USA
e-mail: conrad.stoldt@colorado.edu

F. Cataldo and S. Iglesias-Groth (eds.), *Fulleranes: The Hydrogenated Fullerenes*, Carbon Materials: Chemistry and Physics 2, DOI 10.1007/978-1-4020-9887-1_2,

and Charnley 2000). Such species are likely to not consist entirely of carbon, as the ubiquitous presence of hydrogen in the interstellar medium suggests that some degree of hydrogenation should be expected. The presence of an unidentified infrared emission (Gillett et al. 1973) and the evidence of adsorption near 2,900 cm^{-1} in the line of sight to the galactic center (Soifer et al. 1976), typical of CH stretching vibrations, lend observational support to this expectation. Potential candidates include polycyclic aromatic hydrocarbons, hydrogenated amorphous carbon and fulleranes.

Here, we summarize our laboratory studies of mixtures of hydrides of buckminsterfullerene, prepared by reaction of C_{60} films with an atomic hydrogen flux in ultrahigh vacuum (UHV), and analyzed by means of High Resolution Electron Energy Loss Spectroscopy (HREELS). These studies afford us a broad window of energy loss, spanning from the infrared to the ultraviolet. A precise assessment of the degree of hydrogenation of the films is not conducted, in part because of the sheer difficulty of the task, but also because of the questionable usefulness of such assessment, vis-a-vis the variability expected in the degree of hydrogenation of carbonaceous species in various interstellar conditions. A theoretical analysis by Webster (1992a) suggests that the degree of hydrogenation results from the equilibrium between ultraviolet flux, which promotes hydrogen photodetachment, and interstellar gas density, which promotes attachment. Thus, a varying degree of hydrogenation will follow, among other factors, from variable environmental harshness.

2.2 The Unidentified Infrared Emission Problem

The term "unidentified infrared emission" is used to refer to the long-known emission features of interstellar dusts in the spectral region from just over 3,000 cm^{-1} to below 800 cm^{-1} (Gillett et al. 1973). These features comprise sharp IR bands at 2,920, 1,610, and 880 cm^{-1}, as well as a broader envelope near 1,300 cm^{-1}. In addition, a recurrent mode at 3,050 cm^{-1}, a weak mode near 1,450 cm^{-1}, and a shoulder near 1,150 cm^{-1} are observed. These spectral features can all be attributed to vibrational modes of hydrogenated carbon species, as summarized in Table 2.1. The chemical structure of these species remains the subject of debate. Furthermore, a number of carbon-rich astronomical objects reveal an emission feature in the far-IR at 490 cm^{-1}, of unclear attribution (Kwok et al. 1989).

Any hydrogenated candidate for the unknown IR emission must exhibit sufficient stability to the harsh conditions of the interstellar medium, including bombardment by particles and ultraviolet and gamma radiation. Equally important, for a species to be considered a viable candidate, its spectra must reproduce two important characteristics found in most diffuse medium spectra. These are (a) that the 1,300 cm^{-1} band be more intense than the 880 cm^{-1} band (Williams 1996), and (b) that the spacing between it and the next main emission feature match the "canonical interstellar spacing" of 300 cm^{-1} (Hudgins and Allamandola 1999).

2.3 The Ultraviolet Extinction Curve

From far-UV to near-IR wavelengths, diffuse ISM spectra exhibit a smooth continuum known as the interstellar extinction curve. The most prominent unidentified spectral feature on this curve is the so-called UV absorption bump at 217 nm. Characterized by a constant position and a bandwidth that varies with interstellar conditions (Fitzpatrick and Massa 1986), this feature is believed to result from electronic π–π^* interband transitions in hydrogen containing carbonaceous dust (Henning and Salama 1998). Most theoretical approaches adopt a small, spherical or spheroidal form of graphite as the candidate for the UV bump carrier (Mathis 1996), although most models indicate that the electronic structure of graphite must be modified to satisfy observational constraints (Li and Greenberg 1997). Below the bump, a number of weak, diffuse interstellar bands (DIBs) are detected on the continuum between 400 and 1,200 nm. The nature of the DIBs is still in question, although their narrow linewidths suggest unresolved rotational and vibrational structure from electronic excitations in gas-phase, organic molecules (Herbig 1995; Sarre et al. 1995). The strongest and broadest DIB, found at 443 nm, exhibits a nonvariable wavelength and weakens in dense interstellar clouds, as do all of the DIBs and the UV absorption bump.

The discovery of C_{60} by Kroto and coworkers (1985) was motivated in part by the interstellar dust problem. C_{60} would seem to be an ideal candidate, as it is spherical and graphite-like, it forms spontaneously in harsh environments with carbon dust, and is stable in intense radiation fields, a condition analogous to that found in the diffuse ISM (Kroto and Jura 1992). In fact, the observation of two DIBs at 957.7 and 963.2 nm are tentatively considered the first evidence of C_{60}^+ in interstellar dust (Foing and Ehrenfreund 1997). Moreover, a mixture of hydrides of C_{60} is shown to exhibit spectral features remarkably similar to those seen in the unidentified infrared emission (Stoldt et al. 2001). The UV absorption spectrum of synthetic $C_{60}H_{36}$ was also observed to possess a broad bump at 217.5 nm (Cataldo 2003).

2.4 Experimental Procedures

The C_{60} films described below were prepared in UHV on (2 × 1) reconstructed Si(100) substrates. C_{60} is evaporated from a Knudsen cell held at 410°C for approximately 10 min, yielding a reddish brown multilayer (>10 layer thick) film. A mixture of hydrides of buckminsterfullerene was prepared via the interaction of atomic hydrogen with the uppermost layers of the deposited C_{60} film at 25°C. Atomic hydrogen was generated by the dissociation of H_2 at a spiral tungsten filament heated to 1,600°C (hydrogen exposures are given in terms of molecular hydrogen and reported in Langmuirs (1 Langmuir (L) ≡ 10^{-6} Torr · s). These exposures correspond to a flux of roughly 3×10^{13} hydrogen atoms/(cm^2·langmuir) at the sample surface.

High-resolution electron energy loss spectra were obtained using an LK Technologies model ELS3000 spectrometer. In this technique, an energetic electron beam impinges on a surface and electrons scattered into a given detection angle (in the specular direction, unless otherwise noted) are energy analyzed. For spectra acquisition in the IR region, the primary electron beam energy is 6 eV, at an incident angle of 60° toward the surface normal, with resolution set at 3 meV. For the ultraviolet-visible region, the beam energy is varied between 18 and 50 eV and the instrumental energy resolution is set to 28 meV. The sample temperature is held at −150°C during spectra acquisition. The reader must note that HREELS differs in several important aspects from optical spectroscopies. In particular, the excitation mechanism has two components, one electromagnetic (electric dipole scattering) and one short range (impact scattering). Thus, HREELS spectra are somewhat richer (and more complex) than optical spectra, and usually reveal features that can be observed only by a combination of other techniques (e.g., IR, Raman, and neutron scattering).

2.5 Results

2.5.1 *Infrared Spectra*

The effect of hydrogen exposure on the vibrational spectrum of fulleranes $C_{60}H_x$ is reported by Stoldt et al. (2001). The results are summarized in Fig. 2.1. The vibrational spectrum of C_{60} is displayed in plot (*a*), while plots (*b*)–(*d*) illustrate the effects of increasing doses of hydrogen exposure. New loss features appear upon hydrogen exposure. C-H stretch vibration modes are detected as an intense peak centered at 2,900 cm^{-1}. Furthermore, a broad band develops between 1,425 and 1,060 cm^{-1}, and two smaller bands appear near 2,490 and 1,620 cm^{-1}. Additional peaks grow in intensity at 880 and 480 cm^{-1}, while the intensities of the peaks at 765 and 525 cm^{-1} begin to decline with increasing exposure to hydrogen flux. A number of loss peaks are not associated with observed astronomical spectral features, as discussed in the original report. The peaks at 765, 575, and 525 cm^{-1} are clearly due to vibrational modes of the unexposed C_{60} multilayer beneath the uppermost hydrogenated layer(s). To support this conclusion, off-specular HREEL spectra were obtained. In this geometry, scattering is mainly due to the impact mechanism. This allows one to suppress the effect of long range excitation, and thus, the signal resulting from deep buried layers. The off-specular data also reveals that the 1,220 cm^{-1} mode is not dipole excited and hence, it should not be prominent in IR emission spectra. Likewise, the mode centered at 2,490 cm^{-1}, attributable to a multiple loss (the sum of the modes between 1,450 and 1,150 cm^{-1}), will not appear in an optical spectrum. The remaining loss features found in the spectra of Fig. 2.1, plots (b–d), i.e., the losses measured at 2,900, 1,620, 1,425, 1,310, 1,150, and 880 cm^{-1}, compare favorably both with theoretical calculations and with observational evidence. In particular, note that the HREEL spectra shown in Fig. 2.1, plots (b–d),

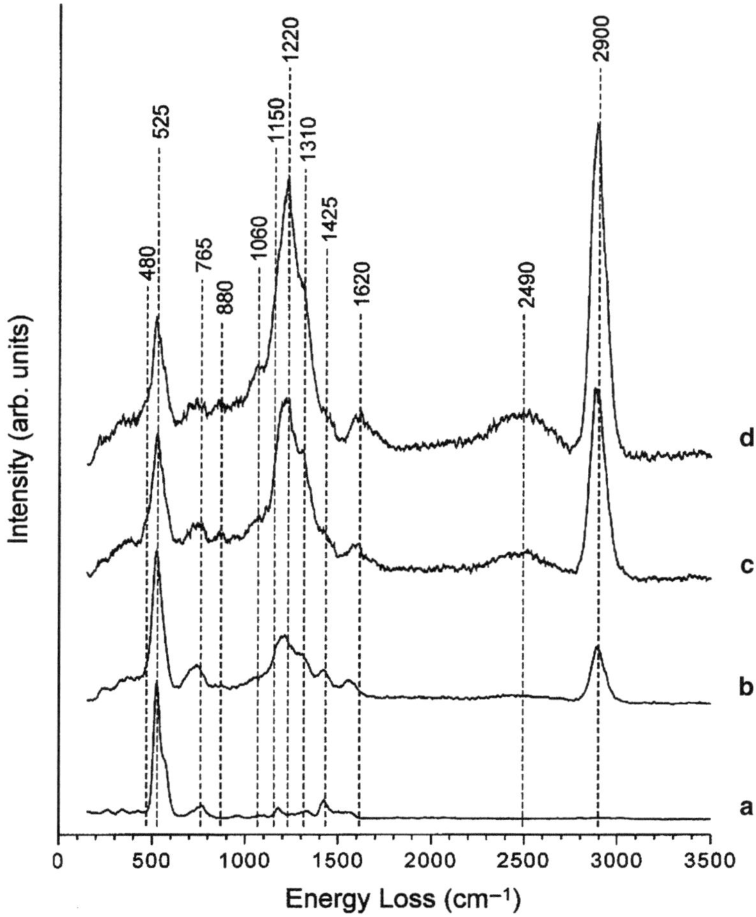

Fig. 2.1 HREEL spectra of C_{60} multilayer films shown as a function of increasing hydrogen exposure. The primary electron beam energy is 6 eV and the sample temperature is −150°C. (**a**) no hydrogen exposure, FWHM = 36.5 cm^{-1}; (**b**) a 45 L hydrogen exposure, FWHM = 34.8 cm^{-1}; (**c**) a 180 L hydrogen exposure, FWHM = 40.4 cm^{-1}; and (**d**) a 1,000 L hydrogen exposure, FWHM = 60.4 cm^{-1}. Spectral features labeled for comparison with Table 2.1 (Reproduced by permission of the AAS from Stoldt et al. 2001).

satisfy both spectral prerequisites mentioned above (canonical interstellar spacing and relative intensity of the 1,310 cm^{-1} band). In this regard, the observation that the position of the 1,310 cm^{-1} band does not shift with degree of hydrogenation is particularly significant, as it lends support to the hypothesis that an admixture of buckminsterfulleranes with broad (and varying) degree of hydrogenation may be responsible for the unidentified emission. Overall, the experimental losses featured in Fig. 2.1 (b–d) and attributable to buckminsterfullerane species are comparable in wavelength to the unidentified IR emission features listed in Table 2.1.

Table 2.1 Unidentified astronomical IR emission frequencies and their typical assignments

Astronomical frequency (wave numbers)	Typical mode assignment
3,050	sp^2 aromatic C–H stretch
2,800–3,000	sp^3 aliphatic C–H stretch
1,620	Aromatic C–C stretch
1,450	Aromatic C–C stretch, aliphatic C–H deformation
1,250–1,300	Blending of aromatic C–C stretch bands
1,150	Aromatic C–H in-plane bend
1,050–1,810	Blending of several weak aromatic C–C stretch bands
885	Aromatic C–H out-of-plane bend
490	Unknown

Lastly, two additional comments are in order. One concerns the shoulder that is seen to grow in intensity with hydrogen exposure near 480 cm^{-1}. This frequency, close to the calculated $2F_{1u}$ normal mode of $C_{60}H_{60}$, is nearly the same frequency of an "unidentified" mode recurring in observations of carbon-rich astronomical objects, and discussed previously. The second comment concerns the identity of the carrier of the IR emission line at 3050 cm^{-1}, an aromatic C–H strech, whose origin was attributed by Webster (1991) to lightly hydrogenated fullerane such as $C_{60}H$ or $C_{60}H_2$. Here, we note that for low hydrogen exposures, HREEL spectra show a broad band centered at 2,900 cm^{-1} and extending up to only 3,000 cm^{-1}. Only with exposures in excess of 2,000 L is the formation of aliphatic CH_2 and CH_3 bending modes detected in HREEL spectra. This observation was interpreted as suggesting that bond cleavage occurs after extended hydrogen bombardment, with generation of hydrogenated aromatic fragments from the broken up C_{60} cage, a likely occurrence also during extensive exposure to atomic hydrogen in the diffuse medium.

2.5.2 Ultraviolet Spectra

The far-UV to near-IR EEL spectrum of the unexposed C_{60} film is exhibited in Fig. 2.2a, in close agreement with spectra reported previously for thick C_{60} films on Si(100) (Gensterblum 1991). The characteristic "camel back" features, which exclude C_{60} as a carrier of the interstellar extinction, are readily observed at 195 and 260 nm. The three peaks at 260, 335 and 420 nm are seen in absorption spectroscopy (AS), and correspond to dipole-allowed, ${}^1A_g$–${}^1T_{1u}$, single-electron π–π^* transitions in the C_{60} molecule (Leach 1992; Hare et al. 1991). The intense peak at 195 nm is the so-called π-plasmon that results from a collective excitation of the molecular π-electron subsystem. The collective nature of this excitation is affirmed by our experimental observation that its intensity is variable upon changing the primary electron energy of the EEL spectrometer (Lucas 1992). Additionally, this band is broadened and blue-shifted when compared to the narrower transition seen in AS

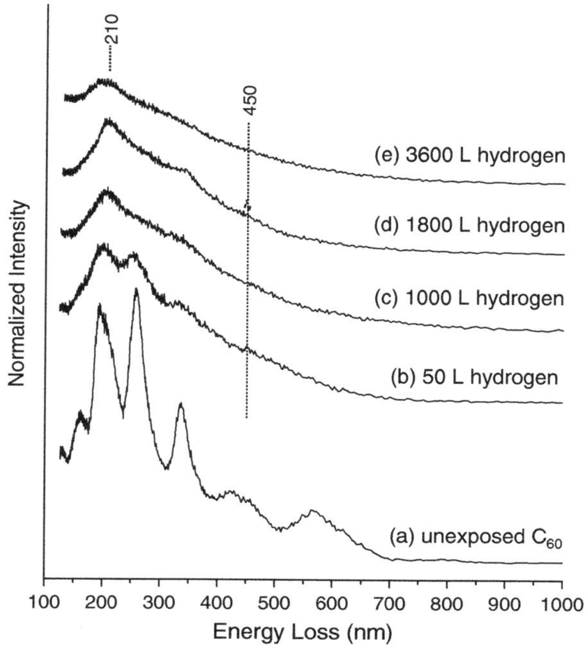

Fig. 2.2 EEL spectra of C_{60} multilayer films shown as a function of increasing hydrogen exposure. The primary electron beam energy is 50 eV and the sample temperature is −150°C. The FWHM of the individual spectra are (**a**) 44.8, (**b**) 34.4, (**c**) 33.1, (**d**) 40.7, and (**e**) 37.2 meV

near 210 nm, owing to the modified response function pertinent to the reflection geometry of EELS experiments.

Looking now at the hydrogenated C_{60} spectra shown in Fig. 2.2b–e, a number of distinct spectral changes are seen. The most significant is the observation of an intense, broad loss peak near 210 nm with low to moderate hydrogen exposure (Fig. 2.2b–d). This strong band is positioned between the two peaks which give the C_{60} spectrum its "camel back" character. The intensity of this loss peak is observed to vary with primary electron energy, indicating that it is associated with excitations of the π-electron system. Further hydrogenation of the C_{60} film results in a decrease in intensity of the 210 nm band, as seen in Fig. 2.2e. The π-plasmon, initially at 195 nm, is observed to shift to 210 nm and decrease in intensity with increasing hydrogen exposure. The astronomical UV bump shows a similar pattern of variability: the intensity is low in dense gas and high in the opposite conditions. As mentioned previously, this mode is blue-shifted in EELS and is detected closer to 220 nm in AS. Also observed is a small loss shoulder near 450 nm in the spectra of Fig. 2.2b–d that loses intensity with increasing hydrogen exposure. The broadest DIB, observed at 443 nm, behaves similarly in that it is most intense in more diffuse interstellar clouds.

The source of the 210 nm UV band in our spectra of partially hydrogenated C_{60} is now discussed in the context of a theory previously detailed by Webster (1992b, 1993a). In Fig. 2.2a, the π-plasmon of unhydrogenated C_{60} yields an intense loss peak near 190 nm that is attributed to the collective oscillations of the global π-electron system. With extended hydrogen exposure, every hydrogen atom that is attached to C_{60} results in the loss of a single π-electron and as a consequence, the spatial extent to which the molecule is conjugated decreases as well. In other words, the global π-electron system is broken up into several isolated local systems surrounded by hydrogen-bearing carbon atoms. When fully hydrogenated, the C_{60} molecule is aliphatic, and will exhibit no π-plasmon (or UV bump). The small local π-systems that remain embedded in C_{60} following prolonged hydrogen exposure have known analogs in molecules such as benzene, and in small odd-alternant radicals such as C_3H_5. Such molecules are known to have transitions at UV or visible wavelengths that may contribute to the astronomical UV absorption bump and the DIBs (Webster 1993b).

A complication that arises with this analysis concerns the nearly 10^{15} possible isomers of hydrogenated C_{60} which will each yield a unique electronic spectrum. Astronomical observations show systematic behavior in the spectra of diffuse molecular clouds (Mathis 1990), suggesting that the dynamic equilibrium which exists between competing astrophysical processes will favor certain molecular species depending on the interstellar environment. As a result, the enormous number of isomers of hydrogenated C_{60} may be significantly reduced to a few stable hydride families having 'magic numbers' of hydrogen atoms favored by symmetry considerations (Webster 1993a). This conjecture is supported by our previous observation that the vibrational spectra of a mixture of hydrides of C_{60} only exhibit the "canonical interstellar spacing" of 300 cm^{-1} after prolonged hydrogen exposure (Stoldt et al. 2001). Low degrees of hydrogenation do not satisfy this requirement of astronomical spectra, indicating that the interstellar environment selects higher hydrides of C_{60}.

In Fig. 2.3a, the EEL spectra of clean and hydrogenated C_{60} are compared with the variable component of the interstellar extinction curve. The two-component model for the interstellar extinction (Webster 1997) has a nonvariable or ubiquitous component attributed to classical-sized dust grains, and a variable component (shown in Fig. 2.3a) which is connected with hydrogenated carbonaceous molecules. The visible spectrum of hydrogenated C_{60} shows good agreement with the variable extinction curve, in particular reproducing the distinct kink near 500 nm, while the EEL spectrum of C_{60} does not. In Fig. 2.3b, the EEL spectrum of hydrogenated C_{60} is compared to the mean interstellar extinction curve. The structure of the two curves is similar: each shows a far-UV extinction at short wavelengths below 200 nm; a strong UV bump between 210 and 220 nm; and a near-featureless extinction from the near-UV to the near-IR.

The main discrepancy between the two curves in Fig. 2.3b is the strong loss shoulder due to the dipole-allowed transitions of C_{60} at 260 and 335 nm in the hydrogenated C_{60} EEL spectrum. Adding hydrogen to C_{60} will reduce the symmetry or eliminate it entirely, simultaneously decreasing the intensity of these transitions. In Fig. 2.2b–e, extended hydrogen exposure does lead to a reduction in the intensity

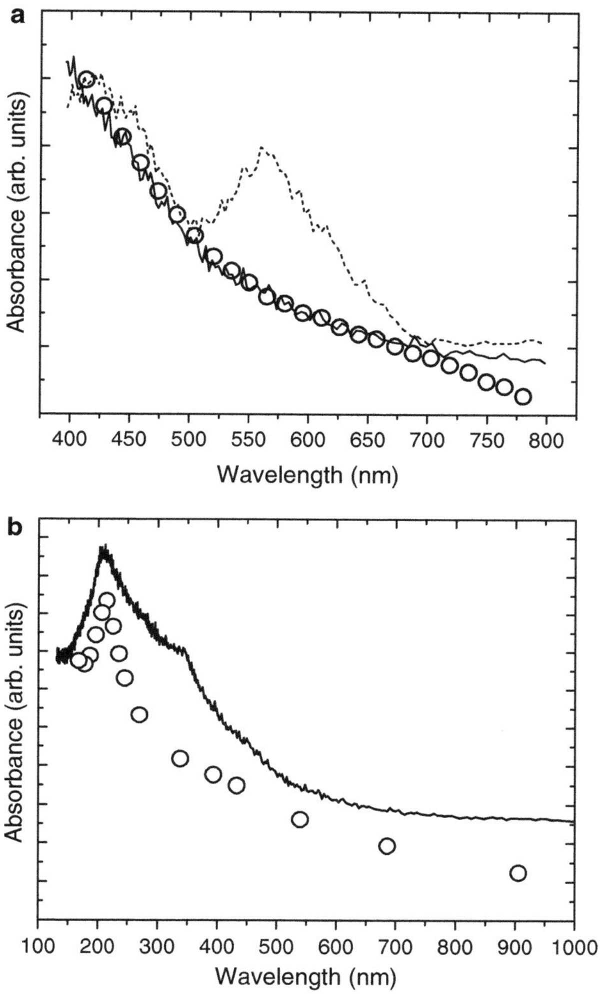

Fig. 2.3 (a) The visible-band EEL spectra of unexposed (*dashed line*) and hydrogenated (*solid line*) C_{60} films shown in Fig. 2.2, spectra labeled (a) and (d), compared with the extinction curve of the variable component (open circles) detailed in Webster (1997). (b) The far-UV to near-IR EEL spectrum of the hydrogenated C_{60} film (*solid line*) in Fig. 2.2 (d) compared to the mean interstellar extinction curve (*open circles*). The vertical scales of the EEL spectra and extinction curves in Fig. 2a and b are incommensurable, and the EEL spectra are arbitrarily scaled to give reasonable qualitative agreement for comparison

of these bands, although they are not entirely removed. By performing experiments on a monolayer C_{60} film, we have confirmed that some contribution to these bands results from excitations of the underlying C_{60} layers which remain not hydrogenated. A second contribution may result from the random nature of energetic hydrogen bombardment, which will yield a wide degree of hydrogenation in any one

HREELS experiment. This should result in some amount of lightly hydrogenated C_{60} molecules in the uppermost layer of the film. Interestingly, the UV-visible spectra of $C_{60}H_2$ and $C_{60}H_4$ (Webster 1993b) are very similar to that of clean C_{60} in regard to the strength of the dipole-allowed transitions, indicating that the addition of a small number of hydrogen atoms does not sufficiently perturb the molecule's electronic structure. Therefore, a wide degree of hydrogenation may be the source of the strong transitions characteristic of pure C_{60} in the hydrogenated films, perhaps intensified by a contribution from optically inactive impact scattering modes. We can also infer that interstellar C_{60} must exhibit both a high average degree of hydrogenation (at least more than four hydrogens) and a smaller range of hydrides lending further support to the 'magic number' scenario.

2.6 Conclusions

Mixtures of fulleranes produced by hydrogenation of solid C_{60} films under atomic H flux have revealed spectral features that bear striking similarity to those observed in the diffuse interstellar medium, both in the far IR and in the UV spectral windows. Of course, one must be cautious not to overextend the interpretation of laboratory data, for a number of reasons: firstly, because electron spectroscopy, the experimental technique used in these studies, differs in several important aspects from the spectroscopic methods employed in observational astronomy, and secondly, because of the specifics of specimen preparation and environmental conditions. In this regard, there is a need to explore the stability of fulleranes to energetic and corpuscular radiation (Cataldo et al. 2009). Nonetheless, our findings lend support to the suggestion of fulleranes as candidates for unidentified emission and absorption features of interstellar and circumstellar media. Whether or not they exist in sufficient abundance is still unclear; however, their spectral features make them undoubtedly an ideal model system for laboratory studies of these fascinating astrophysical phenomena.

References

Cataldo F (2003) Fullerenes Nanotubes Carbon Nanostructure 11:295
Cataldo F, Strazzulla G, Iglesias-Groth S (2009) MNRAS 394:615
Ehrenfreund P, Charnley SB (2000) Annu Rev Astron Astrophys 38:427
Fitzpatrick EL, Massa D (1986) Astrophys J 307:286
Foing BH, Ehrenfreund P (1997) Astron Astrophys 317:L59
Gensterblum G (1991) Phys Rev Lett 67:2171
Gillett FG, Forrest WJ, Merrill KM (1973) Astrophys J 183:87
Hare JP, Kroto HW, Taylor R (1991) Chem Phys Lett 177:394
Henning T, Salama F (1998) Science 282:2204
Herbig GH (1995) Annu Rev Astron Astrophys 33:19

Hudgins DM, Allamandola LJ (1999) Astrophys J 513:L69
Kroto HW, Jura M (1992) Astron Astrophys 263:275
Kroto HW, Heath JR, O'Brien SC, Curl RF, Smalley RE (1985) Nature 318:162
Kwok S, Volk K, Hrivnak BJ (1989) Astrophys J 345:L51
Leach S (1992) Chem Phys 160:451
Li A, Greenberg JM (1997) Astron Astrophys 323:566
Lucas AA (1992) J Phys Chem Solids 53:1415
Mathis JS (1990) Annu Rev Astron Astrophys 28:37
Mathis JS (1996) Astrophys J 472:643
Sarre PJ, Miles JR, Scarrott SM (1995) Science 269:674
Snow TP, Witt AN (1995) Science 270:1455
Soifer BT, Russell RW, Merrill KM (1976) Astrophys J 207:L83
Stoldt CR, Maboudian R, Carraro C (2001) Astrophys J 548:L225
Webster A (1991) Nature 352:412
Webster AS (1992a) Astron Astrophys 257:750
Webster A (1992b) MNRAS 255:41
Webster A (1993a) MNRAS 263:385
Webster A (1993b) MNRAS 265:421
Webster A (1997) MNRAS 288:221
Williams DA (1996) Astrophys Space Sci 237:243

Chapter 3
The Potential Role Played by the Fullerene-Like Structures of Interstellar Carbon Dust in the Formation of Molecular Hydrogen in Space

Franco Cataldo[1] and Susana Iglesias-Groth[2]

Abstract After a general introduction to the problem of formation of molecular hydrogen from atomic hydrogen in the interstellar medium and in the dense molecular clouds in particular, and after the explanation of the key role played by the surfaces on this process, it is proposed that the most suitable carbon surface for the formation of molecular hydrogen (from the radiative association process of atomic hydrogen) can be represented by carbon black rather than by graphite. Furthermore, it is proposed that the fullerene-like structures present in the carbon black graphene sheets are the reaction sites where molecular hydrogen may be formed.

3.1 General Aspects About Molecular Hydrogen Formation in the Interstellar Medium

Hydrogen is the most abundant element of the Universe and constitutes 88.6% of all atoms of cosmos; helium is eight times less abundant than hydrogen: 11.3% of the atoms of cosmos are made by this element. Summing up all the H atoms and He atoms we get that 99.9% of the Universe is made by these two elements (Burbridge et al. 1957; Aller 1961). All the other elements can be considered as impurities. Hydrogen is present in the Universe both under atomic and under molecular form. Molecular hydrogen (H_2) is the simplest and most abundant molecule in our cosmos but we have problems in explaining how molecular hydrogen can be formed for instance in the interstellar clouds. The problem of the formation

[1]Istituto Nazionale di Astrofisica, Osservatorio Astrofisico di Catania, Via S. Sofia 78, 95123 Catania, Italy and Actinium Chemical Research, Via Casilina 1626/A, 00133 Rome, Italy
e-mail: franco.cataldo@fastwebnet.it

[2]Instituto de Astrofisica de Canarias, Via Lactea s/n, E-38200 La Laguna, Tenerife, 2 CSIC, Spain
e-mail: sigroth@iac.es

F. Cataldo and S. Iglesias-Groth (eds.), *Fulleranes: The Hydrogenated Fullerenes*, Carbon Materials: Chemistry and Physics 2, DOI 10.1007/978-1-4020-9887-1_3,

of molecular hydrogen (H_2) in interstellar clouds would not exist if it could be produced in gas phase by the radiative association of two hydrogen atoms:

$$H + H \rightarrow H_2$$

This reaction is highly exothermal since at 0°C involves the liberation of 103.24 Kcal/mol corresponding to 4.48 eV per hydrogen molecule formed. Because of the high recombination energy, even at pressure much higher (e.g. 0.5 torr) than that present in the interstellar medium, the lifetime of molecular hydrogen is about 0.5 s. This means that on average at 0.5 torr only 1 collision after 10^6 collisions between atomic hydrogen is effective in forming molecular hydrogen (MacKay 1973 see also Henning 1998 and Herbst 2001). This occurs because two-centered collisions are ineffective, as there is no process for dissipating the energy of formation of H_2, which in fact re-dissociates into two atoms of hydrogen. For recombination, three-body collisions, H + H + M are needed in order to dissipate the excess of energy and this takes place for instance on surfaces, as already demonstrated in laboratory experiments (MacKay 1973). The efficiency of recombination (ε) is measured by the fraction of collisions which are effective in causing the H_2 formation. It has been reported (MacKay 1973) that metal surfaces are effective in hydrogenation reactions. For example, Ti, Pd, Pt, Au, Cu, Al surfaces are extremely effective in causing the recombination reaction of hydrogen atoms with $\varepsilon \cong 1$–10^{-2}. A similar behavior is offered by graphite (MacKay 1973), although additionally it presents as side reaction the formation of hydrocarbons. Instead on glass and silicon surfaces the value of ε falls to 10^{-4}–10^{-5}, while certain oxides like for instance ZnO present efficiencies comparable to those reported for metals (MacKay 1973).

Obviously, by decreasing the temperature down to 15 K (the typical temperature of a dense molecular cloud) and reducing dramatically the pressure (in the gas phase of a molecular cloud there are from 10^2 to 10^6 particles cm^3 in comparison to our atmosphere that under normal conditions contains 10^{19} molecules/cm^3), with a further constrain of the lack of surfaces, the effectiveness of the recombination reaction between atomic hydrogen becomes negligible in comparison to that discussed above on "laboratory" conditions. In fact, at the temperature of a dense molecular cloud of the interstellar medium (10–20 K), the recombination rate is negligible (Herbst 1995, 2001; Henning 1998).

Of course, also in the space conditions the H_2 molecule formed by the combination of two H atoms, must be in a highly excited state and, in order to avoid fragmentation, recombination energy (4.48 eV) must be released in a very short time (Herbst 1995, 2001). The unique way for an isolated molecule to release such amount of energy is the photon emission. However the radiative relaxation via vibro-rotational transitions is forbidden and the molecule just formed breaks apart in its atomic constituents (Herbst 1995, 2001). This process involving the mentioned radiative relaxation is also known as radiative association; it is much more efficient for larger systems involving molecules, in which the collision complex can live for a long period of time and so can increase the possibility that radiation is emitted.

However, it cannot convert atomic to molecular hydrogen under interstellar conditions. Nor is the three-body associative process possible because the density of a dense molecular cloud involves, say, 10^4 particles/mL; the chance that a third body strikes the H_2 collision complex before it dissociates so as to stabilize it, is zero under considered conditions. There is, however, a finite but exceedingly small possibility that a molecule of hydrogen can be formed in the gas phase in interstellar conditions; the rate constant is in fact extremely low $K = 10^{-31}\ s^{-1}$ not sufficient to explain the amount of molecular hydrogen present in the Universe (Pirronello and Averna 1988).

It is now commonly accepted (Herbst 1995, 2001; Pirronello and Averna 1988, 1997a, b; Averna and Pirronello 1991; Biham et al. 1998, 2001; Katz et al. 1999; Vidali et al. 2001, 2002) that the surfaces of the dust particles present in the interstellar medium must necessarily catalyze the formation of molecular hydrogen and other more complex molecules. In fact, it is estimated that at least 1% of interstellar matter by mass is tied up in dust particles. When a hydrogen atom strikes a grain surface, thanks also to the low temperatures, remains adsorbed with high efficiency. If the hydrogen atom is physically adsorbed, it is weakly bound, it can move along the surface by thermal hopping or by tunneling and diffuse to other binding sites present on the surface where another H atom is adsorbed. The surface chemistry that ensues goes via the Langmuir-Hinshelwood, or diffusive, mechanism (Somoraj 1994; Thomas and Thomas 1997): a diffusing hydrogen atom on a surface eventually finds another such and they recombine to form a molecule by giving off sufficient energy to the dust particle, the temperature of which rises slightly. The nascent molecule can evaporate even at 10 K, or, as has been actually measured in recent experiments of the process, can desorb by utilizing a portion of the reaction exothermicity as kinetic energy.

From the surface chemistry and heterogeneous catalysis it is also known (Somoraj 1994; Thomas and Thomas 1997; Vidali et al. 2002) another mechanism by which a reaction can take place in a surface: the Eley–Rideal mechanism. If we look at the equations and at the conditions, the Eley–Rideal process can be distinguished from the Langmuir–Hinshelwood for the fact that one atom of hydrogen is adsorbed on the surface and can be more or less mobile, but its reaction with another hydrogen atom occurs only when it is struck by an atom coming from the gas phase. Already at first sight the Eley–Rideal mechanism seems to be less adaptable to the space conditions of low density and low temperatures in the gas phase in comparison to the conditions needed for the Langmuir–Hinshelwood mechanism. At low temperatures, sites of very weak binding become populated and this leads to increased recombination by Langmuir–Hinshelwood mechanism (MacKay 1973).

3.2 Surfaces Involved in the Synthesis of Molecular Hydrogen in the Interstellar Medium

To put in evidence the key role played by the surfaces the famous scientist Wolfgang Pauli once said "God created matter; surfaces were invented by the devil".

The surface of the dust particles of the interstellar medium may have different chemical nature. In fact, the interstellar dust is not at all made exclusively by carbon particles. Such particles can or cannot be mixed with silica, silicates or other elements, in the simpler conditions the carbon dust can be made in large part by elemental carbon (Henning 1998; Henning and Salama 1998). However, the interstellar light extinction curve presents a "bump" at 217 nm. Such bump has been attributed to carbonaceous dust and the position and the shift of the absorption maximum is due to the degree of hydrogenation of such dust. (Mennella et al. 1997; see also discussions in Cataldo and Keheyan 2002; Cataldo and Pontier-Johnson 2002; Cataldo 2002a,b).

It could also be produced by fullerenes (Iglesias-Groth 2003, 2004). Recent work (Cataldo 2003; Cataldo et al. 2009a, b) has shown that hydrogenated fullerenes, fulleranes, have potential in modeling the carbon dust, first of all because they show an absorption maximum just at 217 nm (Cataldo et al. 2009a, b). However the real situation is more complex since it is expected that in certain conditions "ices" of various different molecules may condense on the dust particles, coating more or less completely their surface. In these conditions the molecular hydrogen formation must occur on the ices surface.

Concerning ices, it has been discussed that they must be amorphous (Smoluchowski 1983) in the interstellar medium and not crystalline. This implies that the adsorbed H atoms are localized in deep traps so that their wavefunctions have a limited spatial extent. This fact reduces significantly their mobility and hence the interaction with another H atom absorbed on another site is slow as compared to the residence time unless the two atoms happens to be localized near each other. This phenomenon reduces the rate of H_2 formation by several orders of magnitude when compared to the situation on crystalline surfaces. Computational simulations on soft and hard ice model surfaces have shown that for a cross-section of 4,000 nm^2 the reaction probability is 1 (Takahashi et al. 1999). Furthermore, the H_2 formed, due to the high amount of energy liberated is rapidly desorbed in an excited state from the ice mantle in timescales of 500 fs (Takahashi et al. 1999).

Concerning silicate surfaces and carbon surfaces, early works (MacKay 1973) have shown that silica and silicates are less effective in the recombination reaction than metals and carbon surfaces. Furthermore, it has been emphasized (MacKay 1973) the risk that the surface poisoning could lead to erratic values and erroneous conclusions (MacKay 1973). A recent experimental work of Pirronello and colleagues (1999) utilizing an hydrogen isotope beam at 200 K impinging a surface at 5–20 K, has shown that the recombination efficiency is higher on carbon surfaces rather than on olivine (silicate) surface, although the activation energy for the overall process is again higher in the case of carbon. A master equation for the evaluation of H_2 formation on interstellar grains (olivine and carbon) has also been proposed (Biham et al. 2001).

Since there are sufficient evidences about the presence of carbon grains and about the fact that the recombination of H atoms on carbon surface is effective even at low temperatures, we will limit our further discussion to this specific aspect: hydrogen and carbon surfaces.

3.3 Carbon Surfaces and Molecular Hydrogen Formation

Although it has been established that the carbon dust in the space must have any structure other than that of pure graphite and although it has been proposed that the real structure of carbon dust can be described by carbon black at a different degree of order and hydrogenation (Henning 1998; Cataldo et al. 2002; Cataldo 2002a), large part of the old and also contemporary works are dealing instead with graphite surfaces as model of interstellar carbon dust. This probably is due to the fact graphite surfaces may be represented more easily in computational studies and also because it is more easily available for experimental works. However graphite is an extremely simple model which nowadays is overcome by other models such as carbon black (Henning 1998), graphene sheets and a defective graphene sheets with fullerene-like defects. Furthermore a graphene sheet is also a model of very large polycyclic aromatic hydrocarbon which have been postulated as the real carriers of the diffuse interstellar bands (Henning and Salama 1998). Before reviewing the last knowledge on carbon structures and topology, let us discuss briefly the early results on the molecular hydrogen formation on graphite surfaces.

Cohen in 1976 concluded that the basal planes of pure graphite are not suitable substrates for H_2 formation whereas H_2O ice, CO ice and SiO_2 surfaces are. This is in contrast with a recent work (Maijer et al. 2001) where it has been concluded, by 3D quantomechanical calculations, that instead graphite is effective in the atomic hydrogen recombination reaction via an Eley–Rideal mechanism. The 3D calculation shows that H_2 molecule formed is released from the surface in a less vibrational excited state than previously thought with also rotational excitation. Furthermore, it has been shown that the absence of a barrier to reaction leads to a reaction probability close to 1 even for low collision energies (Maijer et al. 2001). Modeling of H_2 formation on graphite surface has shown (Xianwei and Jackson 2002) that the physisorption sites lies at 0.4 nm above the surface with a small binding energy of 8 meV. Simultaneously, localized and stable chemical adsorption sites can be found only when H is placed on the top of the site, with the help of substantial surface reconstruction. The above model has been proved to be useful in describing the reaction between a chemisorbed H atom and a gas phase H atom via the Eley–Rideal mechanism (Xianwei and Jackson 2002). A relatively similar approach to that just described can be found in the work of Jeolaika and Sidis (1999) where atomic hydrogen and graphite interactions were studied in the frame of density functional theory using a coronene-like model of the graphene sheet. Also in this case two adsorption regions were found: a physisorption region lying at 0.3 nm above the surface and a chemisorption region just at 0.15 nm above the surface in correspondence of a carbon atom. It is of importance the fact that the chemisorption site requires also substantial surface relaxation.

Perhaps the most complete set of works both theoretical and experimental is represented by the series of papers of Pirronello and colleagues (Pirronello et al. 1988, 1997a, b; Averna and Pirronello 1991; Biham et al. 1998, 2001; Katz et al. 1999; Vidali et al. 2001, 2002). It is remarkable that among the possible surfaces used by these authors also amorphous carbon was considered and found to be effective in H_2 formation (Pirronello et al. 1999). One of the key conclusions of a series of works (Pirronello et al. 1988, 1997a, b; Averna and Pirronello 1991; Biham et al. 1998, 2001; Katz et al. 1999; Vidali et al. 2001, 2002) was that at 10–15 K the formation of H_2 is appreciable but lower than what it has been estimated originally. At 10 K the mobility of H atoms on the surfaces considered is low and tunneling is not sufficient to assure the required mobility, thus thermal stimulation or selected radiation absorption is advocated to cause a minimum mobility. Furthermore, at low H mobility and coverage the recombination is low and proportional to the square of H adatom concentration on the surface, while at high mobility and coverage the H_2 formation probability is independent on adatom coverage.

In line with the idea that the correct surface analogous for the interstellar carbon dust must be amorphous, it is worth mentioning here the work of Wada and colleagues (1999) on quenched carbon condensate (QCC) as interstellar dust analogue. First of all the QCC is practically carbon black although it is called in a different way. In fact it is formed by the thermal (plasma) decomposition of mixtures of methane and hydrogen. The other extremely important aspect of QCC is its morphology, which is characterized by the presence of both onion-like carbon (a typical feature of carbon black) and, for the less hydrogenated samples, by the presence of polyhedral particles with bent graphene sheet. This fact introduces the argument of the present paper: the fullerene-like structures present on the surface of carbon black are the sites where specifically molecular hydrogen could be formed in the interstellar medium.

3.4 Fullerene-Like Structures on Carbon Dust of the Interstellar Medium: Their Role in Molecular Hydrogen Formation

Following the discovery of the fullerene-like structures on carbon black graphene sheets (Howard et al. 1994; Donnet et al. 2000), one of us has published a series of papers (Cataldo 2000a, 2000b, 2001a, 2001b, 2001c, 2001d, 2001e; Cataldo 2002a, 2002b) dealing with the role played by these structures on the interaction, adsorption and chemisorption of macromolecules on carbon black surface and the resulting macroscopic reinforcing effect observed in polymer-carbon composites (Cataldo et al. 2003; Cataldo 2005). Elsewhere (Cataldo 2002; Cataldo et al. 2003) we have reviewed in great details the reason of this behavior and the results of other researchers.

In other papers (Cataldo and Keheyan 2002; Cataldo and Pontier-Johnson 2002) we have tried to extend the results and the conclusions we have drawn in the field of the interaction between carbon black surfaces, activity and morphology, with radical and macroradicals also in the field of interstellar carbon dust.

As reported by Cataldo (2002b), the electron affinity of a fullerenes homologue series grows from 2.65 eV for C_{60} fullerene to 3.4 eV for fullerenes $C_{>100}$. The largest "fullerene" which can be conceived, can be considered as an open and defective graphene sheet surface made by condensed hexagonal rings but containing pentagonal rings and other kind of defects such as heptagonal rings. As an example, Fig. 3.1a shows a graphene sheet made of condensed hexagonal rings with only a "defective" pentagonal ring. It can be observed that this surface is no more flat and planar as in the case of a normal graphene sheet made exclusively by condensed hexagonal rings, but instead is bent and presents a positive curvature. Two pentagonal "defects" in a graphene sheet (Fig. 3.1b) fully surrounded by hexagonal rings causes the surface to bend more than the previous case. This is a consequence of the Euler's rule: with 12 pentagons we can close a graphene sheet in a fullerene cage (Cataldo 2002b). Examples of bent graphene sheets but with open curvature are presented in Fig. 3.1c, d where two pentagonal rings are abutting, or where an heptagonal defect is present in the sheet. The simultaneous addition of a pentagonal and an heptagonal ring in a graphene sheet, irrespective to the fact that these two defects are condensed or not, causes a simultaneous combination of a positive and negative curvature which in Fig. 3.1e, f is presented as a wave.

All these defects, have been defined as "fullerene-like" structures and defects (Cataldo 2002b) and have been experimentally detected in carbon blacks and soot of different origin (Cataldo et al. 2002b). An example of these defects are reported in Fig. 3.2. Furthermore, in a beautiful work, Hashimoto et al. (2004) has experimentally detected by TEM (Transmission Electron Microscopy) such topological defects in graphene sheets irradiated with electrons and ions. Thus, the fullerene-like structures and defects depicted in Fig. 3.1a–f not only are formed spontaneously during carbon black synthesis, but can be formed also on exposure to corpuscular radiation as clearly shown by Hashimoto et al. (2004) or by the action of high energy photons as indirectly shown with Raman spectroscopy by Cataldo (2000c). The disordered (turbostratic) aggregation of more or less defective graphene sheets leads to the formation of carbon black and this process occurs effectively also in space for instance around late-type carbon-rich stars with a mechanism very similar to that occurring in sooting flames (Henning 1998). The action of cosmic rays and high energy radiation on such interstellar carbon black causes a further enrichment of such fullerene-like structures the sites where precisely we think that the molecular hydrogen is formed with a high efficiency starting from the adsorption of atomic hydrogen.

If the electron affinity of fullerenes grows by increasing the size of the fullerene, it has been reported (Cataldo 2002b) that for a open graphene sheet containing fullerene-like structures it is expected an even higher electron affinity than that observed for the closed-shell fullerenes with electron affinities between 3.4 and 4.7 eV. The latter value being the work function of a fullerene surface. Therefore, if normal fullerenes, like for instance C_{60} fullerene, are considered "free radical sponges" because they react promptly with free radicals forming adducts, it must be expected that a graphene sheet surface, containing fullerene-like defects and having a higher electron affinity than fullerenes must act even more swiftly with free radicals. Furthermore, a graphene sheet from a graphite surface, made by condensed hexagonal rings is already known to be highly inert in many reactions including the reactions with free radicals. Only extremely reactive free radicals such as fluorine radicals are known to react with the graphene sheet under controlled conditions (e.g. low temperatures, dilution with inert gases) to yield graphite fluorides

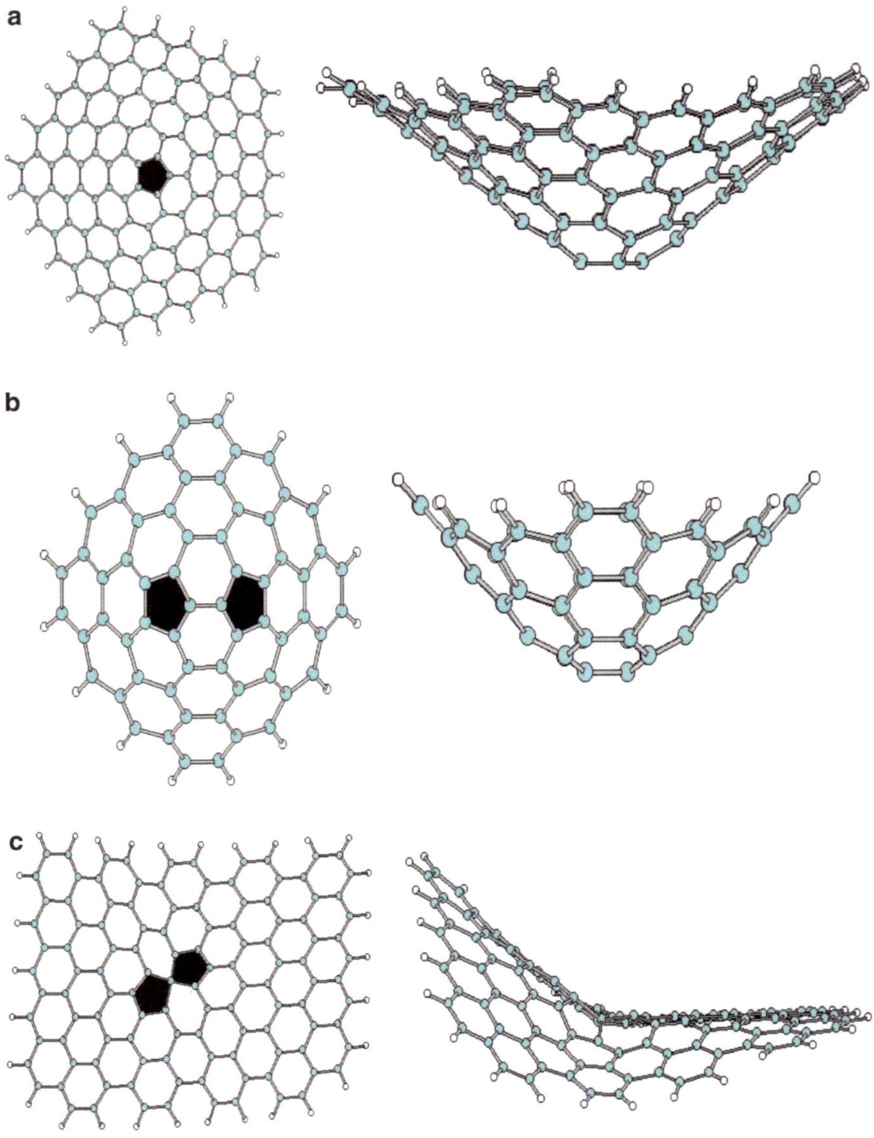

Fig. 3.1 Graphene sheets with some possible fullerene-like defects. (**a**) One pentagonal ring: a positive curvature of the surface. (**b**) Two pentagonal rings: a positive curvature of the surface. (**c**) Two abutting pentagonal rings: an open curvature of the surface

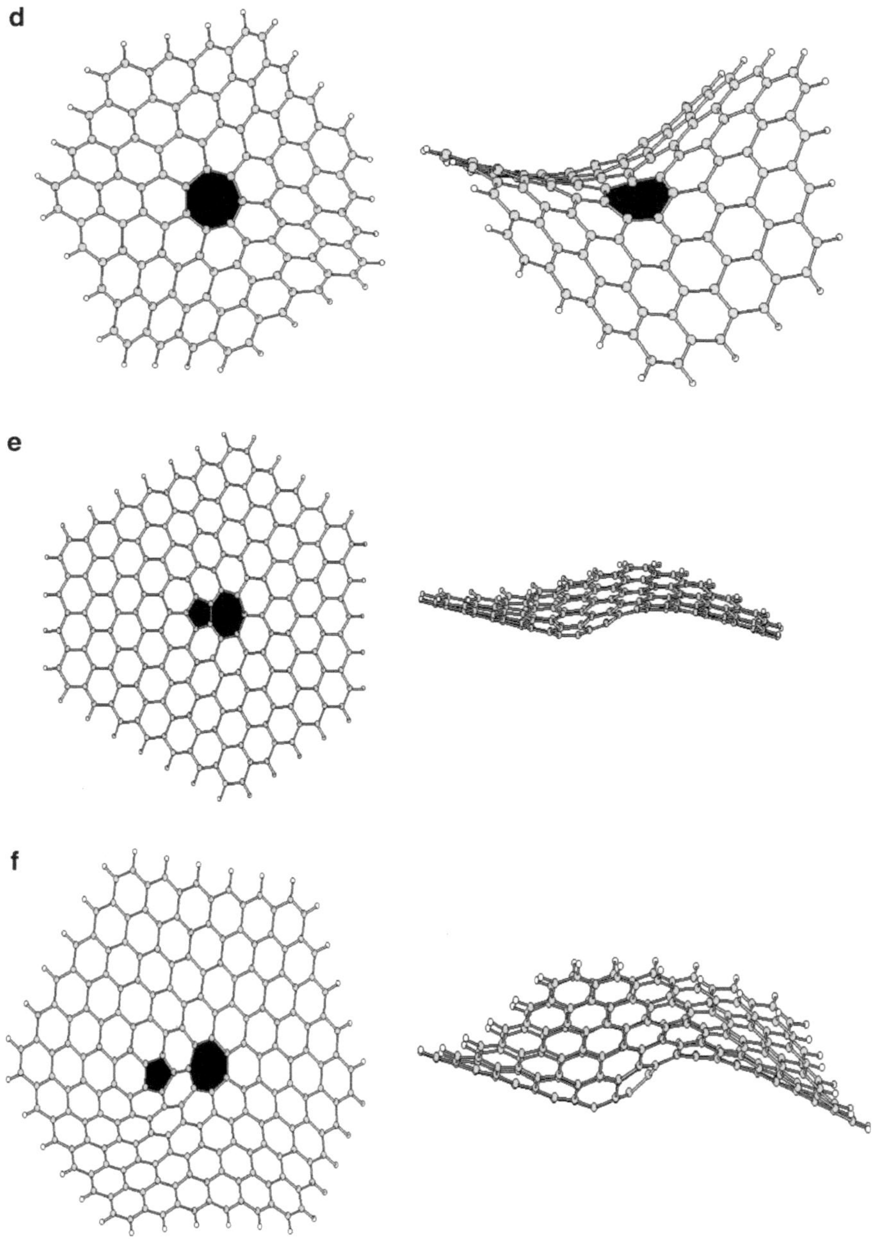

Fig. 3.1 (continued) (**d**) One heptagonal ring: a negative curvature of the surface. (**e**) Fused pentagonal and heptagonal ring: a waving sheet. (**f**) One pentagonal and heptagonal ring annealed by hexagonal rings

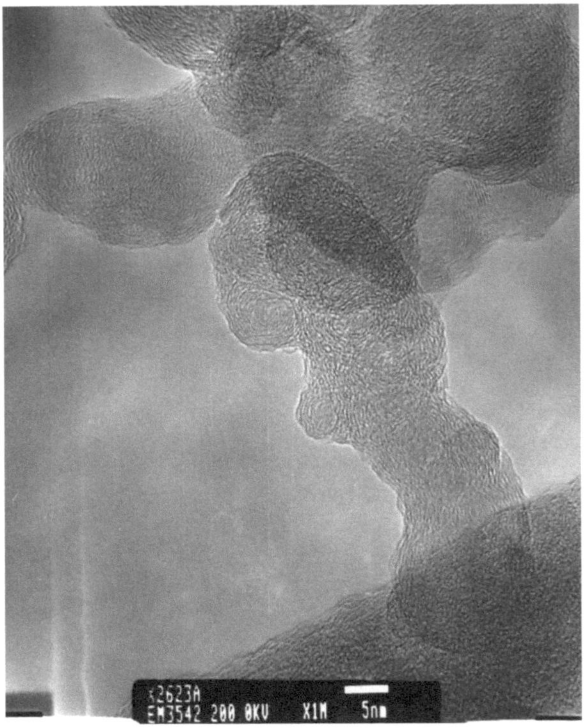

Fig. 3.2 HRTEM (high resolution transmission electron microscopy) image of a furnace carbon black. Curved graphene sheets and surfaces can be easily observed

(Nakajima 1994). In general for graphite, only the edges of the graphene sheets can be considered reasonably reactive. In fact, the edges of a graphene sheet have been simulated by the addition of atomic hydrogen to a naphthalene, a coronene or a circumcoronene anion (with $\Delta H_r = -50$ Kcal/mol), followed by the addition of another atomic hydrogen with the desorption of molecular hydrogen (Bauschlicher and Bakes 2001). This kind of reaction has been found to occur without barriers and to be exothermic with a $\Delta H_r = -53.3$ kcal/mol (Bauschlicher and Bakes 2001). Instead, for a graphene sheet containing fullerene-like structures as defects, in addition to the edges of the sheets, we must expect a significant reactivity just in correspondence of the mentioned defects. Heuristically we can expect that these defects may be the preferential adsorption sites for atomic hydrogen at the low temperatures existing in the interstellar medium and in the molecular clouds. Both physisorption and chemisorption may occur. The former implying a much more mobility of the atomic hydrogen by thermal hopping or tunneling. The chemisorption implies instead the formation of a localized chemical bond between the adsorbate and the surface. In this context, it is worth remembering the reader that atomic hydrogen is a free radical and thus it must react and "stick" easily on a surface containing the fullerene-like defect, just in the correspondence of the defect. By using the group increment and the calculation approach described in the Van Krevelen's book (1990) it can be estimated a $\Delta G_0° =$ -40.8 kcal/mol for the process of chemisorption of one atomic hydrogen in correspondence of a localized double bond present in a fullerene-like defect. A further addition of another atomic hydrogen attached to the adjacent carbon atom where the former has been adsorbed appears even more favorable thermodynamically with an estimated $G_0° = -88.6$ kcal/mol. Of course, in the adjacent site of the first chemisorption we will have a C atom with a dangling bond.

Apart from the favorable thermodynamics, it must be emphasized that the insertion of one or more fullerene-like defects in a graphene sheet surface is beneficial in understanding the exact position of the chemisorption site and the initial physisorption site for an adequate modeling. Furthermore, if we have a chemisorbed hydrogen atom (localized) and another mobile physisorbed atom of hydrogen, it is highly probable that the physisorbed atom will lie in proximity of the chemisorbed atom and the recombination to molecular hydrogen will occur easily with desorption, following the Langmuir–Hinshelwood mechanism. Similarly, if we have a chemisorbed hydrogen atom, another hydrogen atom from the gas phase must hit the chemisorption site in order to cause the formation and the desorption of the molecular hydrogen in agreement with the Eley–Rideal mechanism. Since the fullerene-like defect implies that the graphene sheet must be bent and the defect must be localized just at the corner of the bent surface, again there will be a preferred reaction site for the recombination reaction and the formation of molecular hydrogen.

3.5 On the Reversibility of Hydrogen Chemisorption and Release from Fulleranes

Fullerenes present the special property to react with atomic hydrogen forming hydrogenated derivatives, for instance $C_{60}H_{36}$ and hydrogen can be easily released as H_2 regenerating C_{60} fullerene without cage degradation. The isotope effect in the photolysis of n-hexane has demonstrated that hydrogen is lost from $C_{60}H_{36}$ in a complex reaction which involves also the fullerane oligomerization. However thermal treatment of $C_{60}H_{36}$ at about 500°C or higher temperatures causes the complete and quantitative release of hydrogen regenerating pure fullerene C_{60}. This phenomenon has been studied in detail by Talyzin et al. (2004) by using mass spectrometric analysis and demonstrating that hydrogen is released.

In Fig. 3.3 we show the FT-IR spectrum of pristine $C_{60}H_{36}$ and the resulting spectrum of C_{60} formed by the thermal annealing of $C_{60}H_{36}$ in a thermobalance under N_2 flow. $C_{60}H_{36}$ was heated at 10°C/min to 630°C and then was kept at that temperature for 10 min. Then the sample was cooled down to room temperature rigorously under N_2 flow and the resulting carbonaceous residue in the crucible was analyzed by FT-IR spectroscopy. The typical four pattern band spectrum of C_{60} (at 6.99, 8.45, 17.39 and 19.05 μm) replaced the spectrum of the pristine $C_{60}H_{36}$ shown at the top of Fig. 3.3. The complete dehydrogenation is supported by the FT-IR

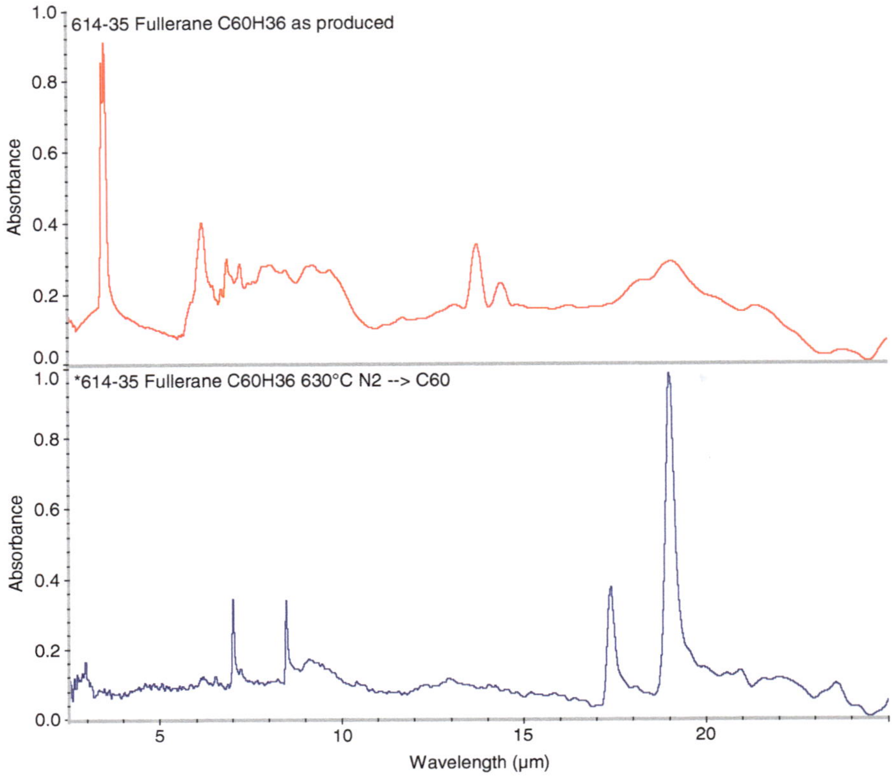

Fig. 3.3 FT-IR spectra (in KBr) of fullerane $C_{60}H_{36}$ (*top spectrum*) and the soot formed by its thermal decomposition at 630°C under inert atmosphere (*bottom spectrum*). The four band pattern spectrum at the bottom of the figure at 6.99, 8.45, 17.39 and 19.05 μm is due to C_{60} fullerene. Thus, fullerane $C_{60}H_{36}$ by thermal processing can form back quite easily fullerene C_{60} which in its turn is able to react with atomic hydrogen forming back hydrogenated fullerene or fullerane

spectra in Fig. 3.3: the C–H stretching bands of pristine $C_{60}H_{36}$ at 3.44 and 3.52 μm are completely absent in the thermally processed sample which consists of C_{60} and carbonaceous matter (Cataldo et al. 2009a).

This experiment has demonstrated that C_{60} cage is impressively resistant both to addition reaction of atomic hydrogen leading to fulleranes and to elimination reaction of molecular hydrogen giving back C_{60}. Such a molecular device should necessary play a key role in the chemistry of molecular hydrogen formation!

3.6 Thermal Decomposition of $C_{70}H_{38}$ and Formation of C_{70}

The reversible and quantitative release of hydrogen from hydrogenated fulleranes depends from the degree of hydrogenation. At very high hydrogen content, the thermal treatment of fulleranes causes the release of H_2 together with other

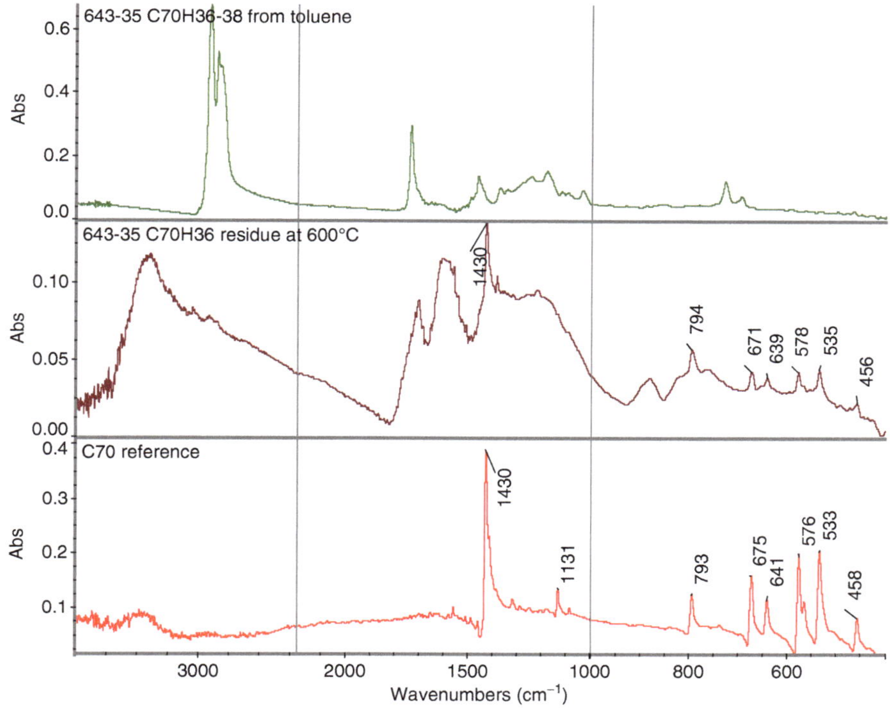

Fig. 3.4 FT-IR in KBr of (*top*) reference $C_{70}H_{38}$; (*middle*) carbonaceous residue obtained at 600°C under N_2 from the thermal decomposition of $C_{70}H_{38}$. Note that C_{70} fullerene has been formed back from $C_{70}H_{38}$. This is testified by the C_{70} band pattern compared with a reference C_{70} infrared spectrum (*bottom*)

hydrocarbons such as methane. Therefore, under such conditions, the process is no more reversible and the fullerene cage breakdown occurs.

We have verified this behaviour on $C_{60}H_{36}$. When $C_{60}H_{36}$ is heated under N_2 to 630°C it decomposes completely forming back C_{60} and releasing H_2. In fact, in our experiment (see Fig. 3.4), C_{60} was recovered from the decomposition of $C_{60}H_{36}$ and it was easily recognized by its characteristic infrared spectrum.

The thermal behaviour of $C_{70}H_{38}$ has already been studied by thermal analysis (Cataldo et al. 2009b) and a remarkable isotope effect has been recorded also in the thermal decomposition of $C_{70}H_{38}$ and $C_{70}D_{38}$. The latter being more stable than the former, as expected (Cataldo et al. 2009b). In the present work, $C_{70}H_{38}$ was heated under N_2 flow at 10°C/min to 600°C and then was cooled under N_2 to room temperature. The FT-IR of the carbonaceous matter recovered is shown in Fig. 3.4 in comparison to a reference spectrum of C_{70} and the infrared spectrum of $C_{70}H_{38}$. It is evident from the spectra of Fig. 3.4 that $C_{70}H_{38}$ is decomposed back to C_{70} by the thermal treatment releasing H_2. In fact, the typical C_{70} band pattern at 1,430, 794, 671, 641, 578, 535 and 456 cm^{-1} is very clearly evident. Thus, not only $C_{60}H_{36}$ gives back C_{60} but also $C_{70}H_{38}$ gives back C_{70} by thermal treatment, although in this last case the reaction does not appear fully reversible because together with C_{70} also a carbonaceous matter is formed. However the conditions selected for the thermal treatment were not optimized.

3.7 Conclusions

It has been shown in an heuristic way that the real site on carbon dust for molecular hydrogen formation in the interstellar medium and in dense molecular clouds could be described as fullerene-like sites. Thus, based on this new concept, the author recommends that in the future calculations, modeling and simulations, instead of adopting simple graphene sheets made exclusively by condensed hexagonal rings (using graphite as model) the researchers should try to use graphene sheets containing fullerene-like defects which not only are more realistic but will reserve more reasonable and fruitful results in comparison to those obtained till now. Furthermore, it is recommended that also in the experimentation for the atomic hydrogen recombination, graphite surfaces should be abandoned in favor of characterized fullerene black and various types of carbon black surfaces.

Acknowledgements The present research work has been supported by grant AYA2007-64748 from the Spanish Ministerio de Ciencia e Innovacion.

References

Aller LH (1961) The abundance of the elements. Interscience, New York
Averna D, Pirronello V (1991) Astronom Astrophys 245:239
Bauschlicher CW, Bakes ELO (2001) Chem Phys 274:11
Biham O, Furman I, Katz N, Pirronello V, Vidali G (1998) Month Notice Roy Astronom Soc 296:869
Biham O, Furman I, Pirronello V, Vidali G (2001) Astrophys J 553:595
Burbridge EM, Burbidge GR, Fowler WA, Hoyle F (1957) Rev Mod Phys 29:547
Cataldo F (2000a) Fullerene Sci Technol 8:105
Cataldo F (2000b) Fullerene Sci Technol 8:153
Cataldo F (2000c) Carbon 38:634
Cataldo F (2001a) Kautschuk Gummi Kunststoffe 54:22
Cataldo F (2001b) Polym Intl 50:828
Cataldo F (2001c) Fullerene Sci Technol 9:409
Cataldo F (2001d) Fullerene Sci Technol 9:497
Cataldo F (2001e) Fullerene Sci Technol 9:515
Cataldo F (2002a) Fullerenes Nanot Carbon Nanostruct 10:155
Cataldo F (2002b) Carbon 40:157
Cataldo F, Keheyan Y (2002) Proceedings of the 2nd European Workshop on Exo/Astrobiology, Graz, Austria, 16–19 September 2002, SP518, p 45–48
Cataldo F, Pontier-Johnson MA (2002) Fullerenes Nanot Carbon Nanostruct 10:1
Cataldo F (2003) Fullerenes Nanot Carbon Nanostruct 11:295
Cataldo F, Abbati G, Santini A, Padella F (2003) Fullerenes Nanot Carbon Nanostruct 11:395

Cataldo F (2005) Macromol Symp 228:91
Cataldo F, Iglesias-Groth S, Manchado A (2009a) Month Not Roy Astron Soc 400:291
Cataldo F, Iglesias-Groth S, Manchado A (2009b) Fullerenes Nanot Carbon Nanostruct 17:401
Cohen SA (1976) Nature 261:215
Donnet JB, Pontier-Johnson MA, Norman DT, Wang TK (2000) Carbon 38:1885
Hashimoto A, Suenaga K, Gloter A, Urita K, Ijima S (2004) Nature 430:870
Henning Th (1998) Chem Soc Rev 27:315
Henning Th, Salama F (1998) Science 282:2204
Herbst E (1995) Chemistry in the interstellar medium. Annu Rev Phys Chem 46:27–53
Herbst E (2001) The chemistry of interstellar space. Chem Soc Rev 30:168
Howard JB, Das Chowdhury K, Vander Sande JB (1994) Nature 370:603
Iglesias-Groth S (2003) Physisorption and photoabsorption of fullerenes. Implications physics and astrophysics, Ph.D. thesis. Universidad de La Laguna, Spain
Iglesias-Groth S (2004) Astrophys J 608:L37
Jeolaika L, Sidis V (1999) Chem Phys Lett 300:157
Katz N, Furman I, Biham O, Pirronello V, Vidali G (1999) Astrophys J 522:305
MacKay KM (1973) In: Bailar JC, Emeleus HJ, Nyholm R, Trotman-Dickenson AF (eds) Comprehensive inorganic chemistry, vol 1, Chapter 1. Pergamon, Oxford
Maijer AJHM, Farebrother AJ, Clary DC, Fischer AJ (2001) J Phys Chem 105:2173
Mennella V, Baratta GA, Colangeli L, Palumbo P, Rotundi A, Bussoletti E, Strazzulla G (1997) Astrophys J 481:545
Mennella V, Brucato JR, Colangeli L (2002) Astrophys J 569:531
Nakajima T (ed) (1994) Fluorine-carbon and fluoride-carbon materials. Chemistry physics and applications. Marcel Dekker, New York
Pirronello V, Averna D (1998) Astronom Astrophys 196:201
Pirronello V, Liu C, Shen L, Vidali G (1997a) Astrophys J Lett 475:L69
Pirronello V, Biham O, Liu C, Shen L, Vidali G (1997b) Astrophys J Lett 483:L131
Pirronello V, Liu C, Roser JE, Vidali G (1999) Astronom Astrophys 344:681
Somoraj GA (1994) Introduction to surface chemistry and catalysis. Wiley-Interscience, New York
Smoluchowski R (1983) J Phys Chem 21:4229
Takahashi J, Masuda K, Masataka N (1999) Month Notice Roy Astronom Soc 306:22
Talyzin A, Shulga YM, Jacob A (2004) Appl Phys A78:1005
Thomas JM, Thomas WJ (1997) Principles and practice of heterogeneous catalysis. Wiley-VCH, Weinheim
Van Krevelen DW (1990) Properties of polymers. Their correlation with chemical structure; their numerical estimation and prediction from addititve group contributions, 3rd edn, Chapter 21. Elsevier, Amsterdam
Vidali G, Roser JE, Weber H, Pirronello V, Manicò G (2001) Proceedings of American Astronomical Society Meeting 199, #11.11
Vidali G, Roser JE, Manicò G, Pirronello V (2002). In: Farid Salama (ed) Investigation of formation of molecular hydrogen on dust grain analogues. NASA Laboratory Astrophysics Workshop, Book of Abstracts of the NASA LAW held in NASA-Ames Research Center, Moffett Field, California, May 1–3, 2002
Wada S, Kaito C, Kimura S, Tokunaga AT (1999) Adv Space Res 24:523
Xianwei S, Jackson B (2002) Surface Sci 496:318

Chapter 4
Thermodynamic Properties of Fullerene Hydrides $C_{60}H_{2n}$ and Equilibria of Their Reactions

Gennady J. Kabo[1], Larisa S. Karpushenkava[1], and Yauheni U. Paulechka[1]

Abstract Thermodynamic properties of fullerene hydrides $C_{60}H_{2n}$ in the gaseous and crystalline states were studied by theoretical methods. Molecular structures and vibrational frequencies were calculated for three isomers of $C_{60}H_2$, four isomers of $C_{60}H_4$, three isomers of $C_{60}H_6$, three isomers of $C_{60}H_{18}$, ten isomers of $C_{60}H_{36}$ and two isomers of $C_{60}H_{60}$ by the density functional theory (DFT) at the B3LYP/6-31G* theory level. The ideal-gas thermodynamic properties were calculated based on those parameters. The formation enthalpies of fullerene hydride $C_{60}H_{2n}$ isomers in the ideal-gas state were derived from homodesmic reactions involving adamantane, cyclohexane, and C_{60} fullerene. The $C_{60}H_{60}$ hydride was demonstrated to be thermodynamically unstable relative to the other hydrides due to very high entropy decrease at its formation from C_{60} and H_2. Using the standard methods of statistical mechanics, the heat capacity and derived thermodynamic properties of crystalline $C_{60}H_{36}$ were calculated at 340–1,000 K that extended the range of experimental measurements. The heat capacity of crystalline hydrofullerenes was modeled and the enthalpies of their sublimation was estimated. The thermodynamic parameters of the hydrogenation of fullerene C_{60} in different states were analyzed. It was shown that C_{60} can act as a hydrogen accumulator.

4.1 Introduction

Shortly after realization of preparative syntheses of fullerenes their high chemical activity was proved and fullerene hydrides were obtained (Goldshleger and Moravskii 1997). The methods of synthesis of fullerene hydrides have been intensively developing because these hydrides are considered to be the base compounds for the development of fullerene chemistry, possess a potential for the use as hydrogen accumulators. The

[1]Chemistry Faculty, Belarusian State University, Minsk, Belarus
e-mail: kabo@bsu.by; karpushenkava@bsu.by; paulechka@bsu.by

F. Cataldo and S. Iglesias-Groth (eds.), *Fulleranes: The Hydrogenated Fullerenes*, Carbon Materials: Chemistry and Physics 2, DOI 10.1007/978-1-4020-9887-1_4,

most effective methods for preparative synthesis of fullerene hydrides can be provisionally divided into three groups (Goldshleger and Moravskii 1997):

1. Direct catalyzed hydrogenation of crystalline fullerite or fullerenes in solution.
2. Hydrogenation by hydrogen in the moment of its formation in the reaction with Zn in acidic solutions or according to Birch (treatment by lithium in liquid ammonium in presence of tert-butyl alcohol).
3. Hydrogenation with the use of H-donating compounds: 9,10-dihydroanthracene (DHA), diimide N_2H_2, hydroboration.

Synthesis of the $C_{60}H_{2n}$ hydrides has been studied the most intensively. It was found that in the synthesis the mixtures of the hydrides with $n = 1, 2, 3, 4, 18, 21–24$ are formed (Goldshleger and Moravskii 1997). Perhydrofullerane $C_{60}H_{60}$ has not been experimentally obtained.

The fullerene hydrides $C_{60}H_{2n}$ are a unique group of hydrocarbons due to special structure of their carbon cage. The C_{60} structure is formed by 20 six-membered and 12 isolated five-memebred cycles (facets) containing two types of bonds (edges): 60 (6–5) "pseudosingle" bonds of 1.45×10^{-10} m length and 30 (6–6) "pseudodouble" bonds of 1.38×10^{-10} m length. Though all carbon atoms in C_{60} are considered to be equivalent (Sidorov et al. 2005), the C_{60} molecules are non-aromatic, and non-uniform distribution of electronic density favours selective formation of the products of nucleofilic substitution (Karaulova and Bagrii 1999).

The synthesized fullerene hydrides vary both in the composition and in the isomeric structures. The number of possible isomers of the $C_{60}H_{2n}$ fullerene hydrides varies over a wide range: from 23 for $C_{60}H_2$ to $>10^{15}$ for $C_{60}H_{30}$ (Balasubramanian 1991). In the synthesis, the mixtures are formed with a limited number of isomers due to thermodynamic and kinetic limitations. However, the synthesis of fullerene hydrides of high purity is still impossible due to the above mentioned peculiarities. The best samples correspond to the isomeric mixtures containing 1–5% of the hydrides of different formula.

Thermal instability of fullerene hydrides makes additional difficulties for investigation of their physicochemical properties, especially at $T > 400$ K.

Thermodynamic properties are necessary for description of chemical equilibria of fullerene hydride synthesis, hydrogen accumulating systems, development of chemistry of exo-substituted fullerene hydrides (Karaulova and Bagrii 1999). Thermodynamic analysis of dehydrogenating ability of fullerenes towards hydrocarbons is of special interest. Fullerenes may act as Pt and Pd substituting catalyzers in the technical processes of dehydrogenation, cracking, reforming (Goldshleger and Moravskii 1997). Fullerene hydrides may be intermediates in chemical transformations of carbon and hydrogen in the interstellar space and thermodynamic analysis is an important tool for determination of pathways of this evolution. We believe that in the near future the thermodynamic properties of fullerene hydrides will be experimentally determined on a solid basis as it was done for other hydrocarbons.

The most effective way for investigation of thermodynamic properties for fullerene hydrides is a combination of the experimental methods, the quantum-chemical calculation of molecular characteristics, the statistical thermodynamic calculation of thermodynamic properties, as well as the comparative and additive procedures. The reliability of quantum-chemical calculations of structures, frequencies of normal vibrations, relative stability of fullerene hydrides must be checked using the experimental vibrational frequencies, structural parameters, ideal-gas thermodynamic properties for the allied compounds, for example, cage hydrocarbons and fullerenes. Thus, the ideal-gas thermodynamic properties were calculated for the C_{60} and C_{70} fullerenes (Diky and Kabo 2000). The resulting entropies were in satisfactory agreement with the values calculated from the experimental heat capacities, parameters of the phase transitions and the saturated vapor pressures. Possibilities of the statistical thermodynamic method for calculation of thermodynamic properties of crystalline fullerene hydrides are substantially limited because the $(C_p - C_v)_{cr}$ values cannot be *a priori* evaluated. The comparative procedures should be justified and applied in such calculations.

In this Chapter the proposed approach is realized for thermodynamic investigation of the $C_{60}H_{2n}$ fullerene hydrides.

4.2 Isomerism of $C_{60}H_{2n}$ Fullerene Hydrides

4.2.1 Isomeric Composition of $C_{60}H_2$

The simplest hydrogen derivative of fullerene C_{60} is the fullerene dihydride $C_{60}H_2$. The main isomer in the mixtures obtained using different synthetic techniques is the product of 1,2-addition to the [6–6] bond (Henderson and Cahill 1993). Its structure was determined from the ^{1}H NMR and ^{13}C NMR spectra of $C_{60}H_2$ and $C_{60}HD$ (Henderson et al. 1994). According to the PM3 semiempirical calculations (Matsuzawa et al. 1992), this isomer is the most stable one. In this work we consider two more $C_{60}H_2$ isomers with close positions of the hydrogen atoms: an isomer of the C_s symmetry with 1,4-addition of hydrogen to the [6,6]-bond and an isomer of the C_1 symmetry hydrogenated to the [6,5]-bond (Fig. 4.1). These isomers can also be observed in the experimental samples (Matsuzawa et al. 1992).

There is no information about the thermodynamic properties of the $C_{60}H_2$ hydrofullerenes. Considering the experimental results necessary for statistical thermodynamic calculations, only the IR spectrum of $C_{60}H_2$ was reported (Ballenweg et al. 1993).

4.2.2 Isomeric Composition of $C_{60}H_4$

The $C_{60}H_4$ fullerane was synthesized and studied in 1994 (Henderson et al. 1994). It was found that the sample contains at least four isomers with different positions of double bonds in the molecule (Henderson et al. 1994; Bergosh et al. 1997; Billups et al. 1997) (Fig. 4.2). According to Henderson et al. (1994), the main isomer is the C_s No. 1 isomer (Fig. 4.2). Its content in the mixture may be as large as 50%.

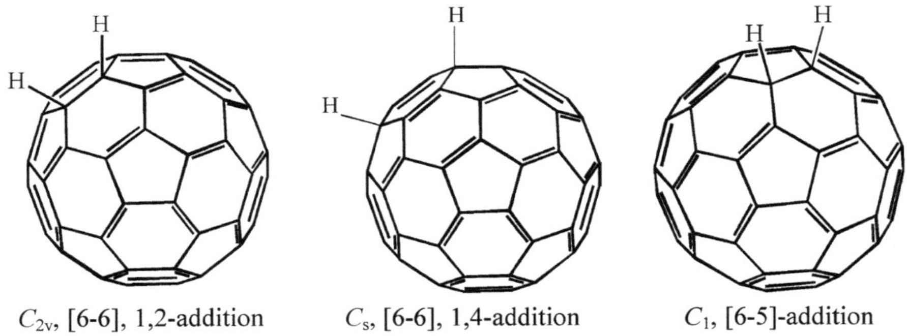

Fig. 4.1 Structure of $C_{60}H_2$ isomers

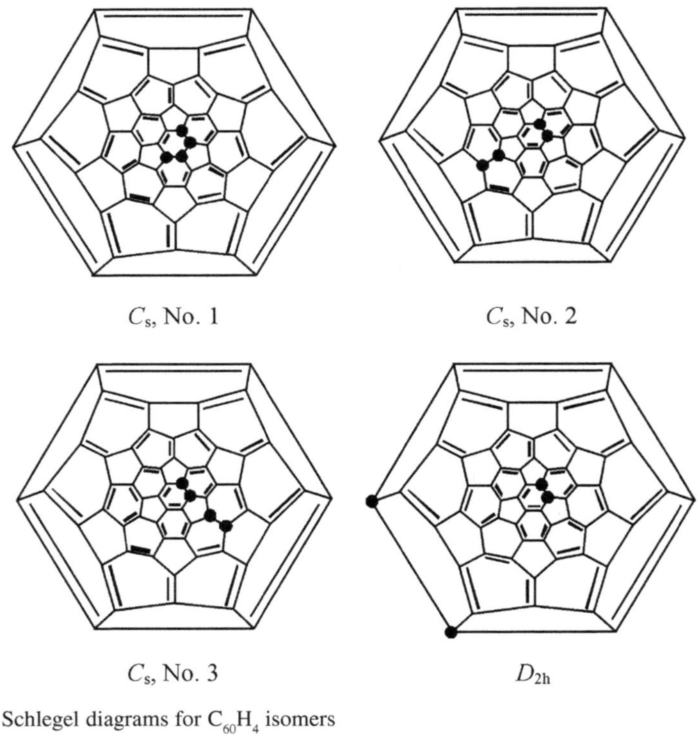

Fig. 4.2 Schlegel diagrams for $C_{60}H_4$ isomers

4.2.3 *Isomeric Composition of $C_{60}H_6$*

The $C_{60}H_6$ hydrofullerane was synthesized by reduction with the Zn/Cu couple in water solution. The compound is a mixture of the D_3 and C_3 isomers with a molar ratio of 6:1 as determined from the spectroscopic data (Nossal et al. 2001; Bergosh et al. 1997) (Fig. 4.3). It should be noted that these isomers are isostructural

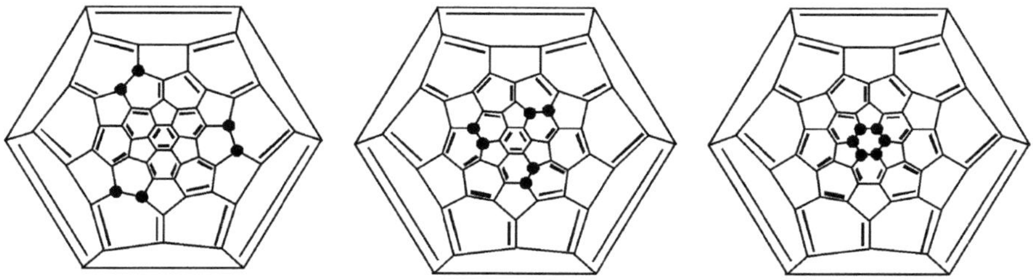

D_3 C_3 C_{3v}

Fig. 4.3 Schlegel diagrams for $C_{60}H_6$ isomers

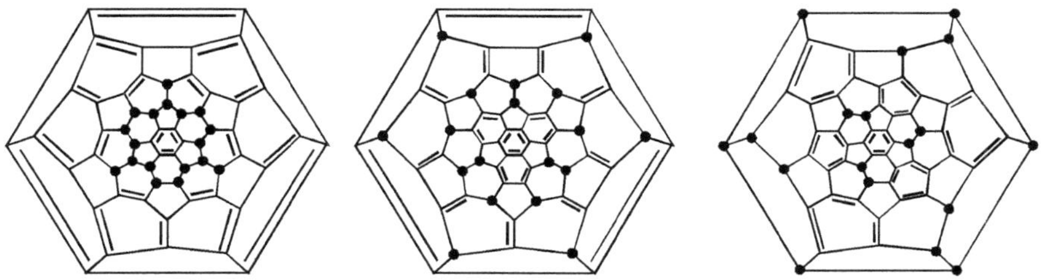

$C_{60}H_{18}$, C_{3v}, 1,2-addition

$C_{60}H_{18}$ C_{3v}, 1,4-addition

$C_{60}H_{18}$, C_3

Fig. 4.4 Schlegel diagrams for $C_{60}H_{18}$ isomers

neither with brominated nor with chlorinated analogues (Sidorov et al. 2005). In this work we also consider the C_{3v} isomer containing a totally hydrogenated benzene ring.

4.2.4 *Isomeric Composition of $C_{60}H_{18}$*

The $C_{60}H_{18}$ hydride was synthesized by hydrogenation of the C_{60} fullerene and by decomposition of $C_{60}H_{36}$. It was demonstrated that hydrogenation of C_{60} by the hydrogen transfer from DHA to fullerene leads either to $C_{60}H_{18}$ or to $C_{60}H_{36}$ depending on the conditions (Rüchardt et al. 1993). It was noted that when the $C_{60}H_{36}$ hydrofullerene is exposed to air for a long time or to vacuum the peaks corresponding to $C_{60}H_{18}$ appear in its mass-spectrum (Lobach et al. 1997). The structure of $C_{60}H_{18}$ was determined by ^1H NMR spectroscopy: this compound consists of the only isomer of the C_{3v} symmetry. This isomer is isostructural to $C_{60}F_{18}$ (Boltalina et al. 1996). All the hydrogen atoms are situated in one hemisphere of the carbon cage (1,2-addition, Fig. 4.4) (Nossal et al. 2001; Darwish et al. 1996). Here we consider one more isomer of the same symmetry, which appeared to be the most stable according to the semiempirical AM1 calculations (Clare and Kepert 1996) (1,4-addition, Fig. 4.4), and the C_3 isomer (Rathna and Chandrasekhar 1993) with more uniform distribution of hydrogen atoms on the fullerene cage.

4.2.5 *Isomeric Composition of $C_{60}H_{36}$*

The question about the isomeric composition of the synthesized samples of the $C_{60}H_{36}$ fullerane have not been completely resolved yet. In this work we consider ten $C_{60}H_{36}$ isomers, which are often formed under the synthetic conditions (Haufler et al. 1990; Rüchardt et al. 1993; Lobach et al. 1997; Billups et al. 1997; Guo and Scuseria 1992; Okotrub et al. 1999; Nossal et al. 2001). Their point groups are T, T_h, D_{3d}, $D_{3d'}$, S_6 (Nos. 91 and 88), C_3 (Nos. 3, 4 and 64) and C_1. The Schlegel diagrams of the considered isomers are demonstrated in Fig. 4.5, the numbering follows (Clare and Kepert 1994, 1999).

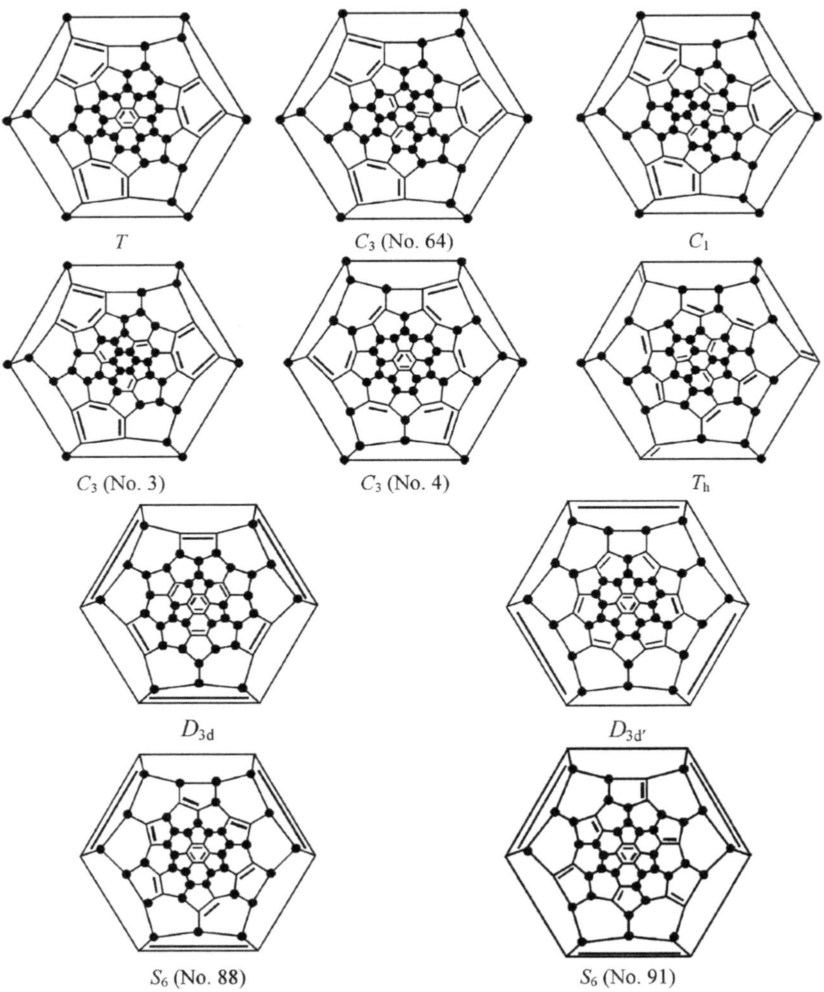

Fig. 4.5 Schlegel diagrams for $C_{60}H_{36}$ isomers

It was found earlier (Rüchardt et al. 1993) that the symmetry of the main isomer in a mixture of the $C_{60}H_{36}$ hydrofullerenes depends on the method of hydrogenation of the initial fullerene. Since all the experimental investigations of thermodynamic properties have been conducted only for the compound obtained by the non-catalytic hydrogen transfer from DHA (Rüchardt et al. 1993), the information about its isomeric composition is extremely important. It was concluded from the NMR spectra (Gakh et al. 2003) that the sample synthesized in this way is a mixture of isomers C_1, C_3 (No. 64) and *T*. High content of the low-symmetry C_1 isomer can be explained by peculiarities of the hydrogenation reaction mechanism (Gakh et al. 2003). It is also interesting to note that the major components of the experimental $C_{60}H_{36}$ sample (C_1 and C_3 No. 64) contain isolated and, as a consequence, more active "pseudodouble" bonds.

4.2.6 Isomeric Composition of $C_{60}H_{60}$

The attempts to synthesize the completely hydrogenated fullerene $C_{60}H_{60}$ have not been successful yet. The authors of theoretical works (Guo and Scuseria 1992; Dodziuk and Nowinski 1996) state that a part of hydrogen atoms must be inside the fullerene cage (T_h isomer). Dodziuk and Nowinski (1996) demonstrated by the molecular mechanics method that the most stable isomer contains ten hydrogen atoms inside the cage, but its symmetry is not clear from the original paper. So, in this paper we consider two isomers: the totally hydrogenated isomer of I_h symmetry and the T_h isomer containing 12 atoms inside the cage (Dodziuk and Nowinski 1996) (Fig. 4.6).

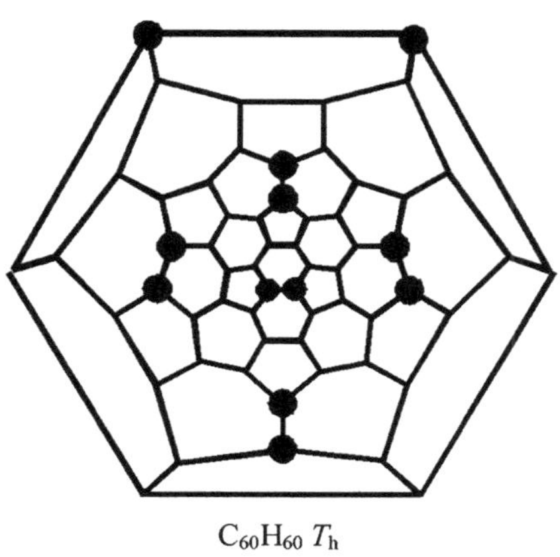

Fig. 4.6 Schlegel diagram for $C_{60}H_{60}$ T_h symmetry (48-exo, 12-endo). Black circles designate the atoms situated inside the fullerene cage

4.3 Experimental Investigation of Thermodynamic Properties for $C_{60}H_{2n}$

Heat capacity (Lebedev et al. 2000), saturated vapor pressure (Dorozhko et al. 2001) and combustion energy (Pimenova et al. 2002) were determined for $C_{60}H_{36}$. In all the cited papers, $C_{60}H_{36}$ was obtained by hydrogenation of C_{60} by hydrogen transfer from DHA. The content of the latter in the samples was determined from the Raman and fluorescent spectra to be in a range of 0.11–2%. The corrections taking into account the DHA impurity was applied only in the measurements of combustion energy (Lebedev et al. 2000). The isomeric composition of the samples in (Lebedev et al. 2000; Dorozhko et al. 2001; Pimenova et al. 2002) was not determined.

Heat capacity measurements in a range of (5–340) K for the sample with the $C_{60}H_{36}$ content of 98% allowed calculation of its thermodynamic properties. It was found that the heat capacity of $C_{60}H_{36}$ is substantially higher than that of C_{60} over the whole interval of the measurements. The phase transition like that in C_{60} at T_{tr} = 260.7 K (Diky and Kabo 2000) was not observed.

Thermodynamic properties of $C_{60}H_{36}$ (Lebedev et al. 2000) are presented in Table 4.1 as well as the $\Delta_f H^o_m$ ($C_{60}H_{36}$, cr) value determined from the combustion experiments for three samples with different DHA content (0.11–0.33 mas.%).

The temperature dependence of the saturated vapor pressure for $C_{60}H_{36}$ was determined by the Knudsen method with detection of mass spectra (Dorozhko et al. 2001). The IR spectroscopy was used for monitoring of $C_{60}H_{18}$ and C_{60} in a vapor phase. It was found from IR spectra of the thermostatted samples that $C_{60}H_{36}$, when exposed to vacuum, is converted into $C_{60}H_{18}$ without formation of intermediate hydrofullerenes and C_{60}. The $P_{sat}(T)$ dependence for $C_{60}H_{36}$ was obtained in the temperature range of 580–680 K:

$$\ln (P/bar) = (14.17 \pm 0.55) - (19.73 \pm 0.33) \times 1000/T. \qquad (4.1)$$

In (Karpushenkava et al. 2007a), the heat capacity difference $\Delta_{cr}^{g}C^{\circ}p,m = (-16.63 - 0.1627 \cdot T)$ J·K·mol^{-1} was obtained for the temperature range of 298–700 K, which allowed us to calculate the sublimation enthalpy for $C_{60}H_{36}$ at 298.15 K.

Dorozhko et al. (2001) estimated saturated vapor pressure for $C_{60}H_{18}$ from the mass-spectral measurements. It was p_{sat}(660 K) = 2.66 × 10^{-8} bar for $C_{60}H_{18}$ formed

Table 4.1 Thermodynamic properties of $C_{60}H_{36}$ (X isomer) (Lebedev et al. 2000; Dorozhko et al. 2001; Pimenova et al. 2002; Karpushenkava et al. 2007a)

	$C_{p,m}$/ J·K^{-1}·mol^{-1}	$\Delta_0^T S^{\circ}_m$/J·K^{-1}·mol^{-1}		$\Delta_f H^o_m$/kJ·mol^{-1}			
T/K	cr	cr	gas	cr	gas	$\Delta_{sub}H^o_m$/ kJ·mol^{-1}	P_{sat}/ bar
298.15	691.0	506.8	–	1,013	1,208	195	–
630	–	–	1,448	–	–	162	4.52·10^{-8}

during thermal decomposition of $C_{60}H_{36}$ and $p_{sat}(660\ K) = 0.44 \times 10^{-8}$ bar for the synthesized $C_{60}H_{18}$ sample. The sublimation enthalpy was also estimated in (Lebedev et al. 2000) to be $\Delta_{sub}H^{\circ}_{m}$ ($C_{60}H_{18}$, 660 K) $\geq$ 186 kJ·mol^{-1}. This value is significantly higher than that for $C_{60}H_{36}$ and is an evidence of its low reliability.

The obtained values of the thermodynamic properties (Table 4.1) are not sufficient for the detailed thermodynamic analysis of equilibria with participation of $C_{60}H_{36}$ in wide temperature ranges. The thermodynamic properties are not available for the other hydrides. So, it seems reasonable to calculate them based on the molecular and spectral data.

The necessary values of the thermodynamic functions for the $C_{60}H_{2n}$ hydrides can be determined from combined calculations by the methods of statistical thermodynamics and quantum chemistry. The experimental $\Delta_0^T S^{\circ}_{m}$(gas, 630 K) and $\Delta_f H^{\circ}_{m}$ (gas, 298.15 K) values (Table 4.1) may be the reliability criteria for the computational methods.

4.4 Thermodynamic Properties of $C_{60}H_{2n}$ in the Ideal-Gas State

The thermodynamic properties of a substance in the state of ideal gas are calculated as the sums of contributions from translation and rotation of a molecule as a whole, vibrations and internal rotation in the molecule, and electronic excitation. For example, for entropy and heat capacity the following equations hold:

$$S^{\circ}_{m}(T) = S^{\circ}_{tr} + S^{\circ}_{rot} + S^{\circ}_{vib} + S^{\circ}_{int.rot} + S^{\circ}_{el} \quad (4.2)$$

$$C^{\circ}_{p,m}(T) = C^{\circ}_{p,\,tr} + C^{\circ}_{p,rot} + C^{\circ}_{p,\,vib} + C^{\circ}_{p,\,int.rot} + C^{\circ}_{p,\,el} \quad (4.3)$$

Equations 4.2 and 4.3 are simplified for $C_{60}H_{2n}$ because internal rotation does not occur in these molecules. As it was demonstrated for the C_{60} fullerene, the electronic contributions to its thermodynamic properties may be neglected at $T <$ 1,000 K (Diky and Kabo 2000). The same is expected for fullerene hydrides.

The set of molecular data required for statistical thermodynamic calculations includes molecular mass, structural parameters for determination of a point group, a symmetry number σ and calculation of a product of principal moments of inertia $I_A \cdot I_B \cdot I_C$, as well as $3n - 6$ frequencies of normal vibrations for an n-atomic molecule.

The molecular parameters for $C_{60}H_{2n}$ can be obtained by the methods of quantum chemistry. The energy of molecules as well as the formation enthalpy is a function of composition and structure of molecules. So, the sequence of computational procedures for calculation of the ideal-gas thermodynamic properties can be presented as the following schemes (Figs. 4.7 and 4.8).

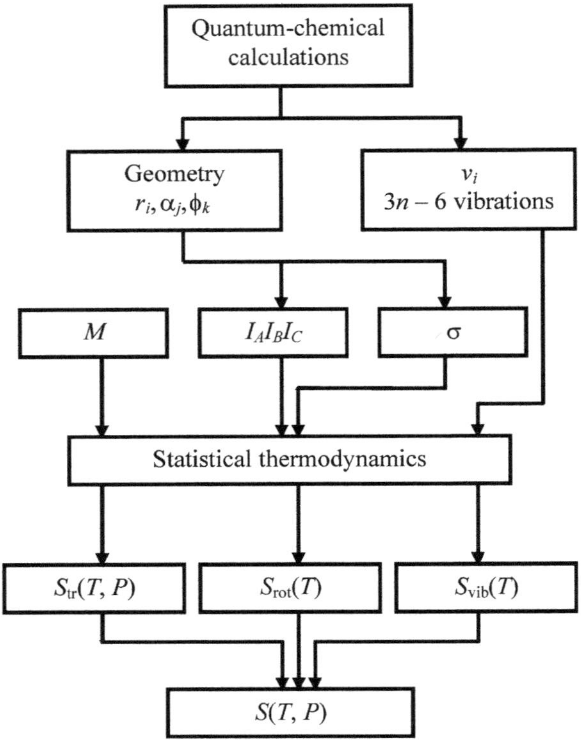

Fig. 4.7 Scheme of statistical thermodynamic calculations of ideal-gas entropy for the compounds without internal rotation

4.4.1 Quantum-Chemical Calculations of Molecular Parameters for $C_{60}H_{2n}$

The molecular structure and vibrational frequencies for all the considered hydrofullerenes were calculated in terms of the density functional theory at the B3LYP/6-31G* theory level using the PC GAMESS/Firefly software (Schmidt et al. 1993; Granovsky 2009). The atomic masses were taken from Loss (2003).

The products of principal moments of inertia $I_A \cdot I_B \cdot I_C$ calculated from the B3LYP/6-31G* geometries, symmetry numbers σ, numbers of chiral isomers for the $C_{60}H_{2n}$ hydrofullerenes are presented in Table 4.2.

Interpretation of vibrational spectra of $C_{60}H_{2n}$ is difficult because of uncertainty in the isomeric composition of the samples, presence of impurities, and low intensity of many absorption bands. The IR spectra of $C_{60}H_2$, presumably a product of 1,2-addition to the 6–6 bond was reported in (Ballenweg et al. 1993). The IR spectrum of $C_{60}H_6$ considered as a mixture of the C_3 and D_3 isomers was presented by Bergosh et al. (1997). The IR and Raman spectra of $C_{60}H_{36}$ and $C_{60}H_{18}$ were studied

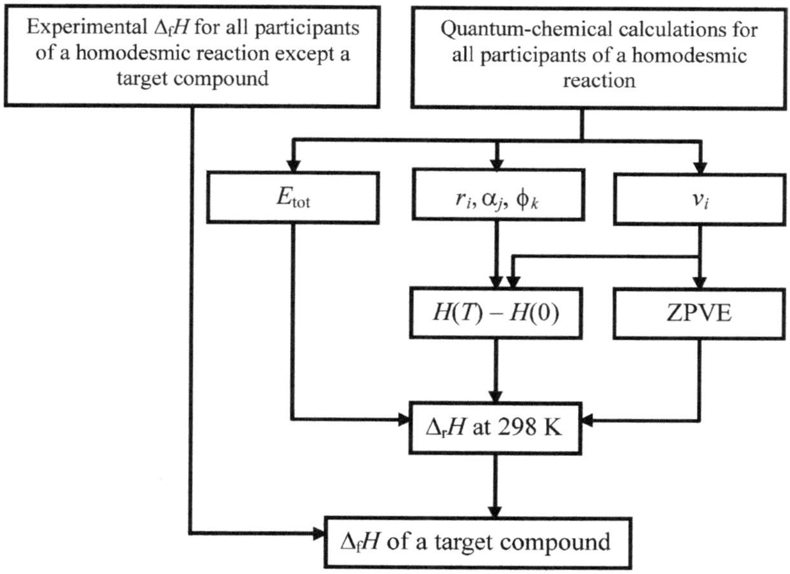

Fig. 4.8 Scheme of $\Delta_f H$ calculation from quantum-chemical calculations using homodesmic reactions

in detail (Bensasson et al. 1997; Bini et al. 1998; Popov et al. 2001; Meletov et al. 2001). The compounds were obtained by hydrogenation either with DHA and other H-donors at 350°C (Bensasson et al. 1997; Bini et al. 1998) or with H_2 at temperatures of 650–700 K and pressure of 3.0 GPa (Meletov et al. 2001). Based on comparison of the vibrational spectra for the samples obtained by different methods with the calculated vibrational frequencies, the authors of the above mentioned studies concluded that the most stable $C_{60}H_{36}$ isomers in the synthesis conditions are the isomers belonging to the S_6, D_{3d} and T point groups. However, it was demonstrated from the 800 MHz ^{1}H NMR spectra for the $C_{60}H_{36}$ sample obtained by hydrogenation of C_{60} with DHA at 350°C (Gakh et al. 2003) that the main components of the reaction mixture are the following isomers: C_1 (60–70%), C_3 No. 64 (25–30%) and T (2–5%). Each of them is a mixture of two chiral forms (Fig. 4.5). The above cited papers (Bensasson et al. 1997; Bini et al. 1998; Popov et al. 2001; Meletov et al. 2001) do not contain the complete set of (96 × 3 – 6) = 282 wavenumbers of normal vibrations for the $C_{60}H_{36}$ isomers required for calculation of the vibrational contributions to thermodynamic properties. These frequencies were calculated for $C_{60}H_{2n}$ and some other hydrocarbons using the B3LYP/6-31G* method in (Karpushenkava et al. 2007a; Karpushenkava and Kabo 2008).

Comparison of the calculated and observed frequencies allowed us to find the scaling factors f_n = (0.985 ± 0.003) for the skeletal C_{60} and CCH bending vibrations and f_n = (0.945 ± 0.005) for the C–H stretching vibrations (Karpushenkava et al. 2007a). The set of 282 wavenumbers for the C_1 isomer of $C_{60}H_{36}$ (Fig. 4.9) is in satisfactory agreement with the experimental spectra (Popov et al. 2001).

Table 4.2 Products of principal moments of inertia $I_A \cdot I_B \cdot I_C$ ($kg^3 \cdot m^6$), symmetry numbers σ, numbers of chiral isomers q for $C_{60}H_{2n}$ molecules

	$C_{60}H_2$			$C_{60}H_4$				$C_{60}H_6$			$C_{60}H_{18}$		
$C_{60}H_{2n}$	C_{2v}	C_s	C_1	C_s (No.1)	C_s (No. 2)	C_s (No. 3)	D_{2h}	D_3	C_3	C_{3v}	C_3	C_{3v} (No.1)	C_{3v} (No. 2)
$I_A \cdot I_B \cdot I_C \cdot 10^{135}$	1.05	1.04	1.05	1.08	1.08	1.08	1.08	1.11	1.12	1.11	1.27	1.29	1.29
σ	2	1	1	1	1	1	4	3	3	3	3	3	3
q	1	1	2	1	1	1	1	2	2	1	2	1	1

	$C_{60}H_{36}$										$C_{60}H_{60}$	
$C_{60}H_{2n}$	T	T_h	D_{3d}	$D_{3d'}$	S_6 (No. 88)	S_6 (No. 91)	C_3 (No. 3)	C_3 (No. 4)	C_3 (No. 64)	C_1	I_h	T_h
$I_A \cdot I_B \cdot I_C \cdot 10^{135}$	1.60	1.64	1.62	1.60	1.60	1.60	1.63	1.63	1.59	1.62	2.15	1.71
σ	12	12	6	6	3	3	3	3	3	1	60	12
q	2	1	1	1	1	1	2	2	2	2	1	1

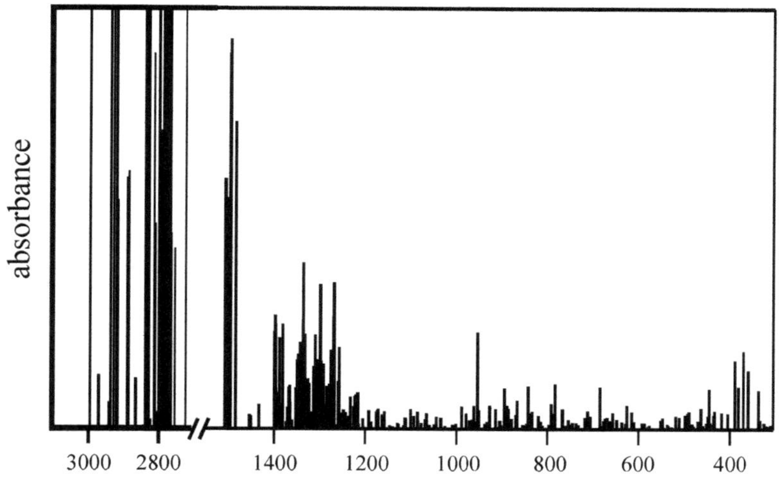

Fig. 4.9 The calculated IR spectrum of the C_1 isomer of $C_{60}H_{36}$. Intensity of C–H stretching bands is decreased by three times

The complete sets of frequencies of normal vibrations for the other $C_{60}H_{2n}$ fulleranes were obtained in a similar way from the calculated B3LYP/6-31G* frequencies.

4.4.2 Thermodynamic Properties of $C_{60}H_{2n}$ in the Ideal-Gas State

The thermodynamic properties of fullerene hydrides in the range 298.15–1,000 K in the state of ideal gas were calculated as described in (Frenkel et al. 1994). The thermodynamic properties of the experimental $C_{60}H_{36}$ sample (hereafter designated as the X isomer) obtained by the hydrogen transfer from DHA were calculated from the molecular properties of the main C_1 isomer. For the X isomer, we accepted the average composition of the sample synthesized by the non-catalytic hydrogen transfer from DHA (Gakh et al. 2003): 68% of C_1, 28% of C_3 (No. 64), 4% of T isomers. All these isomers are chiral. The resulting thermodynamic properties of isomers of the $C_{60}H_{2n}$ fulleranes are presented in Tables 4.3 and 4.4. Table 4.3 contains the data for the main isomers formed in the experimental syntheses. The calculation of entropy for $C_{60}H_{36}$ at 630 K was of special interest since that value could be compared with the experimental one. The calculated value agrees with the experimental value $\Delta_0^T S^o_m$ (630 K) = (1,448 ± 43) J·K^{-1}mol^{-1} within the uncertainties.

4.4.3 Calculation of $\Delta_f H^o_m$ for $C_{60}H_{2n}$ in the Ideal-Gas State

The formation enthalpies for the $C_{60}H_{2n}$ hydrides were calculated from the results of the B3LYP/6-31G* calculations using homodesmic reaction (4.4) with fullerene C_{60}, adamantane $C_{10}H_{16}$ and cyclohexane C_6H_{12}:

Table 4.3 Thermodynamic properties of $C_{60}H_{2n}$ hydrides in the ideal-gas state. $C^{o}_{p,m}$, $\Delta^{T}_{0}S^{o}_{m}$ and $\Delta^{T}_{0}H^{o}_{m} / T$ are in J·K^{-1}·mol^{-1}; $\Delta_{f}H^{o}_{m}$ is in kJ·mol^{-1}

T/K	298	400	500	600	800	1,000	1,500
	$C_{60}H_2$ C_{2v}						
$C^{\circ}_{p,m}$	507.0	719.7	887.7	1,017	1,190	1,293	1,414
$\Delta^{T}_{0}S^{\circ}_{m}$	576.6	756.3	935.7	1,110	1,428	1,706	2,257
$\Delta^{T}_{0}H^{\circ}_{m}/T$	198.7	305.1	405.5	497.2	650.8	769.7	967.6
$\Delta_f H^{\circ}_{m}$	2,471	2,469	2,467	2,465	2,459	2,453	2,432
	$C_{60}H_4$ C_s No. 1						
$C^{\circ}_{p,m}$	514.1	731.6	904.1	1,038	1,217	1,324	1,453
$\Delta^{T}_{0}S^{o}_{m}$	586.8	769.2	951.8	1,129	1,454	1,738	2,304
$\Delta^{T}_{0}H^{o}_{m}/T$	201.4	309.6	411.9	505.6	663.0	785.4	990.0
$\Delta_f H^{o}_{m}$	2,388	2,384	2,380	2,377	2,371	2,364	2,345
	$C_{60}H_6$ D_3						
$C^{\circ}_{p,m}$	522.6	745.1	954.7	1,060	1,245	1,357	1,493
$\Delta^{T}_{0}S^{o}_{m}$	588.8	774.4	927.9	1,141	1,474	1,765	2,345
$\Delta^{T}_{0}H^{o}_{m}/T$	204.8	314.9	419.3	515.1	676.4	802.2	1,015
$\Delta_f H^{o}_{m}$	2,335	2,329	2,324	2,320	2,313	2,306	2,290
	$C_{60}H_{18}$ C_{3v} (1,2-addition)						
$C^{\circ}_{p,m}$	562.3	813.3	1,018	1,180	1,405	1,547	1,729
$\Delta^{T}_{0}S^{o}_{m}$	609.5	810.7	1,015	1,216	1,588	1,918	2,585
$\Delta^{T}_{0}H^{o}_{m}/T$	220.38	340.16	456.0	563.7	748.1	894.6	1,147
$\Delta_f H^{o}_{m}$	1,753	1,734	1,720	1,710	1,695	1,688	1,686
	$C_{60}H_{36}$ C_1						
$C^{\circ}_{p,m}$	621.8	915.4	1,162	1,360	1,645	1,832	2,965
$\Delta^{T}_{0}S^{o}_{m}$	665.0	889.6	1,121	1,351	1,784	2,173	2,090
$\Delta^{T}_{0}H^{o}_{m}/T$	244.4	378.5	511.3	636.9	855.9	1,034	1,348
$\Delta_f H^{o}_{m}$	1,196	1,159	1,131	1,110	1,085	1,077	1,096
	$C_{60}H_{60}$ I_h [a]						
$C^{\circ}_{p,m}$	679.8	1,017	1,314	1,563	1,935	2,189	2,545
$\Delta^{T}_{0}S^{o}_{m}$	653.5	900.8	1,161	1,423	1,927	2,388	3,352
$\Delta^{T}_{0}H^{o}_{m}/T$	264.3	413.4	564.6	710.9	973.3	1,193	1,592
$\Delta_f H^{o}_{m}$	2,063	1,999	1,948	1,910	1,864	1,849	1,888

[a]Not synthesized

$$(60\text{–}2n)/60\ C_{60} + n/2\ C_{10}H_{16} = C_{60}H_{2n} + n/2 C_6H_{12} \qquad (4.4)$$

Since fullerene C_{60} is one of the participants of reaction (4.4) one may suppose that all the conjugation effects in a fullerene cage are properly considered. The values of thermodynamic properties for fullerene C_{60}, cyclohexane and adamantane have been

Table 4.4 Standard thermodynamic functions of $C_{60}H_{2n}$ isomers in the ideal-gas state at $T = 298.15$ K ($P^o = 10^5$ Pa)

Isomer	$C^o_{p,m}$ J·K⁻¹·mol⁻¹	$\Delta_0^T S^o_m$	$\Delta_0^T H^o_m/T$	$\Phi^{o\,a}_m$	$\Delta_f H^o_m$ kJ·mol⁻¹
$C_{60}H_2$					
C_{2v}	507.0	576.6	198.7	377.9	2,471
C_s	507.7	583.1	199.2	383.9	2,502
C_1	506.7	587.3	198.3	389.0	2,554
$C_{60}H_4$					
C_s (No. 1)	514.1	586.8	201.4	385.3	2,388
C_s (No. 2)	514.8	587.4	201.8	385.6	2,712
C_s (No. 3)	504.8	580.6	197.4	383.2	2,725
D_{2h}	515.7	576.7	202.3	374.4	2,715
$C_{60}H_6$					
D_3	522.6	588.8	204.8	378.2	2,335
C_3	522.4	583.1	204.8	387.4	2,336
C_{3v}	520.6	581.7	203.9	379.0	2,343
$C_{60}H_{18}$					
C_{3v} (1,2-add.)	562.3	609.5	220.4	389.1	1,753
C_{3v} (1,4-add.)	570.9	614.5	224.2	390.3	2,023
C_3	564.8	615.8	221.3	388.7	1,947
$C_{60}H_{36}$					
T	625.9	649.5	246.8	402.7	1,049
T_h	624.2	641.7	246.5	395.2	1,362
D_{3d}	619.2	640.7	242.7	398.0	1,352
$D_{3d'}$	622.6	644.5	244.7	399.8	1,180
S_6, No. 88	623.8	652.4	245.7	406.7	1,107
S_6, No. 91	618.5	646.9	242.7	404.2	1,191
C_3, No. 3	622.2	656.2	244.6	411.6	1,243
C_3, No. 4	625.9	659.6	246.7	412.9	1,385
C_3, No. 64	624.0	658.7	245.7	413.0	1,070
C_1	621.8	665.0	244.4	414.9	1,196
X	621.8	659.2	244.4	414.9	1,208
$C_{60}H_{60}$ [b]					
I_h	679.8	653.5	264.3	389.2	2,063
T_h	684.9	660.7	260.9	399.7	1,123

[a] $\Phi^o_m = \Delta_0^T S^o_m - \Delta_0^T H^o_m / T$. $\Delta_0^T S^o_m$ and Φ^o_m are calculated for racemic mixtures

[b] Not synthesized

reliably determined (Frenkel et al. 2009; Kabo et al. 2000; Pedley 1994; Diky and Kabo 2000). The calculated $\Delta_f H^o_m$ values for the $C_{60}H_{2n}$ hydrofullerenes are presented in Table 4.4. The estimated uncertainty of the calculated $\Delta_f H^o_m$ values is ± 100 kJ·mol⁻¹, which is 7–9.5% $\Delta_f H^o_m$ for the $C_{60}H_{36}$ isomers. The experimental value for the $C_{60}H_{36}$ sample $\Delta_f H^o_m$(gas, $C_{60}H_{36}$) = 1,208 ± 28 kJ·mol⁻¹ (Pimenova et al. 2002; Karpushenkava et al. 2007a) was determined with the uncertainty of 2.3%. The $C_{60}H_{36}$ isomers can be arranged in a row according to their thermodynamic stability at 298

K: $T > C_3$ (No. 64) > S_6 (No. 88) > $D_{3d'}$ > S_6 (No. 91) > C_1 > C_3 (No. 3) > D_{3d} > T_h > C_3 (No. 4). The experimental formation enthalpy for $C_{60}H_{36}$ is in the middle of the interval of the calculated values for the $C_{60}H_{36}$ isomers (Table 4.4). The calculated value for the X isomer $\Delta_f H^o_m(X) = 0.68\Delta_f H^o_m(C_1) + 0.28\Delta_f H^o_m(C_3$ No. 64) + $0.04\Delta_f H^o_m(T) = 1155 \pm 100$ kJ·mol^{-1} is in satisfactory agreement with the experimental value.

4.4.4 *Thermodynamics of Hydrogenation Reactions for Hydrocarbons and Fullerene C_{60} in the Gas State*

Thermodynamic parameters of the ideal-gas hydrogenation reactions for hydrocarbons and fullerene C_{60} are presented in Table 4.5. The enthalpies of C_{60} hydrogenation are within (–59 to –86) kJ per mol of H_2 ($\Delta_r H^o_m$/n). These values are closer to similar values for aromatic hydrocarbons (–68 ÷ –69 kJ·mol^{-1}) than to those for alkenes and cycloalkenes (–110 ÷ –119 kJ·mol^{-1}). This fact is probably due to conjugation of double bonds in C_{60} in spite of non-aromaticity of this compound. The symmetryless hydrogenation entropies $\Delta_r SH^{o\sigma}_m$/n are close for hydrocarbons and C_{60} (Table 4.5).

Table 4.5 Thermodynamic characteristics of ideal-gas hydrogenation reactions at T = 298.15 K

Reaction	$\sigma^{product}/\sigma^{reactant}$	$\Delta_r H^o_m$ (kJ·mol^{-1})	$\Delta_r H^o_m$/n	$\Delta_r S^o_m$ (J·K^{-1} mol^{-1})	$\Delta_r S^{o\sigma}_m$	$\Delta_r S^{o\sigma}_m$/n
Fullerene C_{60}						
$C_{60} + H_2 \rightarrow C_{60}H_2$ (C_{2v})	2/60	–59.0	–59.0	–98.2	–126.5	–126.5
$C_{60} + 2\ H_2 \rightarrow C_{60}H_4$ (C_s No. 1)	1/60	–142.0	–71.0	–218.5	–252.5	–126.3
$C_{60} + 3\ H_2 \rightarrow C_{60}H_6$ (D_3)	6/60	–195.0	–65.0	–347.2	–366.3	–122.1
$C_{60} + 9\ H_2 \rightarrow C_{60}H_{18}$ (C_{3v}, 1,2-addition)	3/60	–777.0	–86.3	–1,111	–1,135	–126.1
$C_{60} + 18\ H_2 \rightarrow C_{60}H_{36}$ (X-isomer)	1/60	–1,322	–73.4	–2,226	–2,260	–125.6
Alkenes and cycloalkenes						
cis-butene-2 + $H_2 \rightarrow$ *n*-butane	2/2	–119.4	–119.4	–122.1	–122.1	–122.1
Cyclohexene + $H_2 \rightarrow$ cyclohexane	2/6	–116.0	–116.0	–134.6	–143.7	–143.7
Cyclopentene + $H_2 \rightarrow$ cyclopentane	2/1	–110.4	–110.4	–139.1	–133.3	–133.3
1,3 cyclohexadiene + 2 $H_2 \rightarrow$ cyclohexane	2/6	–229.6	–114.8	–267.4	–276.5	–138.3
Aromatic hydrocarbons						
Benzene +3 $H_2 \rightarrow$ cyclohexane	6/6	–206.0	–68.7	–363.7	–363.7	–121.2
Naphthalene +5 $H_2 \rightarrow$ *trans*-decalin	4/4	–332.7	–66.5	–612.9	–613.0	–122.6
Anthracene +2 $H_2 \rightarrow$ 1,2,3,4-tetrahydroanthracene	1/4	–134.7	–67.4	–228.0	–233.8	–116.9
Anthracene +$H_2 \rightarrow$ 9,10-dihydroanthracene	1/4	–68.00	–68.00	–109.3	–115.1	–115.1

$\Delta_r S^{o\sigma}_m$ is symmetryless reaction entropy. $\Delta_r S^{o\sigma}_m = \Delta_r S_m{}^o + R\ln(\sigma_{product}/\sigma_{reactant})$. Symmetry of the H_2 molecule was not taken into account

4.5 Thermodynamic Properties of $C_{60}H_{2n}$ Crystals

4.5.1 *Heat Capacity and Derived Thermodynamic Properties of $C_{60}H_{2n}$ Crystals*

As it was previously noted, C_p data are available only for $C_{60}H_{36}$ in the temperature range of 5–340 K. The heat capacity of other crystalline $C_{60}H_{2n}$ have not been measured probably due to absence of the required amount of pure samples. However, thermodynamic properties of the $C_{60}H_{2n}$ hydrofullerenes can be calculated as accurately as it is required for calculation of equilibrium composition of their reactions.

Isochoric heat capacity of crystals can be presented as a sum of contributions from lattice vibrations (6 degrees of freedom) and intramolecular vibrations (3n–6 degrees of freedom, where n is a number of atoms in a molecule)

$$C_{\mathrm{v}} = \sum_{i=1}^{6} C_{\mathrm{lat},i} + \sum_{j=1}^{3n-6} C_{\mathrm{int},j} \tag{4.5}$$

Isobaric heat capacity is equal to

$$C_{\mathrm{p}} = \sum_{i=1}^{6} C_{\mathrm{lat},i} + \sum_{j=1}^{3n-6} C_{\mathrm{int},j} + (C_{\mathrm{p}} - C_{\mathrm{v}})_{\mathrm{cr}} \tag{4.6}$$

The strict equation $(C_{\mathrm{p}}-C_{\mathrm{v}})_{\mathrm{cr}} = \frac{TV_{\mathrm{m}}\alpha_{\mathrm{p}}^2}{\kappa_T}$ cannot be used for $C_{60}H_{2n}$ because the values of the thermal expansion coefficient α_{p} and the isothermal compressibility coefficient κ_T are unknown. So, it is reasonable to use the empirical equation (Takahara et al. 1995; Westrum et al. 1968)

$$(C_{\mathrm{p}} - C_{\mathrm{v}}) = AC_{\mathrm{lat}}^2 T \tag{4.7}$$

where A is an empirical parameter normally estimated from the experimental $C_{\mathrm{p}}(T)$ data.

The procedure for heat capacity calculation can be divided into two parts: (a) extrapolation to high-temperature region; (b) calculation down to $T = 0$ K. For $C_{60}H_{36}$ only extrapolation above $T = 340$ K is required.

Heat capacity of the $C_{60}H_{2n}$ crystals was extrapolated to high temperatures using Eqs. 4.6 and 4.7 and the following assumptions:

1. There is no phase transition in crystalline $C_{60}H_{2n}$.
2. The lattice contribution for all the crystals C_{lat} is equal to $6R$ at $T > 150$ K. This assumption is true for most crystals of hydrocarbons with large molar mass.
3. The intramolecular contributions to the heat capacity of the $C_{60}H_{2n}$ crystals are found from the scaled B3LYP/6-31G* vibrational frequencies of the molecules.

Earlier we extrapolated the experimental heat capacity for hydrofullerene $C_{60}H_{36}$ (Fig. 4.10) (Karpushenkava et al. 2007a) and 1,1′-biadamantane $C_{20}H_{30}$ (Karpushenkava et al. 2007b) to the high-temperature region. The following values of the A parameter in Eq. 4.7 were obtained: 7.1×10^{-5} J^{-1} mol for $C_{60}H_{36}$ and

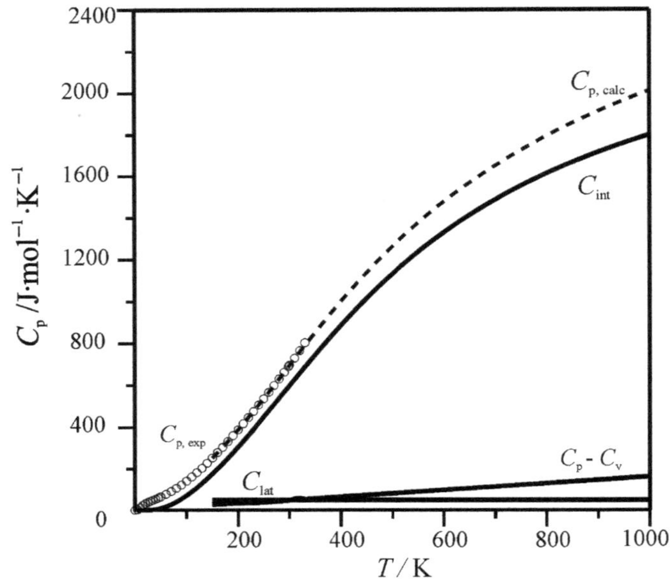

Fig. 4.10 Temperature dependence of heat capacity and different contributions to heat capacity for crystalline $C_{60}H_{36}$. O, experimental values; --------, extrapolation; ———, contributions to heat capacity

Table 4.6 Thermodynamic properties of $C_{60}H_{36}$ in the crystalline state. $C^{o}_{p,m}$, $\Delta_0^T S^{o}_{m}$ and $\Delta_0^T H^{o}_{m}/T$ are in J·K^{-1}·mol^{-1}; $\Delta_f H^{o}_{m}$ is in kJ·mol^{-1})

T/K	200	298.15	300	400	500	600	700	800	900	1,000
$\Delta C^{o}_{p,m}$	387.8	691.0	696.9	1,003	1,266	1,483	1,658	1,803	1,923	2,025
$\Delta_0^T S^{o}_{m}$	297.2	506.8	511.1	754.2	1,007	1,258	1,500	1,731	1,951	2,159
$\Delta_0^T H^{o}_{m}/T$	162.1	285.0	287.5	428.7	570.7	705.3	829.3	942.3	1,045	1,138
$\Delta_f H^{o}_{m}$	1,050	1,013	1,012	984.2	965.7	955.9	954.0	959.0	970.1	986.0

4.1×10^{-5} J^{-1} mol for $C_{20}H_{30}$. The resulting equation for extrapolation of heat capacity of crystalline $C_{60}H_{36}$ to 1,000 K was

$$C_p(C_{60}H_{36}, T) = 6R + C_{int} + 7.1 \cdot 10^{-5} (6R)^2 T \qquad (4.8)$$

The thermodynamic properties of crystalline $C_{60}H_{36}$ calculated from the $C_p = f(T)$ dependence are presented in Table 4.6. The experimental and calculated values at $T = 298.15$ K are compared in Table 4.7.

In this work we used $A = 5 \times 10^{-5}$ J^{-1}·mol for $C_{60}H_2$, $C_{60}H_4$, $C_{60}H_6$ and $C_{60}H_{18}$, and $A = 7.1 \times 10^{-5}$ J^{-1}·mol for $C_{60}H_{60}$. The estimated probable error in A is ± 25%. Since $(C_p - C_v)_{cr}$ contributes less than 10% to the total heat capacity, the error in A will introduce an error of <2% to the total heat capacity of crystalline hydrofullerenes.

The found values at $T = 150$–250 K were used for the C_p extrapolation to $T = 5$ K according to the Kelley–Parks–Huffman method (Kelley et al. 1929). For the $C_{60}H_{2n}$ hydrofullerenes it can be expressed as

Table 4.7 Experimental (Lebedev et al. 2000) and calculated thermodynamic properties of $C_{60}H_{36}$ in the crystalline state at T = 298.15 K, J·K^{-1}·mol^{-1}

Property	Experiment	Calculation
$\Delta C^{o}_{p,m}$	691.0	691.7
$\Delta_0^T S^{o}_{m}$	506.8	507.2
$\Delta_0^T H^{o}_{m}/T$	285.0	285.5

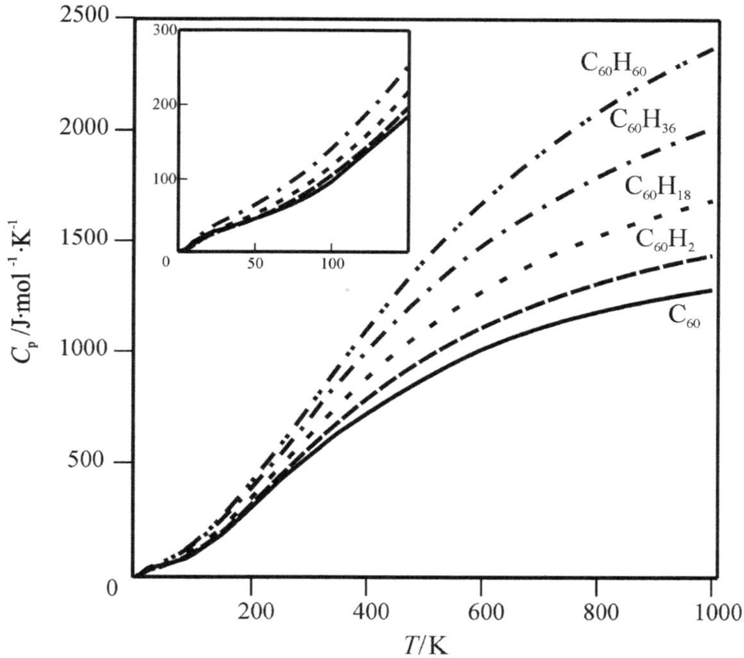

Fig. 4.11 Temperature dependence of heat capacity for fullerene C_{60} and hydrofullerenes C_nH_{2n}

$$C_p(C_{60}H_{2n}) = C_p(C_{60}H_{36})\,(a + bT) \tag{4.9}$$

The $C_{60}H_{36}$ hydrofullerene was used as a reference compound. The a and b coefficients of Eq. 4.9 were found for each hydrofullerene from the temperature dependence of the ratio $C_p(C_{60}H_{2n})/C_p(C_{60}H_{36})$ over the range of 150–250 K by the least-squares method.

The enthalpy and the entropy at 5 K were estimated from the extrapolated heat capacity at this temperature according to the Debye T^3 law. The thermodynamic functions at higher temperatures were calculated from the obtained $C_p(T)$ dependence.

The temperature dependences of heat capacity for fullerene C_{60} and hydrofullerenes $C_{60}H_{2n}$ are presented in Fig. 4.11 and Table 4.8.

Table 4.8 Thermodynamic properties of $C_{60}H_{2n}$ in the crystalline state. $C^{o}_{p,m}$, $\Delta_0^T S^{o}_{m}$ and $\Delta_0^T H^{o}_{m}/T$ are in J·K^{-1}·mol^{-1}; $\Delta_f H^{o}_{m}$ is in kJ·mol^{-1})

T/K	298.15	400	500	600	700	800	900	1,000
$C_{60}H_{2}$, C_{2v}								
$C^{o}_{p,m}$	560.7	786.1	966.6	1,109	1,219	1,307	1,376	1,434
$\Delta_0^T S^{o}_{m}$	395.8	593.0	788.6	977.9	1157	1,326	1,484	1,632
$\Delta_0^T H^{o}_{m}/T$	229.0	342.9	450.3	548.7	636.9	715.3	785.0	847.1
$\Delta_f H^{o}_{m}$	2,283	2,280	2,280	2,280	2,284	2,289	2,296	2,303
$C_{60}H_{4}$, C_{s} (No. 1)								
$C^{o}_{p,m}$	567.8	798.0	983.0	1,129	1,243	1,333	1,406	1,465
$\Delta_0^T S^{o}_{m}$	404.2	604.2	802.9	995.5	1,178	1,351	1,512	1,663
$\Delta_0^T H^{o}_{m}/T$	232.5	348.0	457.2	557.4	647.5	727.8	799.3	863.0
$\Delta_f H^{o}_{m}$	2197	2,199	2,203	2,208	2,214	2,222	2231	2,242
$C_{60}H_{6}$, D_{3}								
$C^{o}_{p,m}$	576.3	811.5	1,001	1,151	1,268	1,361	1,436	1,498
$\Delta_0^T S^{o}_{m}$	412.7	615.9	818.1	1,014	1,201	1,377	1,541	1,696
$\Delta_0^T H^{o}_{m}/T$	236.4	353.7	464.9	567.2	659.3	741.4	814.6	880.0
$\Delta_f H^{o}_{m}$	2,144	2,144	2,146	2,151	2,157	2,164	2,173	2,184
$C_{60}H_{18}$, C_{3v} (1,2-addition)								
$C^{o}_{p,m}$	616.0	879.7	1,097	1,271	1,409	1,521	1,612	1,688
$\Delta_0^T S^{o}_{m}$	437.9	656.6	877.0	1,093	1,300	1,495	1,680	1,854
$\Delta_0^T H^{o}_{m}/T$	251.3	378.5	501.2	615.5	719.4	812.9	896.8	972.2
$\Delta_f H^{o}_{m}$	1,561	1,549	1,542	1,539	1,541	1,546	1,554	1,564
$C_{60}H_{60}$ I_{h} [a]								
$C^{o}_{p,m}$	745.0	1,099	1,412	1,677	1,897	2,082	2,237	2,369
$\Delta_0^T S^{o}_{m}$	533.9	802.6	1,082	1,364	1,639	1,905	2,160	2,402
$\Delta_0^T H^{o}_{m}/T$	303.9	461.6	621.1	775.7	920.6	1,055	1,178	1,290
$\Delta_f H^{o}_{m}$	1869	1,812	1,771	1,743	1,727	1,723	1,728	1,740

[a]Not synthesized

4.5.2 *Formation and Sublimation Enthalpies for $C_{60}H_{2n}$ Crystals*

The formation enthalpies for gaseous fullerene hydrides were calculated above from quantum chemical calculations. Since the experimental sublimation enthalpies are available only for fullerene C_{60}

$\Delta_{sub}H^{o}_{m}$ (C_{60}, 298.15K) = (183.7 ± 5.1) kJ·mol-1 (Diky and Kabo 2000).

and fullerane $C_{60}H_{36}$

$\Delta_{sub}H^{o}_{m}$ ($C_{60}H_{36}$, 298.15K) = (195 ± 10) kJ·mol-1 (Dorozhko et al. 2001) the sublimation enthalpies for the other fulleranes were estimated from an additive procedure:

$$\Delta_{sub}H^{o}_{m}(C_{60}H_{2n}) = \Delta_{sub}H^{o}_{m}{}'(C_{60}) + n\frac{\Delta_{sub}H^{o}_{m}(C_{60}H_{36}) - \Delta_{sub}H^{o}_{m}{}'(C_{60})}{18} \qquad (4.10)$$

Table 4.9 Sublimation enthalpies for $C_{60}H_{2n}$ at $T = 298.15$ K

$C_{60}H_{2n}$	$C_{60}H_2$	$C_{60}H_4$	$C_{60}H_6$	$C_{60}H_{18}$	$C_{60}H_{60}$ [a]
$\Delta_{sub}H^{\circ}_{m}$/ kJ·mol^{-1}	191 ± 5	191 ± 5	191 ± 6	193 ± 10	198 ± 20

[a]Not synthesized

The contribution $\Delta_{sub}H^{\circ}_{m}{}'(C_{60})$ is the sublimation enthalpy for hypothetical rigid crystal. It was calculated as a sum of the sublimation enthalpy of C_{60} at $T = 298.15$ K and the enthalpy of the solid-phase transition (Karpushenkava and Kabo 2008):

$\Delta_{sub}H^{\circ}_{m}{}'(C_{60}) = (191.2 \pm 5.1)$ kJ·mol^{-1}.

The resulting sublimation enthalpies for $C_{60}H_{2n}$ are presented in Table 4.9.

The data of Table 4.3 and 4.9 were used for estimation of the formation enthalpies of $C_{60}H_{2n}$ in the crystalline state (Table 4.8).

4.6 Equilibria of Reactions of $C_{60}H_{2n}$ Hydrides

Hydrogenation of fullerite C_{60} is normally realized at $T = 450$–850 K and $P = 1$ MPa – 2 GPa in presence of catalyzers (Goldshleger and Moravskii 1997; Tarasov et al. 2001). Under these conditions kinetic limitations become less important and the significance of thermodynamic control increases. Additionally, it has been proved that hydrogenation of fullerite C_{60} is reversible and its degree depends on temperature and pressure of hydrogen. So, the analysis of equilibria $C_{60}+nH_2= C_{60}H_{2n}$ may be actual for evaluation of possibilities for the increase of the yield of fullerene hydrides and their working conditions in the hydrogen accumulating systems.

The equilibrium compositions were calculated either from the equilibrium constants or the method of chemical potential minimization (Kabo et al. 1986).

4.6.1 *Gas-Phase Hydrogenation* $C_{60} + nH_2 = C_{60}H_{2n}$

The initial composition for investigation of the equilibria included gaseous C_{60} and H_2. Formation of gaseous $C_{60}H_{2n}$ (n = 1, 2, 3, 9, 18, 30) was allowed. The obtained results are presented in Fig. 4.12.

At low temperatures, only $C_{60}H_{18}$ and $C_{60}H_{36}$ hydrides are stable and the ratio $C_{60}H_{18}/C_{60}H_{36}$ is completely determined by the initial C_{60}:H_2 ratio. If the total pressure is increased, this region widens to higher temperatures. The other hydrides are less thermodynamically stable. Their synthesis in the real conditions (Tarasov et al. 2001) is determined by the kinetic factors preventing disproportionation to $C_{60}H_{18}$ and $C_{60}H_{36}$. Two latter hydrofullerenes, especially $C_{60}H_{36}$, are formed at higher yield.

The effect of temperature and pressure on the equilibrium mole fraction of individual hydrides $C_{60}H_2$ (C_{2v}), $C_{60}H_{18}$ (C_{3v}, 1,2-add.) and $C_{60}H_{36}$ (X) under the conditions of gas-phase decomposition allowing formation of gaseous C_{60} and H_2 is

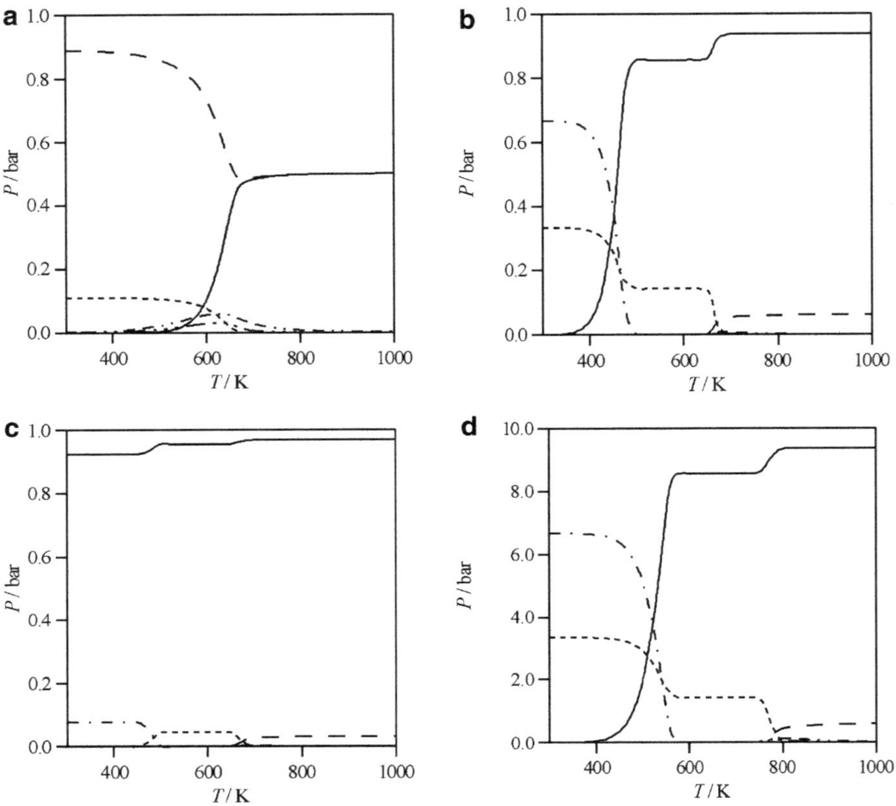

Fig. 4.12 Equilibrium pressures for the system $C_{60} + nH_2$ $C_{60}H_{2n}$ at **a**) $P = 1$ bar and C_{60}:H_2 = 1:1; **b**) $P = 1$ bar and C_{60}:H_2 = 1:15; **c**) $P = 1$ bar and C_{60}:H_2 = 1:30; **d**) $P = 1$ bar and C_{60}:H_2 = 1:20. ——, H_2, _ _, C_{60}, _.._, $C_{60}H_2$, _..._, $C_{60}H_4$, - - -, $C_{60}H_{18}$, _ . _, $C_{60}H_{36}$

demonstrated in Fig. 4.13. Stability of the studied compounds increases in the gas-phase decomposition reaction at $P = 1$ bar in a row $C_{60}H_{36} < C_{60}H_{18} < C_{60}H_2$. However, this order changes when the pressure is increased.

4.6.2 Hydrogenation $C_{60}(cr) + nH_2(g) = C_{60}H_{2n}(cr)$

Direct hydrogenation of C_{60} in solid phase is carried out at pressure 50–80 MPa and temperatures 573–623 K (Goldshleger and Moravskii 1997). The equilibrium pressures of hydrogen in the reactions (Fig. 4.14)

$$C_{60}(cr) + nH_2(g) \rightleftarrows C_{60}H_{2n}\,(cr) \qquad (4.11)$$

were calculated from the corresponding equilibrium constants.

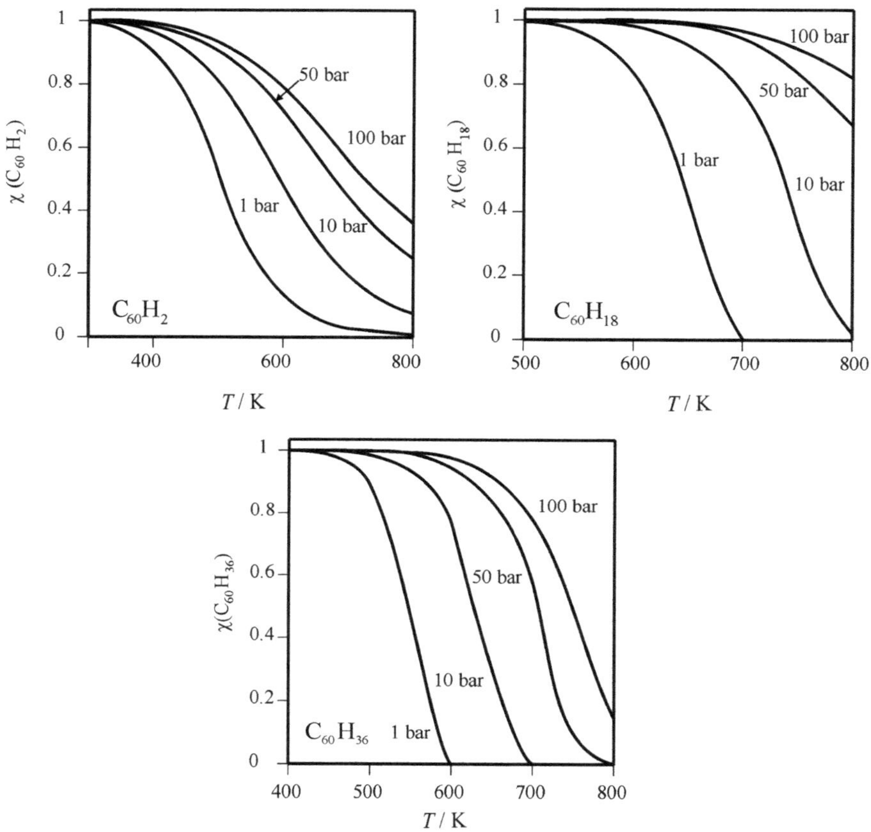

Fig. 4.13 Dependence of equilibrium mole fraction of $C_{60}H_{2n}$ (x) on total pressure and temperature in the reaction of gas-phase $C_{60}H_{2n}$ decomposition

As follows from Fig. 4.14, the hydride $C_{60}H_{18}$ (C_{3v}, 1,2-add.) is stable almost to 700 K, while $C_{60}H_{36}$ (X isomer) starts to decompose at $T > 500$ K. It was noted in (Dorozhko et al. 2001) that exposition to vacuum during P_{sat} measurements favored dehydrogenation of $C_{60}H_{36}$ into $C_{60}H_{18}$.

The $C_{60}H_{36}$ hydride contains 4.76 mass% of hydrogen. It can be used as hydrogen accumulator because in the temperature of 400–600 K the hydrogenation–dehydrogenation cycles are thermodynamically allowable. The thermodynamic parameters and equilibrium hydrogen pressures for the reaction

$$C_{60}(cr) + 18H_2(g) \rightleftarrows C_{60}H_{36}(cr) \quad (4.12)$$

are presented in Table 4.10.

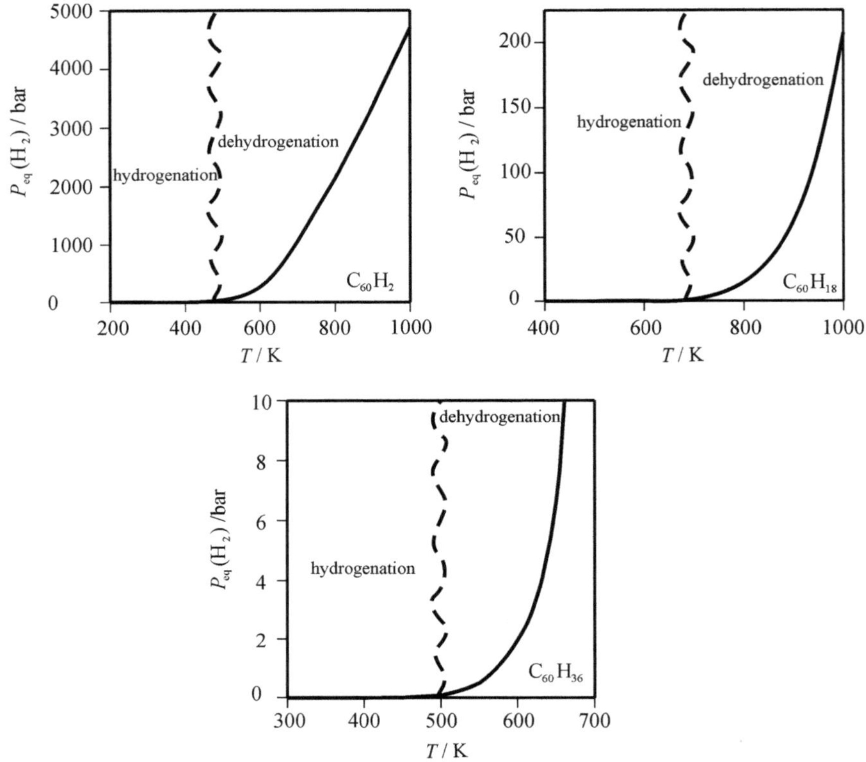

Fig. 4.14 Decomposition pressures of individual hydrofullerenes in the crystalline state

Table 4.10 Equilibrium constants and compositions for solid-state reaction (4.11) (P° = 1bar) for isomer X of $C_{60}H_{36}$

T/K	$\Delta_r H^{\circ}_m$ /kJ·mol^{-1}	$\Delta_r S^{\circ}_m$ /J·K^{-1}·mol^{-1}	$\Delta_r G^{\circ}_m$ /kJ·mol^{-1}	ln$K$	F_{eq} (H$_2$)[a]/bar	P_{eq} (H$_2$)[b]/bar
300	−1,334	−2,275	−651.3	261.1	$4.95 \cdot 10^{-7}$	$4.95 \cdot 10^{-7}$
400	−1,364	−2,362	−418.7	125.9	$9.04 \cdot 10^{-4}$	$9.04 \cdot 10^{-4}$
500	−1,383	−2,405	−180.1	43.32	$8.89 \cdot 10^{-2}$	$8.89 \cdot 10^{-2}$
600	−1,393	−2,424	61.20	−12.27	1.95	1.95
700	−1,395	−2,427	304.0	−52.23	18.0	17.9

[a]F_{eq}, equilibrium fugacity of hydrogen
[b]P_{eq}, equilibrium pressure of hydrogen

4.6.3 Hydrogenation with DHA

Hydrogenation of fullerene with DHA

$$C_{60} + nC_{14}H_{12} \rightleftarrows C_{60}H_{2n} + nC_{14}H_{10} \quad (4.13)$$

is performed at temperatures 570–620 K and approximately 7-fold excess of DHA. Under these conditions, $C_{60}H_{36}$ and $C_{60}H_{18}$ are the main products. The activity coefficients of the compounds in the DHA solution are unknown. However, one would expect that the activity coefficients are close for C_{60} and $C_{60}H_{2n}$ as well as for $C_{14}H_{12}$ and $C_{14}H_{10}$. If this assumption holds, K_x for the reaction in solution will be numerically equal to K_p for the gas-phase reaction at $P = 1$ bar. The gas-phase reaction equilibria can be a good model for the equilibria in solution.

The equilibrium compositions for the gas-phase reaction

$$C_{60}\,(g) + 18C_{14}H_{12}(g) \rightleftarrows C_{60}H_{36}\,(g) + 18C_{14}H_{10}(g) \quad (4.14)$$

are presented in Fig. 4.15 and Table 4.11.

Obviously, fourfold excess of DHA provides the almost complete hydrogenation of C_{60} to $C_{60}H_{36}$ in the temperature range of 300–700 K. The ratio $m(C_{14}H_{12})/m(C_{60}) \approx 18$ is 1.5–2 times less than that applied in the preparative syntheses.

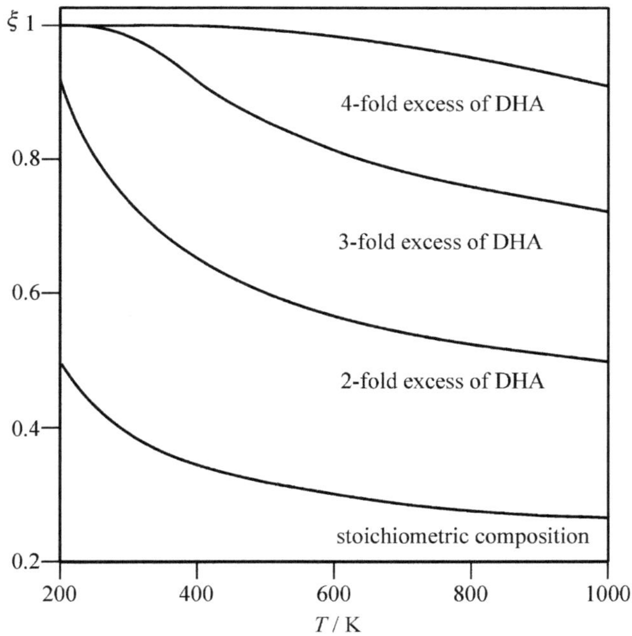

Fig. 4.15 Dependence of degree of C_{60} conversion ξ on temperature and excess of DHA in reaction (4.14)

Table 4.11 Equilibrium constants and compositions for ideal-gas reaction (4.14) at $P^{\circ} = 1$ bar. x is a mol fraction of a component, ξ is a degree of conversion

T/K	$\ln K_p$	ξ	$x(C_{60}H_{36})$	$x(C_{14}H_{10})$	$x(C_{60})$	$x(C_{14}H_{12})$
Stoichiometric composition						
300	−7.77	0.3991	0.0210	0.3781	0.0316	0.5692
400	−12.00	0.3469	0.0183	0.3286	0.0344	0.6188
500	−14.18	0.3217	0.0169	0.3048	0.0357	0.6426
600	−15.59	0.3057	0.0161	0.2896	0.0365	0.6578
700	−17.24	0.2876	0.0151	0.2724	0.0375	0.6749
Fourfold excess of DHA						
300	−7.77	→ 1	0.0137	0.2466	$8.22 \cdot 10^{-8}$	0.7397
400	−12.00	0.9996	0.0137	0.2465	$5.84 \cdot 10^{-6}$	0.7398
500	−14.18	0.9966	0.0137	0.2457	$4.65 \cdot 10^{-5}$	0.7406
600	−15.59	0.9886	0.0135	0.2438	$1.56 \cdot 10^{-4}$	0.7425
700	−17.24	0.9663	0.0132	0.2383	$4.62 \cdot 10^{-4}$	0.7480

Table 4.12 Equilibrium constants and compositions for ideal-gas reaction (4.15) at $P^{\circ} = 1$ bar. x is a mol fraction of a component, ξ is a degree of conversion

T/K	$\ln K_p$	ξ	$x(C_{60}H_{18})$	$x(C_{14}H_{10})$	$x(C_{60})$	$x(C_{14}H_{12})$
300	42.46	0.9859	0.0986	0.8873	0.0014	0.0127
400	28.94	0.9475	0.0948	0.8528	0.0052	0.0472
500	20.90	0.8899	0.0890	0.8009	0.0110	0.0991
600	15.54	0.8256	0.0826	0.7430	0.0174	0.1570
700	11.44	0.7584	0.0758	0.6826	0.0242	0.2174

For the reaction

$$C_{60}\,(g) + 9C_{14}H_{12}(g) \rightleftarrows C_{60}H_{18}\,(g) + 9C_{14}H_{10}(g) \tag{4.15}$$

the satisfactory yield of $C_{60}H_{18}$ is reached already at the stoichiometric ratio of the reagents (Table 4.12).

4.6.4 *Is It Possible to Synthesize $C_{60}H_{60}$?*

It was reported (Shigematsu et al. 1993) that the signal at m/z = 780 was observed in the mass-spectrum of the products of C_{60} hydrogenation at 280°C and $P \sim 130$ bar. The authors (Shigematsu et al. 1993) supposed that it was due to the $C_{60}H_{60}$ hydride. However, the possibility of preparative synthesis of $C_{60}H_{60}$ has not been proved yet. Quantum-chemical calculations show that the formation enthalpies of the exo- (I_h, $\Delta_f H^{\circ}_m = 2{,}063$ kJ·mol^{-1}) and 12-endo-, 48-exo-(T_h, $\Delta_f H^{\circ}_m = 1{,}123$ kJ·mol^{-1}) isomers differ significantly (Table 4.4). Both the isomers are thermodynamically unstable relative to other fullerene hydrides because of a very large entropy decrease in the reaction of C_{60} hydrogenation.

Table 4.13 Thermodynamic characteristics of reaction (4.16) with C_{60}(cr) and $C_{60}H_{60}$(cr)

T/K	$\Delta_r H^o_m$ /kJ·mol^{-1}	$\Delta_r S^o_m$/ J·K^{-1}·mol^{-1}	$\Delta_r G^o_m$ / kJ·mol^{-1}	ln K_p	P_{eq} (H_2)/bar
200	−406.0	−3,509	295.8	−177.9	371.2
250	−441.4	−3,663	474.4	−228.2	1,987
300	−482.2	−3,818	663.2	−265.9	6,976
400	−539.4	−3,984	1,054	−316.9	38,206
500	−581.4	−4,078	1,458	−350.7	117,866
1,000	−608.0	−4,087	3,479	−418.4	1.2 10^{11}

$P_{eq}(H_2)$ is the decomposition pressure for $C_{60}H_{60}$

Table 4.14 Equilibrium constants and compositions for gas-phase reaction (4.16)

		P_o = 100 bar			P_o = 10 kbar		
T/K	ln*K*	*x* (H_2)	*x* (C_{60})	*x* ($C_{60}H_{60}$)	*x* (H_2)	*x* (C_{60})	*x* ($C_{60}H_{60}$)
300	−271.2	0.9677	0.0323	→ 0	0.8758	0.02919	0.09499
500	−354.4	0.9677	0.0323	→ 0	0.9677	0.0323	0

P_o is the total pressure, *x* is the mole fraction of a component

Using the data from Table 4.3, we calculated the equilibrium compositions of reaction (4.16)

$$C_{60} + 30H_2 \rightleftarrows C_{60}H_{60}\,(I_h) \qquad (4.16)$$

for C_{60} and $C_{60}H_{60}$ in the gas and crystalline states (Tables 4.13 and 4.14). One may see that at T = 500 K the equilibrium pressure of hydrogen over solid $C_{60}H_{60}$ (I_h) exceeds 117 × 10^3 bar. Therefore, the conditions of its synthesis are unique. At the gas-phase hydrogenation of C_{60} at T = 300 K and P_o = 10 kbar, the equilibrium partial pressure of $C_{60}H_{60}$ is 950 bar and decreases with temperature. So, the possibility of synthesis of perhydrofullerane $C_{60}H_{60}$ seems to be highly improbable due to the thermodynamic limitations.

4.7 Conclusion

Comprehensive investigation of thermodynamic properties for fullerene hydrides is actual for optimization of the conditions of their synthesis, chemical functionalization, justification of their technical application. The structure, spectra, isomeric compositions for the samples of fullerene hydrides seriously affect their thermodynamic properties. So, the thermodynamic investigations will favor solution of many theoretical problems of chemistry and physics of fullerene hydrides. Taking into account the difficulties in synthesis and separation of individual fullerene hydrides of high purity, one should suppose that the theoretical evaluation of their physicochemical properties is the most effective tool for their investigation.

Based on the results of thermodynamic analysis of the processes with participation of the $C_{60}H_{2n}$ fullerene hydrides the following conclusions can be made.

1. The conditions of syntheses of fullerene hydrides can be corrected with respect to dependence of equilibrium compositions on temperature and pressure (concentration).
2. The thermodynamic analysis allowed us to determine the most favorable parameters of hydrogenation–dehydrogenation cycles in the system C_{60} + 18 H_2 = $C_{60}H_{36}$ under the conditions of thermodynamic control.
3. The failure of many attempts to synthesize $C_{60}H_{60}$ has been explained by high steric energy in the perhydrofullerane molecule (Goldshleger and Moravskii 1997). Here we demonstrated that the main difficulty for this synthesis is a very high entropy decrease in the reaction C_{60} + 30 H_2 = $C_{60}H_{60}$ aggravated by high symmetry of the considered I_h and T_h isomers of the hydride.

References

Balasubramanian K (1991) Chem Phys Lett 182:257

Ballenweg S, Gleiter R, Kratschmer W (1993) Tetrahedron Lett 34:3737

Bensasson RV, Hill TJ, Land EJ, Leach S, McGarvey DJ, Truscott TJ, Ebenhoch J, Gerst M, Ruchardt C (1997) Chem Phys 215:111

Bergosh RG, Meier MS, Cooke JAL, Spielmann HP, Weedon BR (1997) J Org Chem 62:7667

Billups WE, Luo W, Gonzalez A, Arguello D, Alemany L, Marriott T, Saunders M, Jimenez-Vazquez HA, Khong A (1997) Tetrahedron Lett 38:171

Bini R, Ebenhoch J, Fanti M, Fowler PW, Leach S, Orlandi G, Rüchardt Ch, Sandall JPB, Zerbertto F (1998) Chem Phys 232:75

Boltalina OV, Markov VYu, Taylor R, Waugh MP (1996) Chem Commun 2549

Clare BW, Kepert DL (1994) J Mol Struct (Theochem) 315:71

Clare BW, Kepert DL (1996) J Mol Struct Theochem 363:179

Clare BW, Kepert DL (1999) J Mol Struct (Theochem) 466:117

Darwish AD, Avent AG, Taylor R, Walton DRM (1996) J Chem Soc Perkin Trans 2:2051

Diky VV, Kabo GJ (2000) Russ Chem Rev 69:95

Dodziuk H, Nowinski K (1996) Chem Phys Lett 249:406

Dorozhko PA, Lobach AS, Popov AA, Senyavin VM, Korobov MV (2001) Chem Phys Lett 336:39

Frenkel ML, Kabo GJ, March KN, Roganov GN, Wilhoit RC (1994) Thermodynamics of organic compounds in the gas state, vol 1, 2. TRC, Texas

Frenkel et al. (2009) TRC thermodynamics tables – hydrocarbons and non- hydrocarbons. TRC NIST, Boulder

Gakh AA, Romanovich AYu, Bax A (2003) J Am Chem Soc 125:7902

Goldshleger NF, Moravskii AP (1997) Russ Chem Rev 66:323

Granovsky AA (2009) PC GAMESS/Firefly version 7.1.F, www http://classic.chem.msu.su/gran/gamess/index.html

Guo T, Scuseria GE (1992) Chem Phys Lett 191:527

Haufler RE, Conseicao J, Chibante LPF, Chai Y, Byrne WE, Flanagan S, Haley MM, O'Brien SCO, Pan C, Xiao Z, Billups WE, Ciufolini MA, Hauge RH, Margrave JL, Wilson LJ, Curl RF, Smalley RE (1990) J Phys Chem 94:8634

Henderson CC, Cahill PA (1993) Science 259:1885

Henderson CC, Rohlfing CM, Assink RA, Cahill PA (1994) Angew Chem Int Ed Engl 33:786

Kabo GJ, Blokhin AV, Charapennikau MB, Kabo AG, Sevruk VM (2000) Thermochim Acta 345:125
Kabo GJ, Roganov GN, Frenkel ML (1986) Thermodynamics and equilibria of isomers. Universiteskoye, Minsk [in Russian]
Karaulova EN, Bagrii EI (1999) Russ Chem Rev 68:889
Karpushenkava LS, Kabo GJ, Diky VV (2007a) Fullerene Nanot Carbon Nanostruct 15:227
Karpushenkava LS, Kabo GJ, Bazyleva AB, Blokhin AV, Kabo AG, Zaitsau DzH, Pimerzin AA, Sarkisova VS (2007b) Thermochim Acta 459:104
Karpushenkava LS, Kabo GYa (2008) Russian J Phys Chem A 82:1170
Kelley KK, Parks GS, Huffman HM (1929) J Phys Chem 33:1802
Lebedev BV, Bykova TA, Lobach AS (2000) J Therm Anal Calorim 62:257
Lobach AS, Perov AA, Rebrov AI, Roshupkina OS, Tkacheva VA, Stepanov AN (1997) Russ Chem Bull 46:641
Loss RD (2003) Pure Appl Chem 75:1107
Matsuzawa N, Dixon DA, Fukunaga T (1992) J Phys Chem 96:7594
Meletov KP, Assimopoulos S, Tsilika I, Bashkin IO, Kulakov VI, Khasanov SS, Kourouklis GA (2001) Chem Phys 263:379
Nossal J, Saini RK, Alemany LB, Meier M, Billups WE (2001) Eur J Org Chem 4167
Okotrub AV, Bulusheva LG, Asanov IP, Lobach AS, Shulga YuM (1999) J Phys Chem A 103:716
Pedley JB (1994) Thermodynamical data and structures of organic compound. vol 1. TRC Data Series, Boca Raton, FL
Pimenova SM, Melkhanova SV, Kolesov VP (2002) J Phys Chem B 106:2127
Popov AA, Senyavin VM, Granovsky AA, Lobach AC (2001). In: Fullerenes for the new millenium, vol 11. The Electrochemical Society. Pennington, NJ, pp 405
Rathna A, Chandrasekhar J (1993) Chem Phys Lett 206:217
Rüchardt C, Gerst M, Ebenhoch J, Beckhaus H-D, Campbell EEB, Tellgmann R, Schwarz H, Weiske T, Pitter S (1993) Angew Chem Int Ed Engl 32:584
Schmidt MW, Baldridge KK, Boatz JA, Elbert ST, Gordon MS, Jensen JH, Kosoki S, Matsunaga N, Nguyen KA, Su SJ, Windus TL, Dupluis M, Montgomery JA (1993) J Comput Chem 14:1347
Shigematsu K, Abe K, Mitani M, Tanaka K (1993) Chem Express 8:37
Sidorov LN, Yurovskaya MA, Borschevskii AYa, Trushkov IV, Ioffe IN (2005) Fullerenes. Ekzamen, Moscow [in Russian]
Takahara S, Yamamuro O, Matsuo T (1995) J Phys Chem 99:9589
Tarasov BP, Goldshleger NF, Moravsky AP (2001) Russ Chem Rev 70:131
Frenkel M et al (eds) (2009) TRC thermodynamics tables – hydrocarbons and non-hydrocarbons. TRC NIST, Boulder
Westrum EF Jr, Furukawa GT, McCullogh JP (1968) In: Cullough JP, Scott DW (eds) Adiabatic low-temperature calorimetry. Chapter in experimental thermodynamics, 2 vols, vol 1. Butterworth, London, pp 133

Chapter 5
Fullerenes by Direct Reaction with Hydrogen Gas at Elevated Conditions

Alexandr V. Talyzin[1]

Abstract Reaction of solid fullerenes with hydrogen gas occurs with or without catalysts at elevated conditions. Composition of hydrofullerene mixture obtained in this reaction depends strongly on temperature (350–450°C), hydrogen pressure (typically 10–120 bar) and duration of treatment. Saturation of hydrogenation occurs after tens of hours, depending on temperature of reaction. In case of extra strong hydrogenation prolonged reaction leads to formation of fulleranes with composition $C_{60}H_x$ approaching number of hydrogen atoms X = 60. These fulleranes are highly unstable and decompose first with formation of fragmented hydrofullerenes with progressively smaller number of carbon atoms C_{59}, C_{58}, C_{57} etc., followed by collapse of cage structure. Since the collapse occurs at the conditions of high temperature and high hydrogen pressure, all breaking C–C bonds are saturated immediately with hydrogen and new C–H bonds are formed. Therefore, large fragments of fullerane molecules are able to survive and large polycyclic aromatic hydrocarbons (PAH's) formed as a result of cage structure collapse.

5.1 Introduction

Following discovery of fullerenes (Kroto et al. 1985) and methods of their production (Krätschmer et al. 1990) in gram amounts, a lot of research activity was focused on functionalization of C_{60}. It is interesting to note that investigation of fullerene chemistry has started from hydrogenation: synthesis of $C_{60}H_{36}$ (obtained by Birch reduction) was first ever successful chemical reaction tested for C_{60} (Hauffler et al. 1990). Various names were used historically for hydrogenated fullerenes: fullerene hydrides, hydrofullerenes and finally new term fulleranes was proposed and seem to become common. Hydrogenated fullerenes have been a subject of very intense

[1]Department of Physics, Umeå University, SE-901 87, Umeå, Sweden
e-mail: alexandr.talyzin@physics.umu.se

F. Cataldo and S. Iglesias-Groth (eds.), *Fulleranes: The Hydrogenated Fullerenes*, Carbon Materials: Chemistry and Physics 2, DOI 10.1007/978-1-4020-9887-1_5,

research over past 19 years and numerous other methods were successfully employed for preparation of various fulleranes. The list of synthesis methods includes for example hydroboration (Henderson and Cahill1993), hydrozirconation (Ballenwey et al. 1993), solid phase hydrogenation (Jin et al. 1994), transfer hydrogenation (Rüchardt et al. 1993), electrochemical reduction (Cliffel and Bard 1994) reduction by Zn-HCl in toluene solution (Darwish et al. 1995; Cataldo 2003), chemical reduction with diimides (Avent et al. 1994), hydrogen radical-induced hydrogenation (Attalla et al. 1993) and photo-induced electron transfer (Fukuzumi et al. 1997).

The reaction occurs through the addition of hydrogen across fullerene C=C double bonds and the subsequent formation of C–H bonds. It means that only even numbers of hydrogen atoms are found in the neutral fullerane molecules ($C_{60}H_x$, where X is an even number). Hydrogenated fullerenes have been considered as promising materials for hydrogen storage. For example, highly reduced $C_{60}H_{36}$ contains about 5% by weight of hydrogen, which is close to the US DOE target of 6.5% for on-board transport storage systems. The volumetric hydrogen density of $C_{60}H_{36}$ is about 329 g L^{-1}; much higher than the DOE specification of 45 g L^{-1}. However, relatively high temperature of hydrogenation and dehydrogenation (which is partly irreversible (Talyzin et al. 2004a, Luzan et al. 2009)) is the main problem which hinders practical application of fulleranes for hydrogen storage.

Attempts to hydrogenate fullerenes using direct reaction with hydrogen gas are dated back to as early as 1988 when bulk methods of C_{60} production were not yet known (Weiss et al. 1988). However this attempt was not successful since the reaction requires using elevated conditions with relatively high temperature and hydrogen gas pressure. Hydrogenation of ionized C_{60} with molecular hydrogen was first observed by S. Petrie et al. 1992. Possibly first bulk synthesis of fulleranes by reaction of C_{60} powder with hydrogen gas at elevated conditions was performed by Attala et al. at 400°C and 69 bar (Attalla et al. 1993). In this study iodoethane was used to promote ionization of C_{60} since it was believed that only fullerene ions could react with molecular hydrogen. Various catalysts have been used in many subsequent studies to promote hydrogenation of solid fullerite with molecular hydrogen, often reactions were performed at extremely high pressures (see e.g. Jin C et al. 1994 where 650 bar H2 pressure was used). However, later studies demonstrated that reaction of powdered fullerite with molecular hydrogen occurs also in absence of any catalyst. Hydrogenation is efficient even at relatively low hydrogen pressures of 5–10 bar if high temperatures (400–450°C) are used for reaction (Jin et al. 1994; Darwish et al. 1996; Loufty and Wexler 2001; Tarasov et al. 2001; Shulga et al. 2003; Talyzin et al. 2004a,b; 2005a,b; 2006a). The main feature which complicates studies of this reaction is that it typically leads to formation of rather complex mixtures of fulleranes with broad distribution of hydrogen content. However, recent progress in studies of gas phase reaction of hydrogen with fullerenes powder proved that synthesis of pure materials $C_{60}H_{18}$ (Wågberg et al. 2005; Talyzin et al. 2006a) and $C_{70}H_{38}$ (Wågberg et al. 2008) is possible at certain conditions.

The aim of this paper is to provide a general outlook on current state of research related to hydrogenation of fullerenes by reaction with molecular hydrogen and to review recent studies conducted with participation of author in this field during past few years.

5.2 Hydrogenation of Fullerenes by Hydrogen Gas: Conditions and Characterization of Products

Reaction of solid fullerite with hydrogen gas at elevated conditions was studied in several groups. This research was mostly related to search of new materials for hydrogen storage application. Analysis of hydrogenation products performed by A.D.Darwish et al. 1996 revealed two main products: $C_{60}H_{18}$ and $C_{60}H_{36}$. This study presented also first attempt to separate complex fullerane mixtures by HPLC, successful isolation and characterization of $C_{60}H_{18}$. The sample studied by A.D.Darwish et al. 1996 was purchased from MER corporation and conditions of hydrogenation reaction (temperature, H_2 pressure and duration of treatment) were not cited. However, scientists of company which provided the sample (MER Corporation) have been involved in broad research program related to hydrogenation of fullerenes with target on hydrogen storage applications.

Research reports from USA DOE sponsored project performed in MER Corporation are available on-line (Loufty and Wexler 2001) and presented possibly most detail study of hydrogenation reaction to the moment of start of our own research in this field. These studies were later summarized in review paper by Loufty and Wexler 2002. It was found that fullerite powder can be successfully reacted with hydrogen gas even without any catalyst. Hydrogenation was tested in the temperature interval 350–450°C and pressures of hydrogen gas 60–80 bar. Dehydrogenation was reported to require higher temperatures, 500–600°C. Maximal hydrogen content achieved in this study was about 6.1 wt% determined from chemical analysis. This can be compared to maximal theoretical composition of $C_{60}H_{60}$ which would be able to store ~7.7 wt% of hydrogen. However, fulleranes $C_{60}H_x$ with number of hydrogen atoms $X > 36$ are progressively less stable and $C_{60}H_{60}$ was never observed. These results were confirmed in our studies (Talyzin et al. 2004a, b, 2005a, b, 2006a, b) which demonstrated that temperature is the major factor which controls hydrogenation degree of fullerenes, while using high pressures is somewhat less important. Hydrogenation of C_{60} was found to occur even at pressures below 10 bar but is more efficient and more complete at higher pressures, about 120 bar.

It is interesting that degree of C_{60} hydrogenation can be evaluated approximately directly from the change of powder color, without any chemical analysis. Pristine C_{60} powder is black, upon hydrogenation it becomes brown, then brick-red ($C_{60}H_{18}$), yellow (close to $C_{60}H_{36}$) and finally completely white. The change of color is not surprising considering change in hybridization of carbon atoms from mixed sp^2-sp^3 in pristine C_{60} to purely sp^3 for carbon atoms with attached hydrogen. Therefore, in highly hydrogenated C_{60} most of carbon atoms are in sp^3 state (ideal $C_{60}H_{60}$ would be completely sp^3), which results in some "diamond-like" properties. Like with graphite and diamond the difference in hybridization of carbon atoms in the pristine and hydrogenated fullerene molecules results in sharp change of physical properties: highly hydrogenated fullerene is transparent and white in color, it is insulator (pristine C_{60} is semiconductor) and chemically much more inert due to saturation of all suitable for functionalization double C=C bonds. For example, white-colored mixtures of highly hydrogenated fulleranes were found to be extremely resistant to photopolymerization by laser light (Talyzin et al. 2006b). Photopolymerization by laser light is extremely easy for pristine $C_{60,}$ which creates certain problems for recording Raman spectra requiring very low laser power levels and long exposition times. In contrast, Raman spectra of "white" fraction of hydrogenated C_{60} could be recorded with relatively large laser power without any traces of photopolymerization (Talyzin et al. 2006b).

Characterization of samples prepared by hydrogenation of C_{60} is not easy due to very complex composition of produced mixtures of $C_{60}H_x$ with various number of hydrogen atoms X. The final composition of sample, as it was found in our studies, depends very strongly not only on pressure and temperature conditions of reaction but also on duration of treatment. Hydrogenation experiments reported by other groups were performed without any in situ control over reaction process and reaction details could be deduced only from analysis of powder samples performed after completion of experiment. Ability to control the weight of sample in process of hydrogenation and thus to monitor reaction appeared to be a strong advantage of our experiments.

Hydrogenation experiments started by author of this review back in 2002 were performed using Rubotherm balance which allows to measure weight change of sample in process of hydrogenation. The system consisted of sealed reaction chamber and high precision balance which operates through the wall of reaction chamber using magnetic suspension mechanism. Fullerene powder was placed into alumina cell and inserted into reaction chamber. Magnetic suspension allowed the whole assemblage, which includes cell with powder, to levitate inside of reaction chamber and to take weight measurements of sample every 30 s with calibration "zero points" measured every 2–3 min. The data are plotted by software in real time scale which allows to monitor the weight change of sample directly in process of hydrogenation (Talyzin et al. 2004a, b). Prior to experiments it was expected that the weight of sample will be increasing in process of reaction proportionally to the number of hydrogen atoms attached to fullerene molecules and stabilize on certain level. This appeared to be true only for relatively low temperatures. The reaction readily occurs already at 350°C and at that temperature the weight of sample is saturated approximately after 2,500–3,000 min. However, if temperature is increased to 400°C the whole picture is drastically different, see Fig. 5.1. The figure shows a weight change of C_{60} samples in process of prolonged hydrogenation reaction. Gravimetric curve goes through the maximum at ~1,200 min and after that the weight starts to decrease. After ~4,300 min the curve suddenly starts to show some huge noise-like weight variations. The nature of these variations became clear after opening of reaction chamber and examination of hydrogenation products. The fullerene powder had transformed under prolonged hydrogenation into boiling liquid which solidified after cooling into black glass-like solid (Talyzin et al. 2004a). Figure 5.2 shows typical images of samples obtained on various stages of reaction (from samples prepared in different experiments).

Analysis of samples obtained by prolonged hydrogenation was performed first using powder XRD, IR spectroscopy and chemical analysis on C/H composition (Talyzin et al. 2004a) and later using various mass spectrometric methods (Talyzin et al. 2004b, 2005a, b, 2006a, b). Unfortunately, most of the samples obtained by

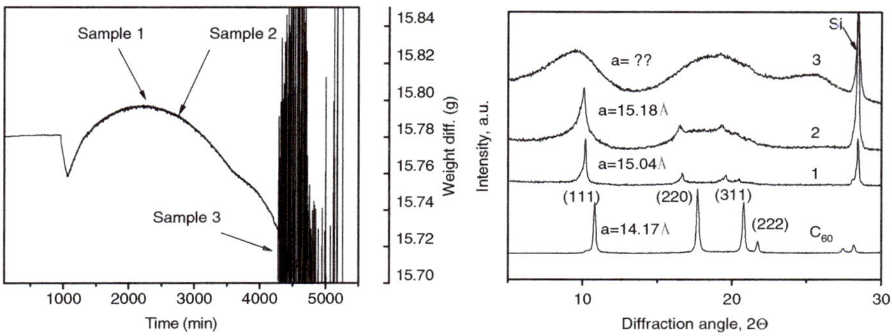

Fig. 5.1 Experimental data on hydrogenation of C_{60} at 120 bar hydrogen pressure and ~673 K. The initial kink of the weight curve at about 1,000 min occurs during an increase of hydrogen pressure and simultaneous increase of temperature which influences the curve due to buoyancy effects. XRD patterns recorded from the samples 1, 2, 3 and compared to the pattern from original C_{60}. Samples 1 and 2 were obtained in one experiment; sample 3 was obtained in a separate experiment (Talyzin et al. 2004a)

Fig. 5.2 Pristine fullerene powder (**1**), medium strong hydrogenation (**2**), hydrogenation to maximal level (up to at least $C_{60}H_{50}$, Talyzin et al. 2006b) (**3**), and glass like sample obtained after hydrogenation experiment shown in Fig. 5.1 (Talyzin et al. 2004a) which consists mostly from products of hydrogenation-induced collapse of fullerene molecules (**4**)

direct reaction with hydrogen gas cannot be studied by Raman spectroscopy with lasers operating in visible range of wavelengths due to very strong luminescence.

Experiment shown in Fig. 5.1 was explained as following:

- On initial stages of hydrogenation more and more hydrogen atoms are attached to the cage forming $C_{60}H_x$ with progressively higher numbers X, thus increasing van der Waals diameter of molecule. This results in expansion of fcc fullerene lattice from initial cell parameter a = 14.17 Å up to a = ~15.2 Å.
- Fulleranes with large number of hydrogen atoms (close to 60) are not stable at conditions of reaction and they collapse with formation of various fragments of hydrogenated cage, thus forming a complex mixture of polycyclic aromatic hydrocarbons, possibly some of them with non-flat geometry due to presence of carbon pentagons.

Mass spectrometric characterization of highly hydrogenated fullerenes showed also that some intermediate products are formed before the cage structure of fulleranes collapses: fragmented fulleranes. The name "fragmented fullerenes" is used when fullerene molecule loses some carbon atoms, thus forming e.g. C_{58}, C_{56}, C_{54} etc. These molecules can be produced from C_{60} in the form of atomic clusters in gas phase and were typically observed only by mass spectrometry (Murry et al. 1993; Hathiramani et al. 2000; Dunser et al. 1997). Fragmented fulleranes are stable and could be obtained in bulk amounts, see next section for more details.

On the first stage of hydrogenation reaction the C/H composition of fulleranes can be approximately determined by XRD from increase of unit cell parameter. The crystal structure of fulleranes powder remains to be fcc like in pristine C_{60} but the cell parameter increases proportionally to the number of attached hydrogen atoms due to increase of van der Waals diameter of molecules. Using chemical analysis on carbon/hydrogen ratio performed on samples hydrogenated at various conditions we constructed the dependence which could be used for approximate evaluation of hydrogen content from known cell parameter determined from powder XRD, see Fig. 5.3

Expansion of the unit cell becomes progressively stronger with an increase of hydrogen content. At around 5 wt% this curve reaches saturation point and prolonged hydrogenation results only in collapse of C_{60} molecules and formation of non-crystalline samples.

The plot shown in Fig. 5.4 shows an interesting trend: the more hydrogen atoms we attach to C_{60} molecule, the bigger the volume they occupy. This effect can be explained by repulsion of hydrogen atoms from each other and weaker C–H bonds in fulleranes with higher hydrogen content.

Hydrogenation degree of C_{60} can also be monitored using IR spectroscopy. Figure 5.5 shows spectra recorded from samples with different degree of hydrogenation. Obviously most of these samples are of quite complex composition. Analysis of these spectra is complicated by absence of published spectra for fulleranes other then $C_{60}H_{18}$ (Darwish et al. 1996) and $C_{60}H_{36}$ (Bini et al. 1998).

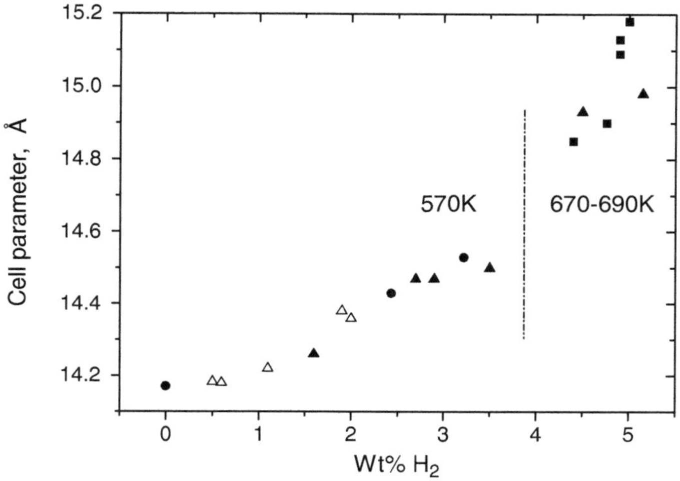

Fig. 5.3 Cell parameter as calculated from XRD (cubic structure) versus hydrogen percentage. Symbols notify different sets of samples. The dependence proved to be the same for hydrogenation experiments performed with with and without Ni and Pd catalysts (From Talyzin et al. 2004a)

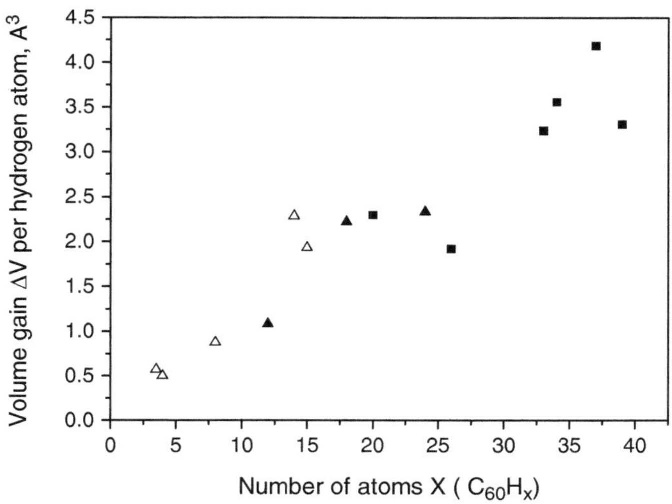

Fig. 5.4 The figure shows dependence of volume gain ΔV versus number of hydrogen atoms in fullerane molecule X. Volume gain per hydrogen atom in $C_{60}H_x$ is calculated per molecule of C_{60} as $\Delta V = (V_{C60Hx} - V_{C60})/X$, V_{C60} is taken as volume per molecule in FCC structure of pristine C_{60} (711 Å^3) (From Talyzin et al. 2004a)

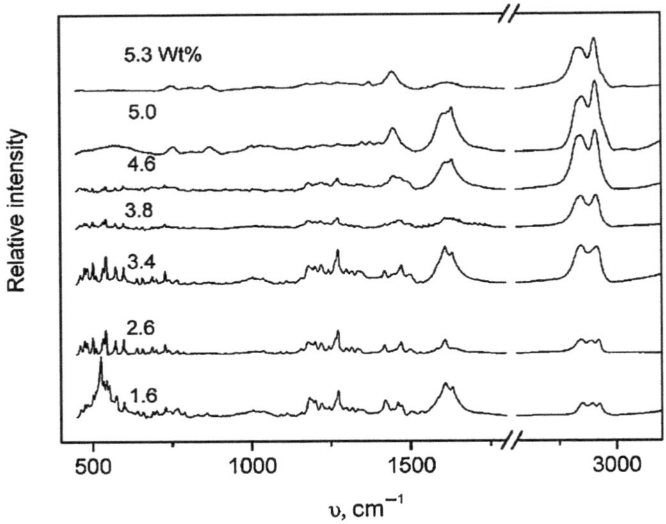

Fig. 5.5 IR spectra recorded from hydrogenated C_{60} samples (mixed fulleranes of various composition) with different average hydrogen content in weight percents (Talyzin et al. 2006a)

The IR spectra of hydrogenated fullerenes show cage vibrations (<1,700 cm^{-1}) and C–H vibrations (2,700–3,000 cm^{-1}). Figure 5.5 shows some clear modifications in both spectral regions upon increased degree of hydrogenation. The region of fullerene C–H vibrations is especially informative: the relative intensity of peaks due to these vibrations becomes increasingly higher proportionally to the degree of hydrogenation. Also, $C_{60}H_{18}$ and $C_{60}H_{36}$, main products of hydrogenation, exhibit clearly distinct peaks in this spectral region (2,847, 2,920 cm^{-1} for $C_{60}H_{18}$ [Darwish et al. 1996] and 2,829, 2,849, 2,911 for $C_{60}H_{36}$ [Bini et al. 1998]), which helps to estimate their presence in hydrogenated samples.

Highly hydrogenated samples show peaks at 2,831, 2,861, 2,913, 2,948 cm^{-1} (for the 5 wt% sample) and some broad weaker peaks centered at ~3,020 and 3,044 cm^{-1} interpreted as signature of PAHs. These large PAH's molecules are suggestively formed due to partial collapse of fulleranes cage structure.

IR spectroscopy becomes even more easy to use for identification of various fulleranes if deuterium is used instead of hydrogen. Experiments with deuterisation of C_{60} proved that it occurs in the same way as with hydrogen (Tarasov et al. 2001; Shulga et al. 2003). However, using deuterium allows to identify peaks due to C–D bonds of fulleranes clearly distinctly from any contamination hydrocarbons which are typically found in almost any fullerene sample due to presence of solvent traces etc.

Summarizing this section, the main parameters which control reaction of solid fullerite with hydrogen gas are temperature, H_2 pressure and duration of treatment. However, some additional parameters also need to be taken into account. For example, our own experience and also reports from other researchers show that freshly synthesized C_{60} can be hydrogenated much easier. Fullerite powder stored on air slowly oxidizes on the surface of grains and some passivation layer of somewhat uncertain composition is formed. It is also assumed that best purity of hydrogen provides better reaction results. For example, using hydrogen from metal hydride batteries was proposed since it is especially clean from typical gaseous contaminations (Tarasov et al. 1998). However, this method may lead to contamination with small metallic particles brought to reaction volume together with gas flow, this kind of contamination should be carefully avoided in studies of ferromagnetic properties reported for some fulleranes.

One more important question is an influence of catalysts on solid–gas hydrogenation reaction of fullerenes. Various catalysts were tested for hydrogenation of fullerenes, for example various metals like Ni, Pd, Pt (Talyzin et al. 2004), intermetallic compounds (Tarasov et al. 1997). Reaction of hydrogen with palladium fulleride (Lobach et al. 1996) should also possibly be related as a Pd catalyzed since it was found that palladium nanoparticles are formed in process of synthesis of this material (Talyzin et al. 2007). Addition of catalyst does not modify reaction significantly except for somewhat lowered temperatures and improved kinetics (Talyzin et al. 2004). It is interesting to note that addition of catalyst in some cases resulted in different crystal structure of powder samples. Body centered cubic structure (bcc) was observed for example in experiments where iodine was used as catalyst (Halle et al. 1993) or in Ni catalyzed reaction (Talyzin et al. 2004). The fcc and bcc structures seem to be observed for fullerane mixtures of very similar average hydrogen content and reasons for different packing are not clear. It could be suggested that it is related to different shapes of fullerane molecules. It is known that addition of hydrogen atoms to the fullerene cage results for some fulleranes in significant change of cage shape, it is especially strong for $C_{60}H_{18}$ where all hydrogen atoms are concentrated on one side of fullerene molecule resulting in a shape which is far from spherical (see e.g. Wågberg et al. 2005). In case if many fulleranes with various hydrogen nembers are present in the sample, the averaged shape of molecules could possibly be close to spherical which results in standard fcc packing. Catalytic hydrogenation may result in more selective formation of certain fulleranes and certain their isomers, which could also lead to more uniform shape of molecules and different packing. Certainly, influence of catalysts on hydrogenation of fullerenes requires farther investigation.

5.3 Mass Spectrometric Characterization of Complex Fullerane Mixtures

Mass spectrometric methods proved to be most efficient and precise in characterization of complex hydrofullerene mixtures. However, it required a lot of efforts to find a proper MS methods and parameters for unambiguous characterization of our samples which are very briefly described below. Main problems of all MS methods applied for characterization of fullerane samples in our studies are following:

- Only soluble part of samples was studied. Hydrofullerenes with few hydrogen atoms seem to be almost insoluble in common organic solvents (Talyzin et al. 2006). Therefore, relative abundance of ions found in mass spectra is not necessarily correlates with relative amounts of molecules in studied samples. Some insoluble molecules could be completely omitted from analysis.
- Mass spectrometric methods are often destructive and strong efforts are required to tune parameters of measurements for correct estimation of sample composition. Fullerane molecules at certain conditions of MS experiments may collapse, undergo fragmentation in gas phase, form dimers. These effects should be analyzed, excluded as much as possible and separated from the data informative about true sample composition.
- High resolution methods are required for highly hydrogenated C_{60} samples. Otherwise, molecules with the same nominal mass but different composition cannot be recognized from each other. However, for high resolution methods the mass of e.g. one carbon atom is clearly distinct from mass of 12 hydrogen atoms and elemental composition of ions can be determined unambiguously.

Initial studies were performed using MALDI MS (Talyzin et al. 2004). These experiments provided an insight into reaction mechanisms and some initial information about composition of samples which were assumed to consist only from $C_{60}H_x$ with various X and in case of "superhydrogenated" samples with addition of large hydrocarbons formed as a result of cage collapse. However, later studies proved that composition of samples is more complex: analysis of main peaks in MALDI MS showed remarkable difference in mass which corresponded often to CH or C_2H_2 mass difference (Talyzin et al. 2004, 2005). It was suggested that highly hydrogenated fullerenes undergo fragmentation of cage with loss of one, two and more carbon atoms off the cage. However, due to the limited resolving power and mass accuracy of MALDI TOF MS, the elemental composition $C_{60}H_x$ species could not be distinguished from $C_{59}H_{(x+12)}$ and $C_{58}H_{(x+24)}$ compounds of the same nominal mass.

Bulk synthesis of hydrogenated fragmented fullerenes with number of carbon atoms in fullerene cage less than C_{60} was proved in next studies (Talyzin et al. 2005a, b, 2006). The limitations of MALDI MS used in first experiments were overcame by use of high resolution field desorption Fourier transform ion cyclotron resonance mass spectrometry (FD FT-ICR MS) (Talyzin et al. 2005b) and high resolution atmospheric pressure photoionization Fourier transform ion cyclotron resonance mass spectrometry (APPI FTICR MS). The later proved to be the best and least destructive method for analysis of complex fullerane mixtures. Several samples with progressively stronger hydrogenation degree were analyzed by APPI FTICR MS which yielded rather complex spectra consisting in some cases from hundreds of peaks due to ions with different mass. Using advantages of high resolution, each peak in these spectra could be assigned to ions of specific C/H composition.

Examples of mass spectra with assignment of most abundant ions are shown in Fig. 5.6 (Talyzin et al. 2006). Analysis of this figure shows that hydrogenation reaction results on the first stage in formation of $C_{60}H_x$ with progressively higher numbers of hydrogen atoms X. Notably, some compositions seem to be more abundant

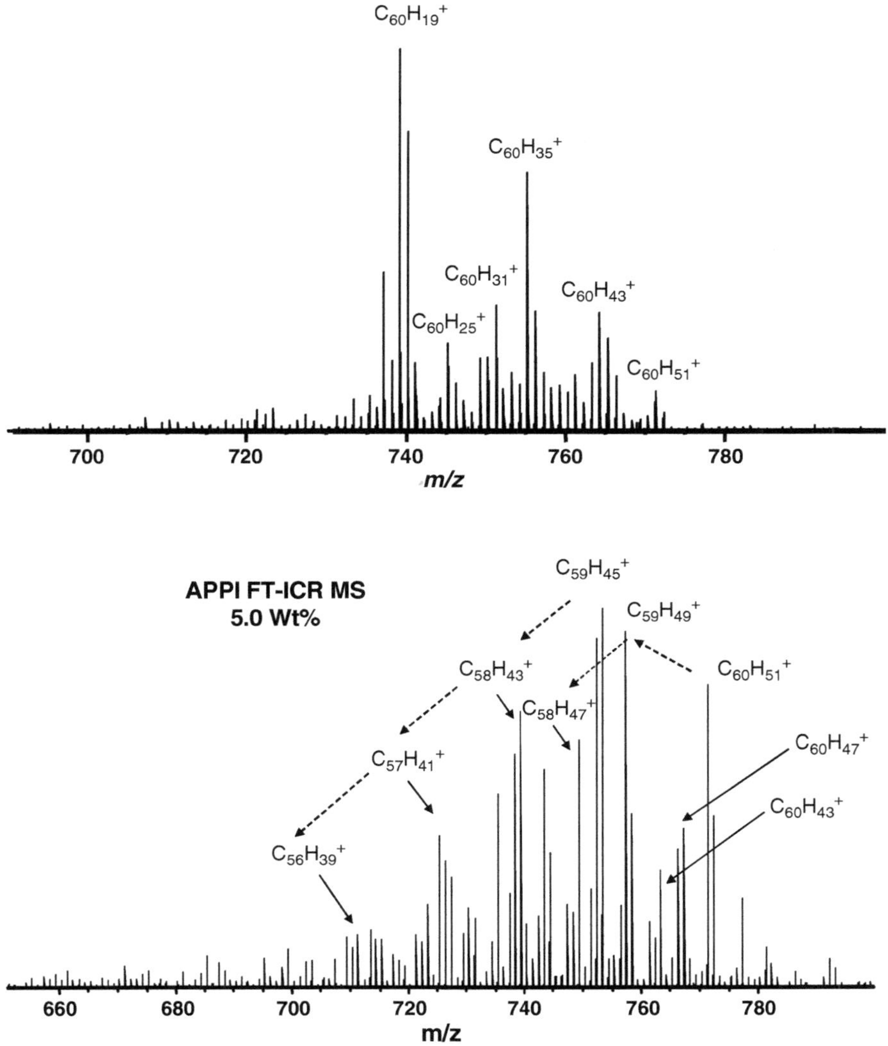

Fig. 5.6 APPI FT-ICR mass spectra from the samples with different hydrogenation degree. Top panel for 3.8 wt% sample and bottom panel 5.0 wt% of hydrogen according to chemical analysis of powders (Talyzin et al. 2006a)

(possibly due to higher stability) than other: most abundant peaks are likely to originate from protonated or deprotonated $C_{60}H_{18}$, $C_{60}H_{36}$, $C_{60}H_{42}$ and $C_{60}H_{50}$. When hydrogenation of fullerene molecules approaches saturation limit ($C_{60}H_{60}$) the fragmentation process starts. Highly hydrogenated hydrofullerenes are not stable at the conditions of reaction (673 K and 100 bar H_2) and eject CH_2 units thus forming $C_{59}H_{x-2}$, $C_{58}H_{x-4}$, … sets of molecules. Two sets of fragmentation chain reactions are marked in the Fig. 5.6. It should be noted that we believe that this fragmentation does not occur in process of MS experiments but corresponds to real composition of studied samples and is explained by fragmentation of hydrofullerenes in process of “superhydrogenation”. Mass spectrometric methods detected also very broad distribution of ions with lower masses, in the region 350–550 amu and for most abundant ions with carbon number/hydrogen number ratio of ~2/1. These ions originate from fragments of hydrofullerene molecules formed as a result of cage collapse and cracking of fullerenes on several pieces.

5.4 High Pressure (GPa) Methods of Fullerane Synthesis

As it was noted above, pressure seem to be less important parameter compared to temperature for outcome of hydrogenation reaction. This conclusion concerns “conventional” range of reaction pressures. Using classical high pressure vessels for reaction of fullerenes with hydrogen has limitations to pressures of some tens or possibly few hundred bars. However, pressures of several GPa (tens of kilobar) do provide some significant difference in hydrogenation of fullerenes. Specific of high pressure experiments is that they typically operate with relatively small volumes of samples and require specific equipment. All components of reaction need to be pre-loaded before compression, which is difficult for gaseous hydrogen. Therefore, thermal decomposition of AlH_3 was used as hydrogen source in first experiments with hydrogenation of fullerene at GPa pressures (Bashkin et al. 1998; Meletov et al. 2001, 2008; Antonov et al. 2002). Thermal decomposition of AlH_3 results in release of hydrogen which then reacts with fullerite powder. It is somewhat questionable how to consider the hydrogen in this reaction-as molecular or as atomic? Possibly this method of hydrogenation provides both atomic and molecular hydrogen for reaction. In our experiments we used $LiAlH_4$ as a source of hydrogen for reaction with C_{60} at 2 GPa pressure and temperatures 500–550°C (Talyzin et al. 2006). The aim of experiments was to achieve as high as possible hydrogenation degree of C_{60} without inducing cage collapse.

The hydrogenation reaction at 723 K was found to be slow, whereas a temperature increase to 800 K resulted in complete collapse of the fullerene cage structure with formation of some amorphous carbon material. Three experiments performed at 2 GPa, 773 K and heating periods of 30, 60, and 90 min resulted in efficient hydrogenation of C_{60}. The sample treated for 90 min consisted of some white powder (see Fig. 5.2, image 3), obtained in amount of ~10 mg in best experiments. Samples prepared with shorter heating period (30–60 min) were less homogeneous: white close to the interface between the reagents and yellow, red or black remote from the source of hydrogen source region of the capsule. Figure 5.7 shows some inhomogeneous samples confirming that hydrogen diffuses through the C_{60} powder as a gas, mostly along inter-grain surfaces, see Fig. 5.7. It is interesting that Raman spectra could easily be recorded from homogeneously white grains while high luminescence prevented recording for grains of other colors (less hydrogenated). Moreover, maximal power levels could be used for recording spectra without any

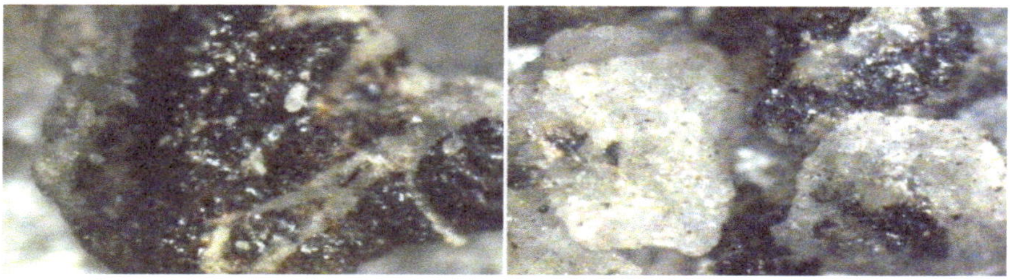

Fig. 5.7 Hydrogenated C_{60} obtained at 2 GPa and 773 K. White grains correspond to maximal degree of hydrogenation

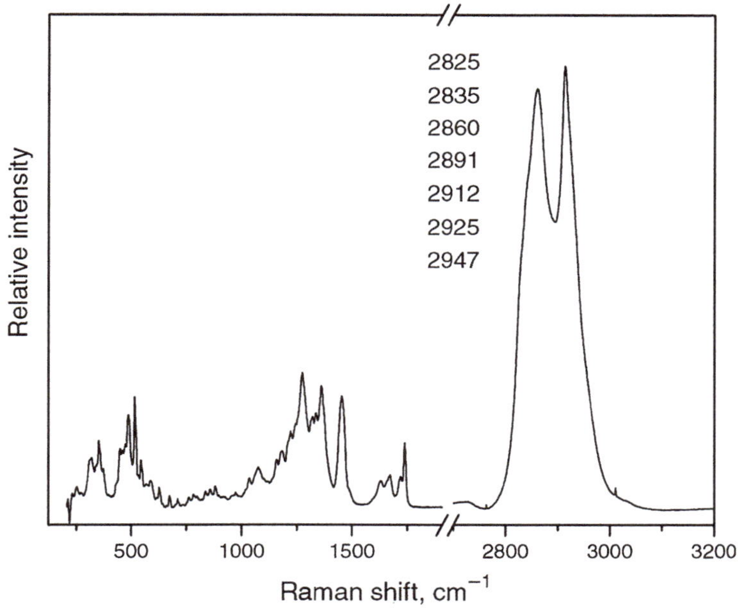

Fig. 5.8 Raman spectrum recorded from the white hydrogenation product of C_{60}. Positions of peaks due to C–H vibrations determined by fitting are shown by numbers (Talyzin et al. 2006b)

signs of laser induced phoropolymerization, see Fig. 5.8. Extremely high relative intensity of peaks from C–H vibrations is an evidence of high degree of C_{60} hydrogenation, in other hand it is obvious from analysis of this spectrum that C_{60} cages are still unbroken.

Powder XRD pattern recorded from this sample was indexed as a bcc structure with cell parameter, a = 11.99(2) Å, corresponding to a cell volume per fullerene molecule of about 862 Å^3. That a-value is somewhat larger than reported previously for pure $C_{60}H_{36}$, a = 11.785 Å (Halle et al. 1993) and a = 11.83 Å (Meletov et al. 2001) which indicated presence of some higher hydrogenated hydrofullerenes.

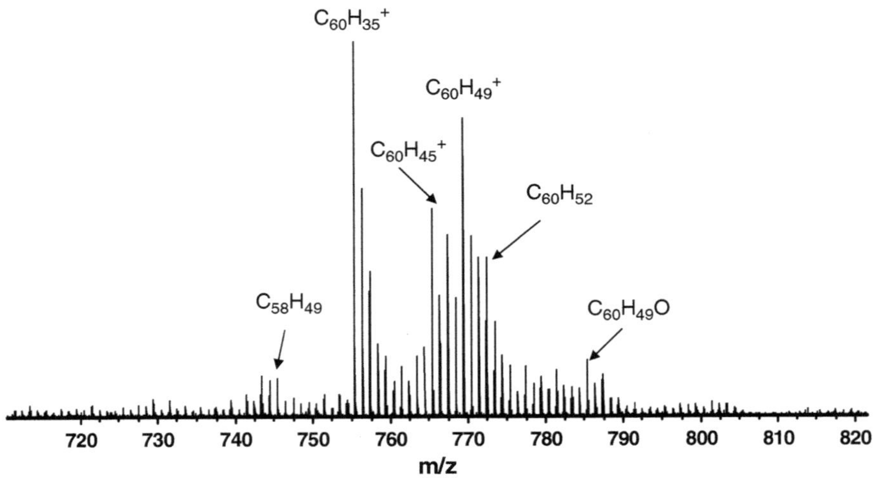

Fig. 5.9 APPI FT-ICR mass spectrum of "white" fraction of hydrogenated C_{60}. (Talyzin et al. 2006a)

Identification of fulleranes by Raman spectroscopy and XRD is difficult due to complexity of obtained mixture. Precise determination of composition for white colored fullerane samples was performed using APPI FT-ICR mass spectrometry.

Figure 5.9 shows high resolution mass spectrum of highly reduced C_{60} which allows to identify each peak with certain composition of $C_{60}H_x$ molecules. As seen from Fig. 5.1, a set of hydrofullerene ions with 45–52 hydrogen atoms appears in relatively high abundance. The spectrum allows to conclude that $C_{60}H_{36}$ and $C_{60}H_{50}$ are the major components of the mixture. The amount of hydrofullerenes $C_{60}H_x$ with x > 36 could possibly be estimated approximately on the level of tens of percents which makes isolation of individual species possible. Further work with samples of that kind requires application of advanced HPLC methods for separation of fullerane mixtures. In our observation the samples are rather stable in solid state but degrade rapidly if dissolved and exposed to the air.

The high pressure hydrogenation differs from other methods in one major point. Under the P–T conditions of the experiment, C_{60} polymerizes into a tetragonal phase in absence of hydrogen (see e.g. Talyzin et al. 2004). Moreover, polymerization occurs at temperatures below those used for hydrogenation experiments. Therefore, the high pressure (2 GPa) method involves hydrogenation of initially polymeric C_{60}. Detail analysis of Raman spectra recorded in various parts of inhomogeneously hydrogenated samples allowed to find an evidence for presence of hydrogenated polymeric C_{60} (Talyzin et al. 2006).

Summarizing all available information about high pressure C_{60} hydrogenation experiments it can be concluded that reaction occurs in several steps. The temperature increase at 2 GPa results first in polymerization of C_{60}, upon further temperature increase hydrogen start to evolve from $LiAlH_4$ and some hydrogenated C_{60} polymers are formed. On the next stage reaction with hydrogen breaks all polymeric C–C bonds which connect C_{60} molecules and complete depolymerisation with formation of new C–H bonds occurs.

Theoretically, the sequence of hydrogen atom attachment in the hydrogenation process is determined by the first reaction site. In pristine C_{60}, all carbon atoms have the same reaction probability. As soon as one hydrogen atom is attached, the possible positions for addition of next atoms become non-equivalent and thus the first H attached atom determines the following chain for subsequent hydrogenation, explaining why $C_{60}H_{36}$ is often found when different hydrogenation methods are used. In high pressure experiments situation is different: intermolecular C–C bonds in C_{60} polymers are weaker than cage C–C bonds. Therefore, it is likely that hydrogenation starts from breaking all polymeric bonds and proceeds after that on monomeric molecules. It can be concluded that using polymers of C_{60} (known in great variety of phases types of intermolecular connections) for hydrogenation reactions presents an opportunity to select initial points of hydrogenation of C_{60} cage: these will be the points where polymeric connections were initially situated. Therefore, hydrogenation path can be designed in different ways by selecting specific polymeric phase of C_{60} and thus defining initial sites of hydrogen attachments to fullerene molecule.

5.5 Future Outlook

Despite long history of research related to hydrogenation of fullerenes by gas phase reaction, this field still presents a lot of challenges and some broad directions for further research. It is convenient to divide the field on three separate subfields in relation to hydrogenation degree of fullerenes.

5.5.1 *Fulleranes with Composition $C_{60}H_x$ and Number of Hydrogen atoms X Below 60*

These are "classical" fulleranes. Mass spectrometric analysis proved that number of hydrogen atoms in fulleranes varies in whole possible region 1–60. However, so far only one fullerane with composition $C_{60}H_{18}$ was isolated (from solid–gas reaction products) and structure was solved by NMR methods (Wågberg et al. 2005). This became possible because of remarkable stability of this material which resulted in direct synthesis without additional purification. Further progress in isolation and characterization of various fulleranes requires using HPLC separation of complex hydrofullerene mixtures. It was demonstrated that fulleranes can be efficiently separated by chromatography (Darwish et al. 1996; Busci et al. 1998). The main problem in HPLC separation is typically large number of various fullerane species (with different member of hydrogen atoms), low solubility in common organic solvents and low stability of highly hydrogenated fullerenes in solution. Fulleranes are rather stable in solid state but after dissolving they slowly degrade if exposed to air. The degradation is more rapid in our observation for fulleranes with number of hydrogen atoms over 36. It should be noted that similar directions for future research could be drawn for $C_{70}H_x$ materials. Only one of them, $C_{70}H_{38}$, was so far prepared in relatively pure form, which allowed solving structure by NMR methods (Wågberg et al. 2008) while many other compositions were observed by mass spectrometric methods in our unpublished experiments (Talyzin AV and Tsybin YO 2007b). It is also possible that variation of hydrogenation parameters or selection of specific catalysts could result in direct synthesis of pure fulleranes of other compositions. It is interesting also to explore dimeric and polymeric hydrogenated fullerenes (Talyzin et al. 2004, 2006; Vasil'ev et al. 2005)

5.5.2 Fragmented Fullerenes, e.g. $C_{59}H_x$, $C_{58}H_x$ etc.

These materials are not yet isolated, but their existence in bulk form seems to be proved by mass spectrometry. Isolation of these molecules in pure form is rather challenging due to lower stability. It is an open question if the structure of fragmented fullerenes is "true" closed cage with all carbon atoms linked by the same number of carbon–carbon bonds or it has a hole with some carbon terminated by additional hydrogen atoms and thus connected to only two neighboring carbons. How many carbons can be taken away from $C_{60}H_x$ without complete collapse of cage molecule is another interesting question. Fullerenes with number of carbon atoms below 60 were not so far isolated in bulk amounts due to their relative instability. Super-hydrogenation with stepwise removal of carbon atoms from the $C_{60}H_x$ cage could be proposed as possible direction for synthesis of smaller hydrogenated fullerenes. If this step is successful, hydrogen could possibly be removed from the cage thus resulting in synthesis of pure small fullerenes.

5.5.3 Fullerene Fragments

Collapse of fulleranes under conditions reactive for carbon and hydrogen is found to result in formation of large fullerene fragments (Talyzin et al. 2005, 2006). In principle, it is interesting to separate these complex mixtures of hydrocarbons which is likely to result in isolation of some unusual PAH's molecules and not-planar fragments of fullerene cage with both hexagon and pentagon carbon rings. This work has not yet started and proposes an interesting field for new research.

5.5.4 Hydrogenation of Peapods

Very recently author of this review successfully hydrogenated fullerenes inside of single walled carbon nanotubes (so called peapods). Evidence of hydrogenation was provided by NMR studies and Raman spectroscopy (Abou-Hamad et al. 2009).

So far, direct evidence of C_{60} hydrogenation inside of nanotubes from HRTEM imaging is absent. It is required as a decisive demonstration of possibility for hydrogen to penetrate inside of peapods but could possibly be challenging experimentally. Chemical reaction within nanospace of carbon nanotubes is possibly only first example of interesting nanoscale chemistry and fullerene hydrogenation in other exotic environments will possibly be successfully demonstrated in future. It is quite likely that hydrogenation in confined space results in formation of fulleranes with different molecular structures.

References

Abou-Hamad E, Kim Y, Talyzin AV, Goze-Bac C, Luzzi DE, Rubio A, Wågberg T (2009) Hydrogenation of C60 in Peapods: physical chemistry in nano vessels. J Phys Chem C 113:8583–8587

Antonov VE, Bashkin IO, Khasanov SS, Moravsky AP, Morozov YG, Shulga YM, Ossipyan YA, Ponyatovsky EG (2002) Magnetic ordering in hydrofullerite $C_{60}H_{24}$. J Alloy Compd 330–332:365–368

Attalla MI, Vassallo AM, Tattam BN, Hanna JV (1993) Preparation of hydrofullerenes by radical induced hydrogenation. J Phys Chem 97:6329–6321

Avent AG, Darwish AD, Heimbach DK, Kroto HW, Meidine MF, Parsons JP, Remars C, Roers R, Ohashi O, Taylor R, Walton DRM (1994) Formation of hydrides of fullerene-c-60 and fullerene-c-70. J Chem Soc Perkin Trans 2:15–22

Ballenwey S, Gleiter R, Krätshmer W (1993) Hydrogenation of buckminsterfullerene c-60 via hydrozirconation - a new way to organofullerenes. Tetrahedron Lett 34:3737–3740

Bashkin IO, Kolesnikov AI, Antonov VE, Ponyatovsky EG, Kobzev AP, Muzychka AY, Moravsky AP, Wagner FE, Grosse G (1998) Vibrational spectra of C-60 hydrofullerite prepared under high hydrogen pressure. Molecular crystals and liquid crystals science and technology section c. Mol Mater 10:265–270

Bini R, Ebenhoch J, Fanti M, Fowler PW, Leach S, Orlandi G, Ruchardt C, Sandall JPB, Zerbetto F (1998) The vibrational spectroscopy of $C_{60}H_{36}$: An experimental and theoretical study. Chem Phys 232:75–94

Busci I, Szabo P, Aniszfeld R, Surya Prakash GK, Olah GA (1998) HPLC separation of hydrogenated derivatives of buckminsterfullerene. Chromatographia 48:59–64

Cataldo F (2003) Fullerane, the hydrogenated C-60 fullerene: Properties and astrochemical considerations. Fullerenes Nanot Carbon Nanostruct 11:295–316

Cliffel DE, Bard AJ (1994) Electrochemical studies of the protonation of c-60- and c-60(2-). J Phys Chem 98:8140–8143

Darwish AD, Abdul-Sada AK, Langley J, Kroto HW, Taylor R, Walton DRM (1995) Polyhydrogenation of [60]-fullerenes and [70]-fullerenes. J Chem Soc Perkin Trans 2:2359–2365

Darwish AD, Avent AG, Taylor R, Walton DRM (1996) Structural characterization of $C_{60}H_{18}$; C_{3v} symmetry crown. J Chem Soc Perkin Trans 2:2051–2054

Dunser B, Echt O, Scheier P, Mark TD (1997) Sequential reaction channels of metastable C-60(4+). Phys Rev Lett 79:3861–3864

Fukuzumi S, Suenobu T, Kawamura S, Shida A, Mikami K (1997) Selective two-electron reduction of C60 by 10-methyl-9,10-dihydroacridine via photoinduced electron transfer. Chem Commun 291–292

Halle LE, McKenzie DR, Attalla MI, Vassallo AM, Davis RL, Dunlop JB, Cockayne DJH (1993) The structure of $C_{60}H_{36}$. J Phys Chem 97:5741–5744

Hathiramani D, Aichele K, Arnold W, Huber K, Salzborn E, Scheier P (2000) Electron-impact induced fragmentation of fullerene ions. Phys Rev Lett 85:3604–3607

Hauffler RE, Conceicao J, Chibante PF, Chai Y, Byrne NE, Flanagan F, Haley MM, OBrien SC, Pan C, Xiao Z, Billups WE, Ciufolini MA, Hauge RH, Margrave JL, Wilson LJ, Curl RF, Smalley REJ (1990) Efficient production of C60 (buckminsterfullerene), $C_{60}H_{36}$, and the solvated buckide ion. J Phys Chem 94:8634–8636

Henderson CC, Cahill P (1993) $C_{60}H_2$-synthesis of the simplest C_{60} hydrocarbon derivative. Science 259:1885–1887

Jin C, Hettich R, Compton R, Joyce D, Blencoe J, Burch T (1994) Direct solid-phase hydrogenation of fullerenes. J Phys Chem 98:4215–4217

Krätschmer W, Lamb LD, Fostiropoulos K, Huffmann DR (1990) Solid C60: a new form of carbon. Nature 347:354–358

Kroto HW, Heath JR, OBrien SC, Curl RF, Smalley RE (1985) C60: Buckminsterfullerene. Nature 318:162–163

Lobach S, Tarasov BP, Shul'ga YuM, Perov AA, Stepanov AN (1996) Reaction of D2 with palladium fulleride C60Pd4.9. Russ Chem Bull 45:464–465

Loufty RO, Wexler EM (2001) Procedings of 2001 DOE hydrogen Prog Rev. NREL/CP-570-30535

Loutfy RO, Wexler EM (2002) Gas phase hydrogenation of fullerenes. Perspectives of Fullerene Nanotechnology. Springer, Berlin, pp 281–287

Luzan SM, Cataldo F, Tsybin YO, Talyzin AV (2009) Thermal decomposition of $C_{60}H_{18}$. J Phys Chem C 113:3133–13138

Meletov KP, Assimopoulos S, Tsilika I, Bashkin IO, Kulakov VI, Khasanov SS, Kourouklis GA (2001) Isotopic and isomeric effects in high-pressure hydrogenated fullerenes studied by Raman spectroscopy. Chem Phys 263:379–388

Meletov KP, Bashkin IO, Shestakov VV, Tartakovskii II, Maksimov AA, Arvanitidis J, Christofilos D, Kourouklis GA (2008) Comparative Raman study of the $C_{60}H_{36}$ and $C_{60}H_{60}$ fullerene hydrides. Fullerenes Nanot Carbon Nanostruct 16:593–596

Murry RL, Strout DL, Odom GK, Scuseria GE (1993) Role of sp(3) carbon and 7-membered rings in fullerene annealing and fragmentation. Nature 366:665–667

Petrie S, Javahery G, Wang G, Bohme DK (1992) Derivatization of the fullerene dications c-60(2+) and c-70(2+) by ion molecule reactions in the gas-phase. J Am Chem Soc 114:6268–6269

Rüchardt C, Gerst M, Ebenhoch J, Beckkaus HD, Campbell EE, Tellgmann R, Schwarz H, Weiske T, Pitter S (1993) Bimolecular radical formation through h-transfer.3. transfer hydrogenation and deuteration of buckminsterfullerene c-60 by 9, 10-dihydroanthracene and 9, 9', 10, 10'[d(4)] dihydroanthracene. Angew Chem Int Ed 32:584–586

Shulga YM, Tarasov BP, Fokin VN, Martynenko VM, Schur DV, Volkov GA, Rubtsov VI, Krasochka GA, Chapusheva NV, Shevchenko VV (2003) Deuterofullerides. Carbon 41:1365–1368

Talyzin AV, Shulga YM, Jacob A (2004a) Comparative study of hydrofullerides $C_{60}H_x$ synthesized by direct and catalytic hydrogenation. Appl Phys A Mater Sci Process 78:1005–1010

Talyzin AV, Sundqvist B, Shulga YM, Peera AA, Imus P, Billups WE (2004b) Gentle fragmentation of C_{60} by strong hydrogenation: a route to synthesizing new materials. Chem Phys Lett 400:112–116

Talyzin AV, Tsybin YO, Peera AA, Schaub TM, Marshall AG, Sundqvist B, Mauron P, Zuttel A, Billups WE (2005a) Synthesis of C59H*x* and C58H*x* fullerenes stabilized by hydrogen. J Phys Chem B 109:5403–5405

Talyzin AV, Tsybin YO, Schaub TM, Mauron P, Shulga YM, Zuttel A, Sundqvist B, Marshall AG (2005b) Composition of hydrofullerene mixtures produced by C_{60} reaction with hydrogen gas revealed by high-resolution mass spectrometry. J Phys Chem B 109:12742–12747

Talyzin AV, Tsybin YO, Purcell JM, Schaub TM, Shulga YM, Noreus D, Sato T, Dzwilewski A, Sundqvist B, Marshall AG (2006a) Reaction of hydrogen gas with C_{60} at elevated pressure and temperature: hydrogenation and cage fragmentation. J Phys Chem A 110:8528–8534

Talyzin AV, Dzwilewski A, Sundqvist B, Tsybin YO, Purcell JM, Marshall AG, Shulga YM, McCammon C, Dubrovinsky L (2006b) Hydrogenation of C_{60} at 2 GPa pressure and high temperature. Chem Phys 325:445–451

Talyzin AV, Dzwilewski A, Pudelko M (2007a) Formation of palladium fullerides and their thermal decomposition into palladium nanoparticles. Carbon 45:2564–2569

Talyzin AV, and Tsybin YO (2007b) Unpublished.

Tarasov BP, Fokin VN, Moravskii AP, Shulga YM (1997) Hydrogenation of fullerites in the presence of intermetallic compounds or metals. Russ Chem Bull 46:649–652

Tarasov BP, Fokin VN, Fokina EE, Rumynskaya ZA, Volkova LS, Moravskii AP, Shul'ga YuM (1998) Synthesis of fullerene hydrides by reaction of fullerite with hydrogen released from hydrides of intermetallic compounds. Russ J Gen Chem 68:1515–1519

Tarasov BP, Shul'ga YM, Fokin VN, Vasilets VN, Shul'ga NY, Schur DV, Yartys VA (2001) Deuterofulleride $C_{60}D_{24}$ studied by XRD, IR and XPS. J Alloys Compd 314:296–300

Vasil'ev YV, Kotsiris SG, Bashkin IO, Antonov VE, Moravsky AP, Drewello T (2005) Bulk production of a strong covalently linked (C60Hx)(2) dimer. J Phys Chem B 109:11875–11879

Wågberg T, Johnels D, Peera A, Hedenström M, Schulga YM, Tsybin YM, Purcell JM, Marshall AG, Noreus D, Sato T, Talyzin AV (2005) Selective synthesis of the C_{3v} isomer of $C_{60}H_{18}$. Org Lett 7:5557–5559

Wågberg T, Hedenström M, Talyzin AV, Sethson I, Tsybin YO, Purcell JM, Marshall AG, Noréus D, Johnels D (2008) Synthesis and structural characterization of $C_{70}H_{38}$. Angew Chem Int Ed 47:2796–2841

Weiss FD, Elkind JL, O'Brien SC, Curl RF, Smalley RE (1988) Photophysics of metal complexes of spheroidal carbon shells. J Am Chem Soc 110:4464–4465

Chapter 6
Chemical Methods to Prepare [60]Fulleranes

Jonathan B. Briggs[1] **and Glen P. Miller**[1]

Abstract Most known chemical methods to prepare hydrogenated [60]fullerenes or [60]fulleranes are described with an emphasis on those methods that are of historical significance and/or convenient, scalable and high yielding. Separate sections are dedicated to $C_{60}H_2$, $C_{60}H_{4-16}$, $C_{60}H_{18}$, $C_{60}H_{36}$, and $C_{60}H_{>36}$. The simplest of all [60]fullerane derivatives, 1,2-$C_{60}H_2$, is prepared in good isolated yield using either a $NaBH_4$ reduction (59% isolated) or a Zn(Cu) couple reduction (66% isolated). A few $C_{60}H_4$ (e.g., 1,2,3,4-$C_{60}H_4$) and $C_{60}H_6$ (e.g., 1,2,33,41,42,50-$C_{60}H_6$) isomers have been formed and identified using structure sensitive analytical techniques but evidence for [60]fulleranes of formula $C_{60}H_{8-16}$ is based largely upon structure insensitive mass spectrometry data. Crown shaped C_{3v} $C_{60}H_{18}$ is the only [60]fullerane species that can be produced in high yield using multiple synthetic methods. It is a stable, easily isolable species of limited solubility that persists for years. Of the synthetic methodologies available, the polyamine chemistries [conventional heating or microwave irradiation] are best because the yields are excellent, the reactions are scalable, only standard glassware is required, the work-up is simple, and no purification is required. Birch reductions, transfer hydrogenations and Zn-acid conditions all lead to mixtures of $C_{60}H_{36}$ isomers in good yield. Of these methods, the transfer hydrogenation reaction gives the cleanest product slate with formation of three isomers possessing C_1 symmetry (60–70%), C_3 symmetry (25–30%) and T symmetry (2–5%), respectively. The formation of [60]fulleranes with more than 36 hydrogens is well documented but the products have generally escaped characterization by structure sensitive analytical techniques. No single isomer of formula $C_{60}H_{>36}$ has ever been isolated.

[1]Department of Chemistry and Materials Science Program, University of New Hampshire, Durham, NH, USA
e-mail: glen.miller@unh.edu

F. Cataldo and S. Iglesias-Groth (eds.), *Fulleranes: The Hydrogenated Fullerenes*, Carbon Materials: Chemistry and Physics 2, DOI 10.1007/978-1-4020-9887-1_6,

6.1 Introduction

Hydrogenated [60]fullerenes or [60]fulleranes were first prepared by Haufler and co-workers using a Birch reduction procedure (Haufler et al. 1990). In doing so, they demonstrated the reactive nature of [60]fullerene and initiated a decade of chemical investigations that produced many intriguing compounds. In addition to the Birch reduction, multiple other methods have been used to hydrogenate [60]fullerenes including borane reductions, Zn-acid reductions, hydrozirconation–hydrolysis, diimide reduction, Zn(Cu) couple reductions, transfer hydrogenations, polyamine hydrogenations, and numerous others. In these manners, a range of [60]fullerane molecules have been prepared and reported. Many of these methods produce complex mixtures of [60]fulleranes that can be difficult to separate. In these cases, time-intensive chromatographic methods including high pressure liquid chromatography (HPLC) are required to isolate small quantities of purified [60]fullerane. A few methods produce a small set of [60] fullerane isomers (e.g., transfer hydrogenation leading to three isomers of $C_{60}H_{36}$) or a single [60]fullerane species (e.g., polyamine hydrogenation leading to C_{3v} $C_{60}H_{18}$) in regioselective fashion. These methods are advantageous in that they minimize or completely avoid time-intensive separations. In this chapter, we will summarize most known chemical methods to prepare [60]fulleranes with emphasis on those methods that are of historical significance and/or convenient, scalable and high yielding. Separate sections are dedicated to $C_{60}H_2$, [60]fulleranes possessing more than 2 but less than 18 hydrogens, $C_{60}H_{18}$, $C_{60}H_{36}$, and [60]fulleranes possessing more than 36 hydrogens.

6.2 $C_{60}H_2$

Of all the hydrocarbon derivatives of [60]fullerene, $C_{60}H_2$, first prepared by Henderson and Cahill using a hydroboration–hydrolysis procedure, is the simplest. Multiple other methods have been demonstrated for the formation of $C_{60}H_2$ including two (i.e., a $NaBH_4$ reduction and a Zn(Cu) couple reduction) that produce $C_{60}H_2$ in good isolated yield.

Henderson and Cahill demonstrated hydroboration–hydrolysis by adding 1–3 equivalents of BH_3-THF to [60]fullerene (10–100 mg scale) in dry toluene followed by hydrolysis with either acetic acid or water (Henderson and Cahill 1993). In this way, the smallest stable fullerane was isolated from unreacted [60]fullerene and trace [70]fullerene in 10–30% yield using HPLC. Of 23 possible isomers of $C_{60}H_2$, Henderson and Cahill argued convincingly for the regioselective formation of a single $C_{60}H_2$ isomer in which the hydrogen atoms added in a 1,2 fashion across a 6,6 junction (i.e., C_{2v} symmetric 1,2-$C_{60}H_2$). Their argument included data from ^1H NMR spectroscopy (i.e., a single singlet with a [solvent dependent] chemical shift of approximately 6 ppm), UV–vis spectrophotometry, mass spectrometry, and most convincingly an analysis of H–D coupling which yielded a J_{DH} coupling constant of 2.4 Hz, consistent with vicinal addition. To prepare the 1,2-C_{60}HD derivative, Henderson and Cahill hydrolyzed the intermediate borane using D_2O. Although the yield of 1,2-$C_{60}H_2$ produced via hydroboration–hydrolysis is modest, the method is quite useful for small scale work given that the product slate does not contain large quantities of more highly hydrogenated [60]fullerenes that are difficult to separate. Thus, Henderson and co-workers reported low yields (<2%) of $C_{60}H_4$ isomers under the same or very similar reaction conditions (Henderson et al. 1994).

Several other methods have been utilized to prepare 1,2-$C_{60}H_2$ (Fig. 6.1a and b) including a Zn-acid reduction by Meier and co-workers (1994) and a hydrozirconation–hydrolysis by Ballenweg and co-workers (1993). In the Zn-acid reduction,

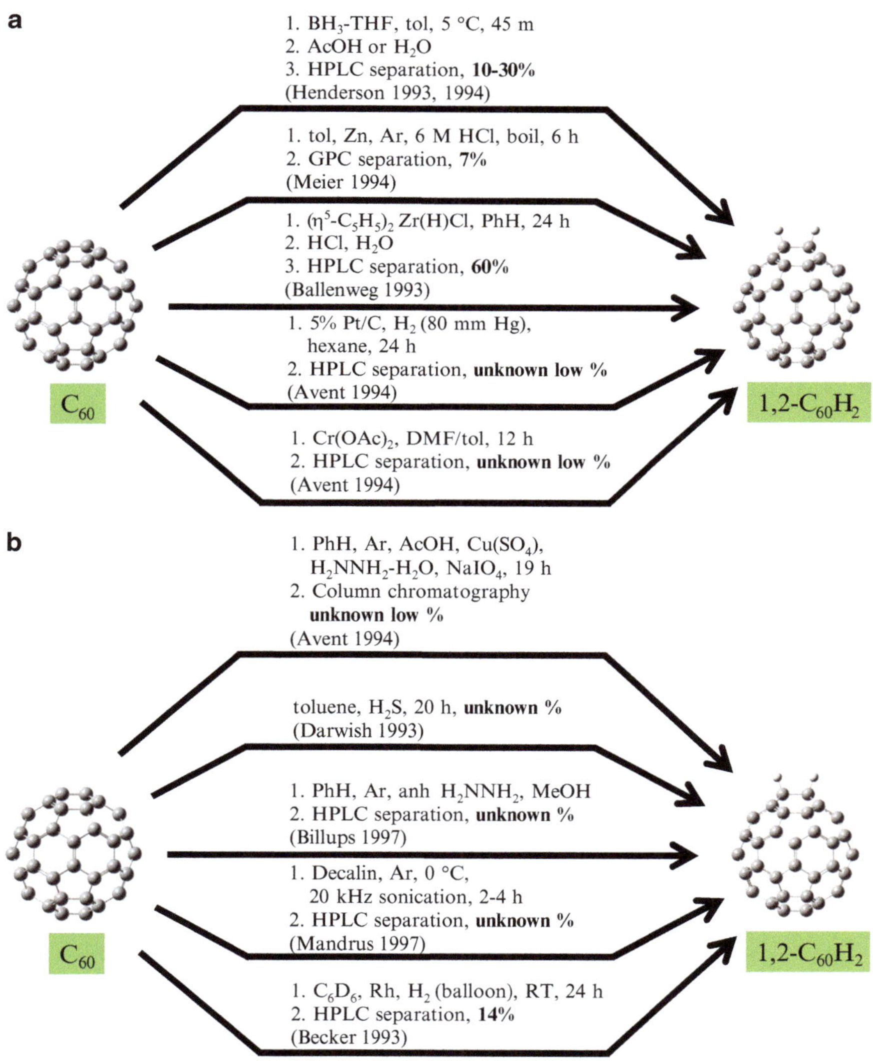

Fig. 6.1 Ten methods to convert C_{60} to 1,2-$C_{60}H_2$ as illustrated in **a** and **b**

20 mg of [60]fullerene was treated with Zn dust and 6 M HCl in boiling toluene to produce several different products including 1,2-$C_{60}H_2$ (7%) and multiple isomers of $C_{60}H_4$. The hydrozirconation procedure was run on a slightly larger scale (40 mg [60]fullerene) and involved the addition of two equivalents of Schwartz's reagent, (η^5-C_5H_5)$_2$Zr(H)Cl, to [60]fullerene in dry benzene followed by hydrolysis with aqueous HCl. Several products formed and were isolated using preparative HPLC including 1,2-$C_{60}H_2$ (60%), $C_{60}H_4$ and $C_{60}H_6$. Almost no details were provided concerning the latter two fulleranes.

Avent and co-workers (1994) attempted to produce fulleranes using three distinct methods: low pressure (80 mm Hg) catalytic hydrogenation with H_2/Pt/C, reduction with $Cr(OAc)_2$, and reduction with diimide. According to ^{1}H NMR analysis, the first two methods produced fullerane products but the yields were low and, at least in the case of catalytic hydrogenation, complicated by the formation of degradation/oxidation by-products. The diimide reduction leads to higher yields of fulleranes but with little selectivity, at least under the reaction conditions reported. Thus, when 20 mg of [60]fullerene was reacted with as much as a 1,000-fold excess of diimide produced in situ via the oxidation of hydrazine hydrate, a complex mixture of fulleranes resulted including multiple $C_{60}H_4$ isomers, more highly hydrogenated fulleranes, 1,2-$C_{60}H_2$ and, reportedly, 1,4-$C_{60}H_2$. Cahill and Henderson report that the 1,4-$C_{60}H_2$ isomer is several kcal/mol less stable than the 1,2-$C_{60}H_2$ isomer according to semi-empirical calculations (Henderson and Cahill 1993). None of the fullerane products produced by diimide reduction were isolated.

Still other methods to produce 1,2-$C_{60}H_2$ include reduction with H_2S as reported by Darwish and co-workers (1993), reduction with anhydrous hydrazine in benzene/methanol as reported by Billups and co-workers (Nossal et al. 2001a; Billups et al. 1997), transfer hydrogenation from decalin as reported by Mandrus and co-workers (1997), and a Rh catalyzed hydrogenation as reported by Becker and co-workers (1993; Fig. 6.1b).

Arguably, two methods to produce 1,2-$C_{60}H_2$ are better than all others (Fig. 6.2). The first is a $NaBH_4$ reduction in ethanol as reported by Wang and co-workers (2005). Thus, 36 mg of [60]fullerene was dissolved in 35 mL of toluene to which 0.5 equivalents of $NaBH_4$ in ethanol was added. After 45 min at 60°C, the crude product

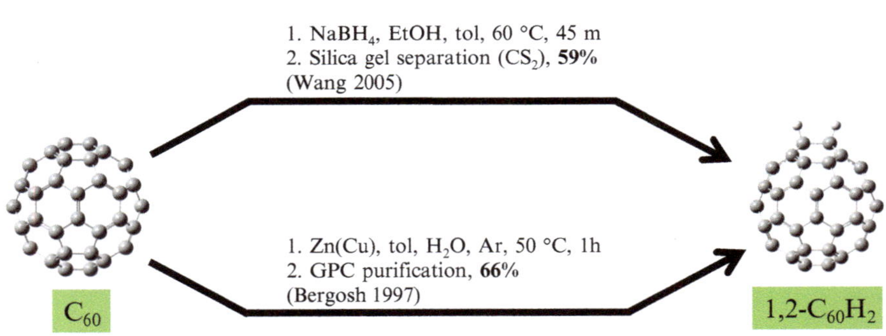

Fig. 6.2 Arguably, the two best synthetic methods to produce 1,2-$C_{60}H_2$

analyzed for C_{60} (20%), 1,2-$C_{60}H_2$ (66%), and more highly hydrogenated fulleranes (14%). Most useful, 1,2-$C_{60}H_2$ was isolated in 59% yield using a silica gel column. The second method is a Zn(Cu) couple reduction performed in wet toluene as reported by Bergosh and co-workers (1997). Thus, 255 mg of [60]fullerene was dissolved in 250 mL of toluene under an Ar purge. Freshly prepared Zn(Cu) couple from 5.1 g of Zn dust was added with 2.5 mL of H_2O. Following 1 h of heating (50°C) and stirring, the crude product analyzed for 1,2-$C_{60}H_2$ (78% by HPLC analysis) and several $C_{60}H_4$ isomers (12% by HPLC). Following GPC purification (four columns in series), 1,2-$C_{60}H_2$ was isolated in 66% yield. Of these two methods, the $NaBH_4$ reduction is the simpler reaction to perform and purification on a silica gel column is attractive. However, the Zn(Cu) couple reduction produces a slightly higher yield of 1,2-$C_{60}H_2$ on a larger scale. It is in fact one of a very small number of fullerene hydrogenation reactions that has been successfully demonstrated to work on a 100+ mg scale.

Interestingly, 1,2-$C_{60}H_2$ shows a remarkably high acidity with a pK_a^1 equal to 4.7 (more acidic than *t*-BuC_{60}H with a pK_a^1 of 5.7 (Fagan et al. 1992)) and a pK_a^2 equal to 16 in DMSO (Niyazymbetov et al. 1994). The large acidities of the two hydrogens on 1,2-$C_{60}H_2$ illustrate the relatively stable nature of the mono- and di-anions of [60]fullerene produced upon single and double deprotonation, consistent with a plethora of electrochemical studies on [60]fullerene. The electrochemical (cyclic voltammetry) reduction of 1,2-$C_{60}H_2$ itself is only slightly more difficult than that of [60]fullerene in benzonitrile (Guarr et al. 1993). When sweep rates greater than 1,000 mV/s were utilized, three reversible reductions were observed between −1 and −2 V (vs. Fc^+/Fc). However, at slower sweep rates, dehydrogenation of the $C_{60}H_2^{n-}$ anions was observed to be competitive with return oxidation.

Regarding the stability of 1,2-$C_{60}H_2$ and other fulleranes, concerns of possible degradation have been raised by several groups and subsequently reinforced by Nossal and co-workers (2001a) in a widely read 2001 review. Consequently, some workers in the field consider *all* fulleranes to be fragile species. While it is often easiest to paint with a broad brush, many fulleranes like, for example, several isomers of $C_{60}H_{36}$ are relatively stable species that can be protected from degradation with only modest effort. Other fulleranes like, for example, C_{3v} $C_{60}H_{18}$, are downright robust species that are difficult to degrade. Any discussion concerning the stability of fulleranes should be handled on a case by case basis. Ideally, these discussions would include substantive mechanistic insights including, for example, the role of impurities (e.g., metal, base) that may be carried over from the corresponding hydrogenation reaction conditions.

6.3 [60]Fulleranes with More Than 2 But Less Than 18 Hydrogens: $C_{60}H_{2<n<18}$

Although distinct fullerane regioisomers of formula $C_{60}H_4$ and $C_{60}H_6$ have been formed and identified using structure sensitive analytical techniques like ^{1}H and ^{13}C NMR spectroscopy, evidence for [60]fulleranes of formula $C_{60}H_{8-16}$ is based largely upon structure insensitive mass spectrometry data. As a consequence, considerably less is known about these latter species.

The synthesis of $C_{60}H_4$ isomers is complicated by a lack of regioselectivity during their formation. Assuming that addition always occurs in a 1,2 fashion, eight unique isomers of $C_{60}H_4$ are possible (Matsuzawa et al. 1992) and a variety of methods have been utilized to produce some quantity of all of them (Figs. 6.3 and 6.4). The methods utilized include the zinc-acid reduction (Meier et al. 1994), the diimide reduction (Avent et al. 1994), hydroboration–hydrolysis (Henderson et al. 1994), hydrozirconation–hydrolysis (Ballenweg et al. 1993), anhydrous hydrazine (reported to form all eight isomers of $C_{60}H_4$ (Nossal et al. 2001a) and the Zn(Cu) couple reduction in wet toluene (Bergosh et al. 1997). Of the eight possible $C_{60}H_4$ isomers, 1,2,3,4-$C_{60}H_4$ is reported to be a major product in several cases (Henderson et al. 1994; Avent et al. 1994; Nossal et al. 2001a) but the reactions are generally low yielding and far from selective (Fig. 6.3). The Zn(Cu) couple reduction is unique in that only three $C_{60}H_4$ isomers (1:1:0.3) form in a combined 45% yield (plus 18% $C_{60}H_6$) when the reduction is stopped after 1.5 h and none of them is 1,2,3,4-$C_{60}H_4$ (Fig. 6.4).

The Zn(Cu) couple reduction shows even greater selectivity when the reaction is run for slightly longer times. Initially, two isomers of $C_{60}H_6$ were observed to form in 30–40% yield (6:1 ratio) when the reduction was stopped after 3–4 h (Bergosh et al. 1997; Meier et al. 1996). The two isomers of $C_{60}H_6$ were separable by HPLC and the major product was identified (Meier et al. 1996) as 1,2,33,41,42,50-$C_{60}H_6$ (Fig. 6.4). Wang and co-workers subsequently discovered a third (minor) $C_{60}H_6$ isomer (Wang et al. 2000) in a reaction that utilized $^3He@C_{60}$ as reactant but was

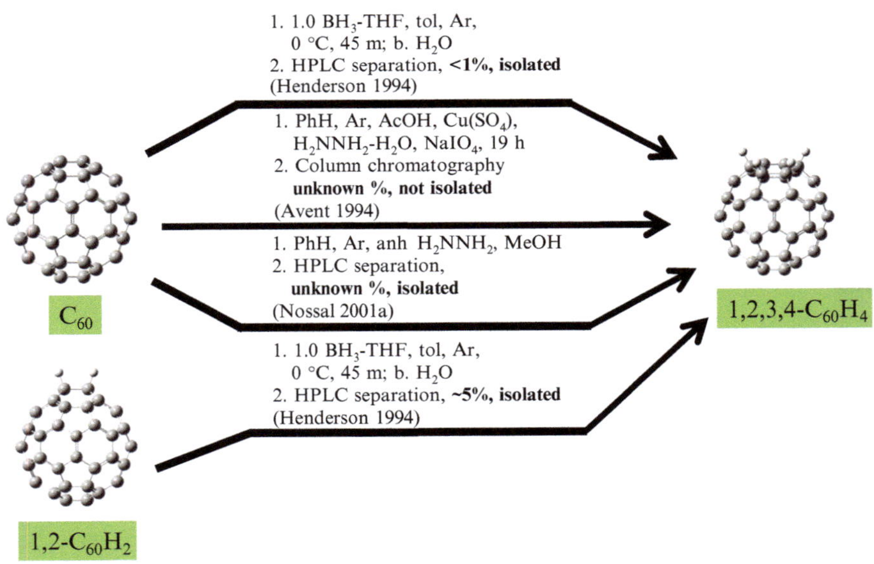

Fig. 6.3 Methods to form 1,2,3,4-$C_{60}H_4$

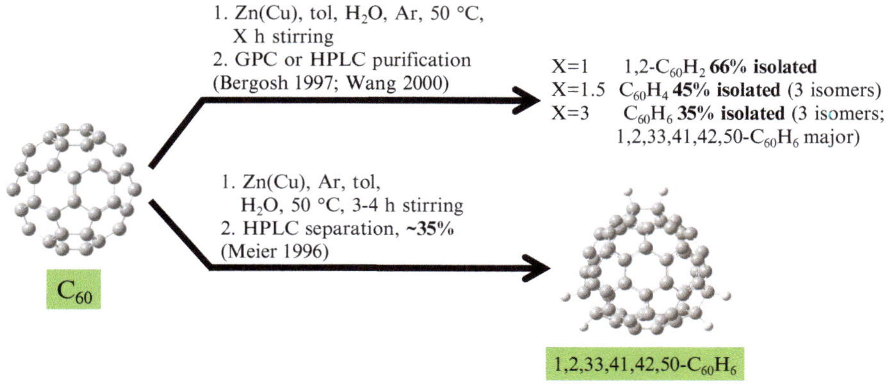

Fig. 6.4 Zn(Cu) couple reduction of [60]fullerene produces three isomers of $C_{60}H_4$ and three isomers of $C_{60}H_6$ including 1,2,33,41,42,50-$C_{60}H_6$ that has been isolated

otherwise identical to previous Zn(Cu) couple reductions (Bergosh et al. 1997; Meier et al. 1996). The third isomer was identified following a ^{3}He NMR analysis of the crude product. A ^{13}C enriched 1,2,33,41,42,50-$C_{60}H_6$ was produced (along with ^{13}C enriched 1,2-$C_{60}H_2$) (Meier et al. 2002) enabling unambiguous ^{13}C NMR chemical shift assignments. Interestingly, the carbon atoms with largest downfield chemical shift are those immediately adjacent to sites of functionalization (i.e., sp^2 hybridized carbons closest to sp^3 hybridized C–H carbons), in contrast to expectations based on pyramidalization angles.

The three sites of 1,2-addition in 1,2,33,41,42,50-$C_{60}H_6$ are as far removed from one another as possible. As this isomer is likely a kinetically controlled product, Meier and co-workers considered a sequential pathway involving reduction of the presumed precursor 1,2,41,42-$C_{60}H_4$ (Meier et al. 1996). However, 1,2,33,41,42,50-$C_{60}H_6$ could also form directly from C_{60}^{3-} (even in wet toluene) according to a pathway that is driven by the dissolving metal-like conditions of the reaction. In this case, minimization of charge repulsion in C_{60}^{3-} would dictate the regioselectivity of addition. An interesting experiment that would rule out one of these two unique pathways involves subjecting an isolated 1,2,41,42-$C_{60}H_4$ molecule to an otherwise identical Zn(Cu) couple reduction (i.e., kinetically controlled reaction conditions). According to Meier's prediction, 1,2,33,41,42,50-$C_{60}H_6$ would form cleanly from 1,2,41,42-$C_{60}H_4$. However, if C_{60}^{3-} is a critical intermediate on the pathway to 1,2,33,41,42,50-$C_{60}H_6$, then a Zn(Cu) couple reduction of 1,2,41,42-$C_{60}H_4$ would likely give a different (or no) product.

Several other methods (Meier et al. 1994; Ballenweg et al. 1993; Avent et al. 1994; Billups et al. 1997; Wang et al. 2005) also produce some (usually small) quantity of $C_{60}H_6$ isomers but the products have not been carefully characterized.

Although the existence of $C_{60}H_8$, $C_{60}H_{10}$, $C_{60}H_{12}$, $C_{60}H_{14}$ and $C_{60}H_{16}$ as intermediates to more highly hydrogenated species cannot be questioned, no single isomer with any of these molecular formulas has ever been isolated. Some of these species may exhibit reduced stability as compared to other fulleranes, but a lack of regioselectivity in their formation is likely an equally culpable factor. Evidence for their formation can be found in mass spectrometry studies where fragmentation processes complicate the analysis. For example, Jin and co-workers reported the direct, high temperature (300–350 °C), high pressure (500–850 bar H_2) hydrogenation of solid phase [60] fullerene (Jin et al. 1994) to produce a series of fulleranes from $C_{60}H_2$ up to and including $C_{60}H_{18}$. Unfortunately, their characterization was limited to laser desorption mass spectrometry (LD-MS) with the most abundant mass peaks consistent with fulleranes of formula $C_{60}H_{8-16}$. The absence of corroborating characterizations by other analytical methods leaves one to question whether or not the most abundant fulleranes detected by LD-MS had formed by fragmentation of higher molecular weight fulleranes like, e.g., $C_{60}H_{18}$.

6.4 $C_{60}H_{18}$

Crown shaped C_{3v} $C_{60}H_{18}$ is the only [60]fullerane species that can be produced in high yield using multiple synthetic methods. It is a stable, easily isolable species of limited solubility that persists for years.

Numerous methods of synthesis have produced product slates that contain $C_{60}H_{18}$ species including direct hydrogenation using 80 atm of H_2 (Sui et al. 1996), hydrogenation in a hot filament chemical vapor modification (CVM) reactor using 20 torr of H_2 (Vieira et al. 2001, 2004), and a transfer hydrogenation using 9,10-dihydroanthracene (Rüchardt et al. 1993). In the first and last of these cases, characterization of products was accomplished by mass spectrometry and little is known about their detailed structures. In all cases where a $C_{60}H_{18}$ species has been isolated and/or characterized by structure sensitive methods (including hydrogenation in a hot filament CVM reactor), it has been shown to possess a C_{3v} crown structure (Fig. 6.5).

Darwish and co-workers first isolated C_{3v} $C_{60}H_{18}$ from a commercial product slate provided by MER Corporation using an HPLC method (Darwish et al. 1996). The fullerane was produced by MER using a poorly defined high temperature, high pressure hydrogenation of solid [60]fullerene. Characterization of the product, while compelling, consisted only of structure insensitive IR spectroscopy and mass spectrometry in addition to ^1H NMR spectroscopy. Briggs and co-workers subsequently produced the same compound in higher yield and without the need for chromatographic purification using a simple, inexpensive polyamine hydrogenation procedure (Briggs et al. 2005). Their characterization included mass spectrometry, UV–vis spectrophotometry, ^1H and ^{13}C NMR spectroscopies plus 2D COSY and phase-sensitive COSY spectroscopy, all of which were consistent with the proposed C_{3v} $C_{60}H_{18}$. The COSY spectrum in particular provided compelling evidence for the C_{3v} crown structure as the weak $^5J_{HH}$ coupling [3 Hz, consistent with $^5J_{HH}$ couplings on other fulleranes (Avent et al. 1994)] between distant protons on each of three repeating units was observed (Fig. 6.6).

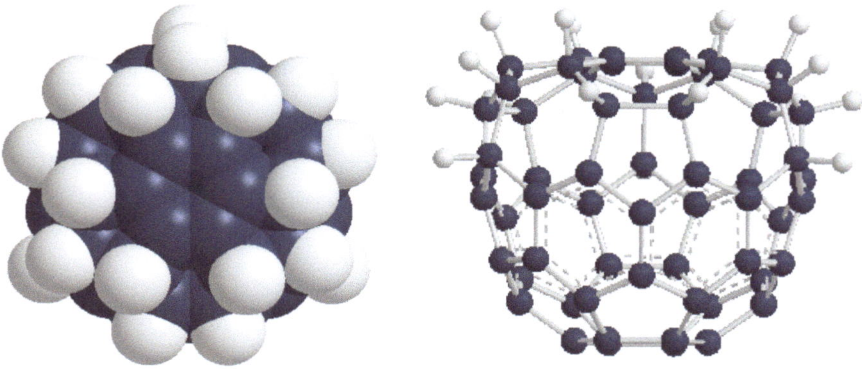

Fig. 6.5 C_{3v} crown structure of a stable, isolable $C_{60}H_{18}$ molecule that is produced in high yield by direct hydrogenation of solid C_{60} (Darwish et al. 1996; Wågberg et al. 2005) and most conveniently by a polyamine hydrogenation at 200 °C in standard glassware (Briggs et al. 2005)

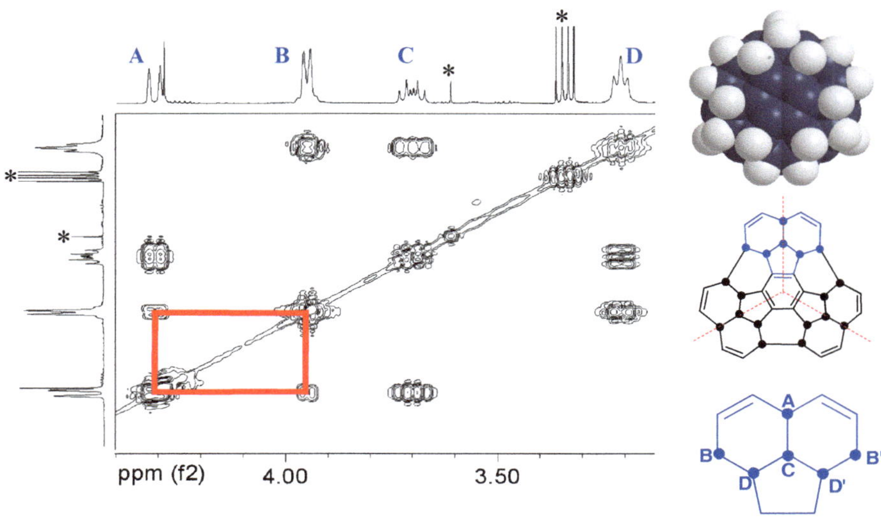

Fig. 6.6 COSY spectrum of C_{3v} $C_{60}H_{18}$ produced in high yield using diethylenetriamine as both solvent and hydrogenation reagent (Briggs et al. 2005). The weak cross-peak between protons A and B corresponds to a $^5J_{HH}$ of 3 Hz. Asterisks indicate residual solvent (*quartet*) and a minor impurity (*singlet*)

The polyamine hydrogenation of [60]fullerene to produce C_{3v} $C_{60}H_{18}$ involves addition of [60]fullerene to a standard round bottomed flask containing diethylenetriamine (DETA) with 10–50 mg of [60]fullerene utilized per every milliliter of DETA. The solution is boiled under N_2 in the dark for 2–20 h. Upon cooling, the bright orange product partially precipitates from solution. Further precipitation is

Table 6.1 [60]Fullerene hydrogenation reactions leading to C_{3v} $C_{60}H_{18}$ run in neat, boiling diethylenetriamine (DETA) under N_2 in the dark

Mass C_{60} (g)	mL DETA	Time (h)	Mass $C_{60}H_{18}$ (g)	% Yield
0.10	7	14	0.083	81
1.0	50	16	0.919	90
5.0	100	20	4.674	91
10.0	200	20	9.413	92

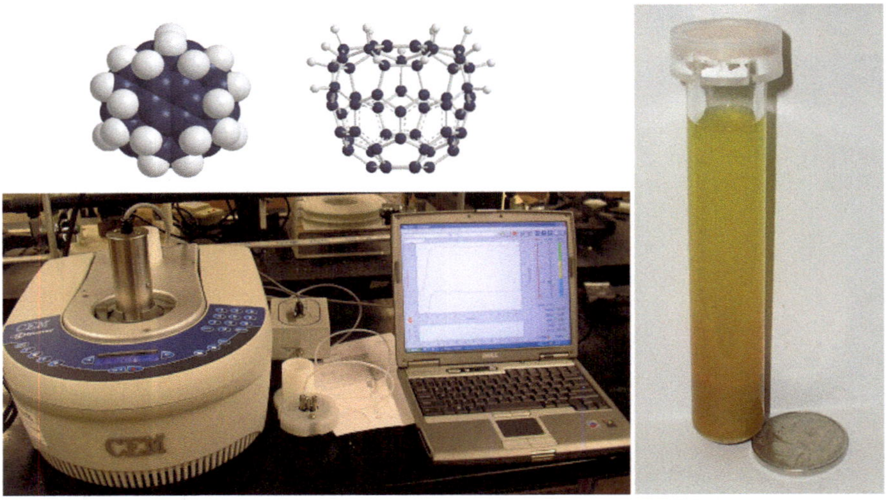

Fig. 6.7 Microwave irradiation (300 W, 250 °C) reduces the time required to produce C_{3v} $C_{60}H_{18}$ in high yield using polyamine from several hours to 10 min (Kintigh et al. 2007). The orange solid that settles to the bottom of the microwave reactor tube is 99%+ C_{3v} $C_{60}H_{18}$

facilitated by addition of ethanol. The orange product is filtered and washed with copious ethanol followed by drying in air or under vacuum. Most impressive, the reaction is completely scalable (Table 6.1) and could easily be run on a kilogram scale, likely without any loss in yield.

Moreover, with access to a microwave reactor (Fig. 6.7), reaction time can be cut from several hours to several minutes (Kintigh et al. 2007).

The mechanism of polyamine hydrogenation (Fig. 6.8) is believed to involve successive electron transfer (from polyamine to fullerene) – proton transfer (from polyamine radical cation to fullerene radical anion) steps (Briggs et al. 2005; Kintigh et al. 2007). At or near room temperature, aliphatic amines and polyamines are known to hydroaminate [60]fullerene (Miller 2006), likely also involving preliminary electron transfer – proton transfer steps followed by free radical coupling of C and N based radicals (Fig. 6.8). At elevated temperatures in polyamine solution, however, this latter free radical coupling step becomes uncompetitive with

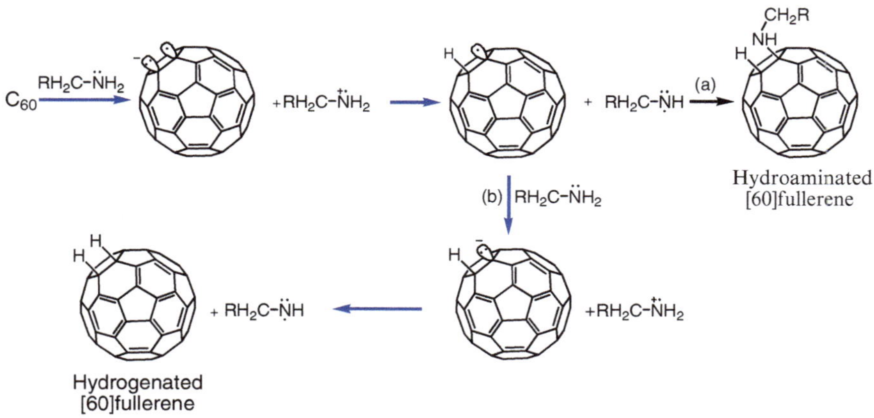

Fig. 6.8 Proposed mechanism for the polyamine hydrogenation of [60]fullerene and its relationship to hydroamination (Briggs et al. 2005; Kintigh et al. 2007)

additional electron transfer. This could be due to faster electron transfers at elevated temperatures or perhaps reversible C–N coupling. The result is tantamount to a mechanistic hijacking in which hydroamination is completely shut down in favor of hydrogenation. Consistent with the path proposed in Fig. 6.8, reaction with a deuterium labeled DETA (i.e., N,N,N′,N″,N″-diethylenetriamine-d_5) provided convincing evidence that all H(D) transferred to [60]fullerene originated from NH(D) bonds on DETA (Kintigh et al. 2007). It is interesting to note that large aliphatic monoamines with boiling points similar to diethylenetriamine also act as hydrogenation agents but the yields of fullerane produced in this fashion are quite low and the reactions are complicated by hydroamination product formation (Briggs et al. 2005). Thus, polyamines are unique in their ability to promote hydrogenations on fullerene suggesting a rate enhancing cooperative effect involving more than one nitrogen atom.

Kim and co-workers performed a density functional theory (DFT) study of the reaction between diethylenetriamine and [60]fullerene (Kim et al. 2008) and suggested a first step involving hydrogen transfer from physisorbed diethylenetriamine to [60]fullerene in order to produce a set of neutral free radical intermediates that subsequently couple to give hydroamination product. Dismissing the fact that intermolecular hydrogen atom transfers from closed shell donors to closed shell substrates are rare, the calculations were performed for molecules in a vacuum rather than in polar liquid media, a major change compared to the experimental reaction conditions. Electron transfer steps are well known to be facilitated by polar media (Maroncelli et al. 1998) that can stabilize ionic intermediates through solvation. It seems likely that the introduction of polar media would alter the course of the DFT calculations.

Shortly after demonstration of the polyamine hydrogenation of [60]fullerene to produce C_{3v} $C_{60}H_{18}$, Wågberg and co-workers similarly produced high yields of

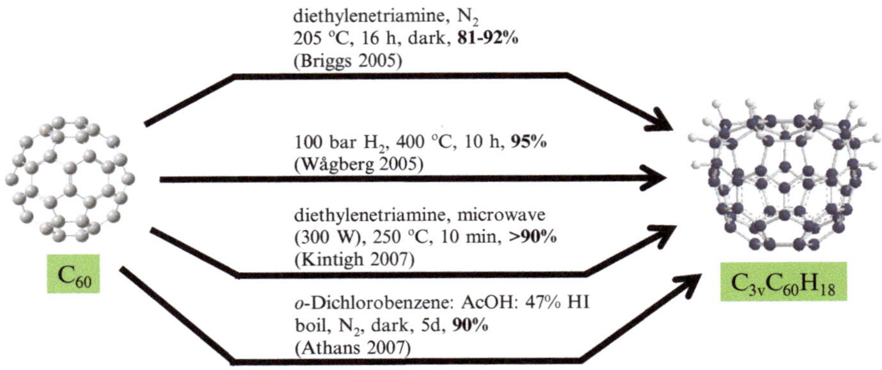

Fig. 6.9 Synthetic methods to produce C_{3v} $C_{60}H_{18}$ in high yield

C_{3v} $C_{60}H_{18}$ using a high temperature (400 °C), high pressure (100 bar H_2) hydrogenation of solid [60]fullerene (Wågberg et al. 2005), similar to the methods employed by Jin and co-workers (1994) as well as MER corporation (Darwish et al. 1996). Likewise, Athans and co-workers reported an HI-AcOH reduction of [60]fullerene to produce C_{3v} $C_{60}H_{18}$ in 90% isolated yield (Athans et al. 2007). Thus, four distinct synthetic methods are known to produce C_{3v} $C_{60}H_{18}$ in high yield (Fig. 6.9). Of these four methods, the polyamine chemistries [conventional heating (Briggs et al. 2005) or microwave irradiation (Kintigh et al. 2007)] are considered best because the yields are excellent, the reactions are scalable, only standard glassware is required, the work-up is trivial, and no purification of C_{3v} $C_{60}H_{18}$ is required. For laboratories already set up to run reactions at high temperature and high pressure, the direct hydrogenation (Wågberg et al. 2005) may be preferred. The HI-AcOH reaction (Athans et al. 2007) is less desirable because it is slow and not easily scaled.

It is important to note that despite reports of facile oxidation and spontaneous decomposition in CH_2Cl_2 solution (Darwish et al. 1996), C_{3v} $C_{60}H_{18}$ is a robust fullerane. In our hands, it has been stored as a solid for over 2 years (vial wrapped in foil, capped but otherwise not protected from air, 5 °C) without any significant decomposition. Moreover, although we do not store C_{3v} $C_{60}H_{18}$ in solution, we have never observed it to decompose in any solvent, rapidly or otherwise. Considering its stability and the ease with which it is prepared on a bulk scale, C_{3v} $C_{60}H_{18}$ represents a useful and as yet unexplored synthetic building block for more advanced structures and materials.

6.5 $C_{60}H_{36}$

Just as there are very few reports of fulleranes with more than 6 but less than 18 hydrogens, there are few substantive reports of fulleranes with more than 18 but less than 36 hydrogens and none of those compounds has ever been carefully characterized. There are, however, numerous reports of $C_{60}H_{36}$ isomers prepared using Birch reduction, transfer hydrogenation and Zn-acid conditions. It appears that $C_{60}H_{18}$ and $C_{60}H_{36}$ represent thermodynamic minima on the [60]fullerane energetic landscape. Nonetheless, there are substantial differences between $C_{60}H_{18}$ and $C_{60}H_{36}$. Thus, while only one $C_{60}H_{18}$ regioisomer is known (i.e., C_{3v} $C_{60}H_{18}$), three $C_{60}H_{36}$ regioisomers (C_1, C_3 and T) prepared via transfer hydrogenation are well documented. Transfer hydrogenation, Birch reduction and Zn-acid procedures do not produce the same set of $C_{60}H_{36}$ regioisomers. Regardless of method of preparation, no single $C_{60}H_{36}$ regioisomer has been isolated.

Haufler and co-workers were the first to hydrogenate [60]fullerene using a Birch reduction procedure (Li, $NH_3(l)$, *t*-BuOH) that, although well known in the organic chemistry community, was not well described in their seminal paper (Haufler et al. 1990). Nonetheless, it is surmised that [60]fullerene was added to a solution of Li^0 in liquid NH_3 at low temperature (likely −78 °C) in an inert atmosphere followed by addition of the weak Brønsted acid *t*-butanol with stirring for an undisclosed period of time. Reduction of [60]fullerene produces an anion or set of anions (likely multiple reductions leading to C_{60}^{n-} anions) which are protonated by *t*-butanol to produce [60] fullerane and Li^+ *t*-BuO^-. In order to produce a highly hydrogenated [60]fullerane like $C_{60}H_{36}$, multiple reduction-protonation steps must occur in succession. Electron Impact Mass Spectrometry (EI-MS) using thermal evaporation (direct insertion probe) of the off-white colored crude product indicated two species, $C_{60}H_{36}$ and $C_{60}H_{18}$, the latter appearing only after extensive heating (360 °C) at 3×10^{-7} torr and likely due to the thermal degradation of a more highly hydrogenated species. The ^1H NMR spectrum exhibited a very broad multiplet between 4.2 and 2.5 ppm. Haufler and co-workers correctly surmised that this broad multiplet was evidence for formation of numerous [60]fullerane isomers and they proposed a structural model for a set of $C_{60}H_{36}$ isomers in which one isolated double bond existed on each of the 12 pentagon rings, the relative orientation of double bonds varying from one isomer to the next. The formation of a [60]fullerane with isolated double bonds seemed reasonable given that Birch reductions of simple aromatic species (e.g., benzene, naphthalene) produce formally hydrogenated compounds with isolated double bonds (e.g., 1,4-cyclohexadiene and 1,4,5,8-tetrahydronaphthalene).

Shortly after Haufler and co-workers' seminal paper (Haufler et al. 1990), Banks and co-workers reported that an analogous Birch reduction procedure gave a complex mixture of pale-yellow product "containing $C_{60}H_{18}$ to $C_{60}H_{36}$ with a skewed distribution centered on $C_{60}H_{32}$" (Banks et al. 1993). Unfortunately, their characterization of Birch reduction products was limited to laser desorption mass spectrometry (LD-MS) techniques, and as is always the case in such circumstances, one must question whether or not those techniques favored fragments rather than products. Indeed, Banks and co-workers recognized this possibility, at least as it applied to the EI-MS characterization reported by Haufler and co-workers: "the radically different mass spectra obtained under the rather severe sample introduction conditions (of Haufler and co-workers) required for EI mass spectroscopic studies can be attributed to thermal decomposition of the polyhydrofullerenes" (Banks et al. 1993). This argument is troubling, however, as thermal decomposition should lead to fullerane with less hydrogen, not more. Note that Banks and co-workers reported an LD-MS distribution centered on $C_{60}H_{32}$ while Haufler and co-workers reported $C_{60}H_{36}$ using EI-MS. Indeed, Vasil'ev and co-workers subsequently reported (Vasil'ev et al. 2000) that products obtained from their Birch reduction gave rise to mass spectral signals consistent with $C_{60}H_{36}$ as evidenced by matrix assisted laser desorption (MALDI) MS with 9-nitroanthracene as matrix.

The well trained synthetic chemist will recognize that absent isolated yields of $C_{60}H_{36}$ and corroborating data from structure sensitive analytical techniques, it is difficult to accept stand-alone mass spectrometry studies with anything more than a grain of salt. Moreover, it should not escape the reader's attention that Haufler and co-workers reported an off-white colored product while Banks and co-workers reported a pale-yellow product (Vasil'ev and co-workers make no mention of the color of their Birch reduction product). None of the papers gives more than a sentence of detail concerning the experimental conditions of their respective Birch reductions. Prolonged exposure of reduced fullerene/fullerane to *t*-BuOH would lead to a more highly hydrogenated product that would tend toward off-white rather than yellow. It is possible, even likely, that the three reactions cited above (Fig. 6.10) gave unique product slates.

Other researchers have also studied the Birch reduction of [60]fullerene (Fig. 6.10) and have also concluded that multiple isomers of $C_{60}H_{36}$ are formed (Nossal et al. 2001b; Zhang et al. 2004b; Wang and Zhang 2006). Thus, Nossal and co-workers reported that after 8–10 h of reaction at −33 °C, the crude product slate obtained from the Birch reduction of [60]fullerene could be separated into several bands by HPLC. The major bands consisted entirely of several isomers of $C_{60}H_{36}$ with two minor bands containing combinations of $C_{60}H_{38}$ and $C_{60}H_{40}$ isomers, and $C_{60}H_{36}$ and $C_{60}H_{32}$ isomers, respectively. ^{1}H NMR, ^{13}C NMR and ^{3}He NMR (obtained on reactions that utilized He@C_{60} as reactant) spectra provided corroborating, compelling evidence that multiple isomers existed in each HPLC band. Attempts were made to identify specific isomers by matching experimental and DFT computed ^{3}He NMR chemical shift values, but these were unconvincing.

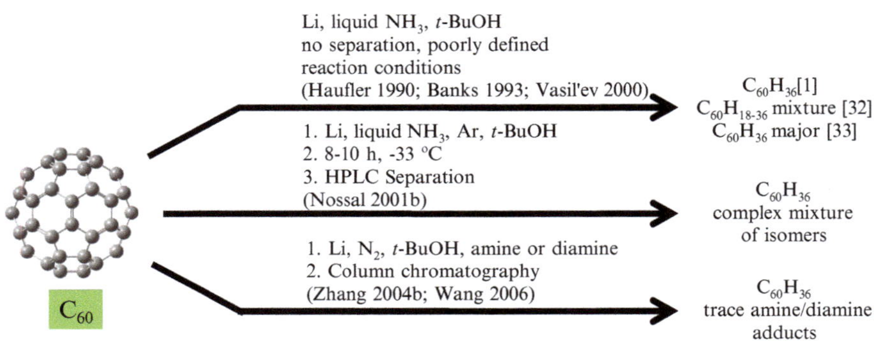

Fig. 6.10 Birch and Benkeser reductions of [60]fullerene leading primarily to multiple isomers of $C_{60}H_{36}$

Zhang and co-workers also reported (Zhang et al. 2004b) $C_{60}H_{36}$ products (multiple isomers) plus small quantities of hydroamination product resulting from a Benkeser reduction (a modified Birch reduction in which an amine or diamine is utilized in place of ammonia) of [60]fullerene. Subsequent thermal degradation of the column chromatography purified $C_{60}H_{36}$ mixture in the presence of transition metal catalysts (Wang and Zhang 2006) produced $C_{60}H_{18}$, strongly suggesting that Haufler and co-workers' EI-MS detection of $C_{60}H_{18}$ following thermal evaporation (direct insertion probe) of their Birch reduction product (Haufler et al. 1990) was due to decomposition of more highly hydrogenated [60]fullerane.

A different, convenient method to produce $C_{60}H_{36}$ involves transfer hydrogenation from 9,10-dihydroanthracene to [60]fullerene. The method was first established by Rüchardt and co-workers (1993) and subsequently repeated by many others. Thus, upon heating a mixture of [60]fullerene and 9,10-dihydroanthracene (120 equivalents) in a sealed glass ampoule (N_2 atmosphere) to 350 °C for several minutes, the initially dark violet melt turns to brown, red, orange, yellow and finally white. Following sublimation of anthracene and its derivatives, a white [60]fullerane that analyzes for several $C_{60}H_{36}$ isomers was isolated. With longer heating times (24 h), the reaction produces a [60]fullerane that analyzes for $C_{60}H_{18}$ (along with [60]fullerene impurity), presumably via dehydrogenation of initially formed $C_{60}H_{36}$. It is important to note that contrary to literature assumptions (Gakh et al. 2003), the $C_{60}H_{18}$ product(s) produced in this manner has not been carefully characterized and it is unclear if it possesses the same C_{3v} crown structure as reported using polyamine (Briggs et al. 2005; Kintigh et al. 2007), direct hydrogenation (Darwish et al. 1996; Wågberg et al. 2005) and HI-AcOH (Athans et al. 2007) chemistries. Conversely, the $C_{60}H_{36}$ products produced via transfer hydrogenation have been characterized by numerous groups. Nossal and co-workers (2001b) concluded that the main $C_{60}H_{36}$ product produced via transfer hydrogenation is only a minor constituent of the $C_{60}H_{36}$ product slate produced via Birch reduction. This is not surprising given that the low temperature Birch reduction conditions should favor kinetically controlled products while the high temperature transfer hydrogenation conditions should favor thermodynamically controlled products. Nossal and co-workers also provided compelling ^{3}He NMR evidence for two $C_{60}H_{36}$ products (3:1 ratio) and suggested the possibility (among other suggestions) that those two products could correspond to C_1 and C_3 symmetric isomers of $C_{60}H_{36}$. The most compelling characterization of $C_{60}H_{36}$ isomers produced via transfer hydrogenation came from Gakh and co-workers (2003) who identified three isomers of C_1 symmetry (60–70%), C_3 symmetry (25–30%) and T symmetry (2–5%), respectively, using 2D NMR methods (^{1}H-^{13}C HSQC, ^{1}H-^{13}C COSY-HSQC, and ^{1}H-^{13}C TOCSY-HSQC). These three isomers are illustrated in Fig. 6.11.

Despite the lucid descriptions of the Birch reduction (Haufler et al. 1990) and transfer hydrogenation (Rüchardt et al. 1993) products, there is occasional confusion in the literature with some authors (mostly engineers and physicists) incorrectly surmising that a single $C_{60}H_{36}$ species forms in these reactions. The careful [60]fullerane aficionado will recognize that one $C_{60}H_{18}$ (C_{3v}) and three $C_{60}H_{36}$ (C_1, C_3 and T) isomers are all well documented.

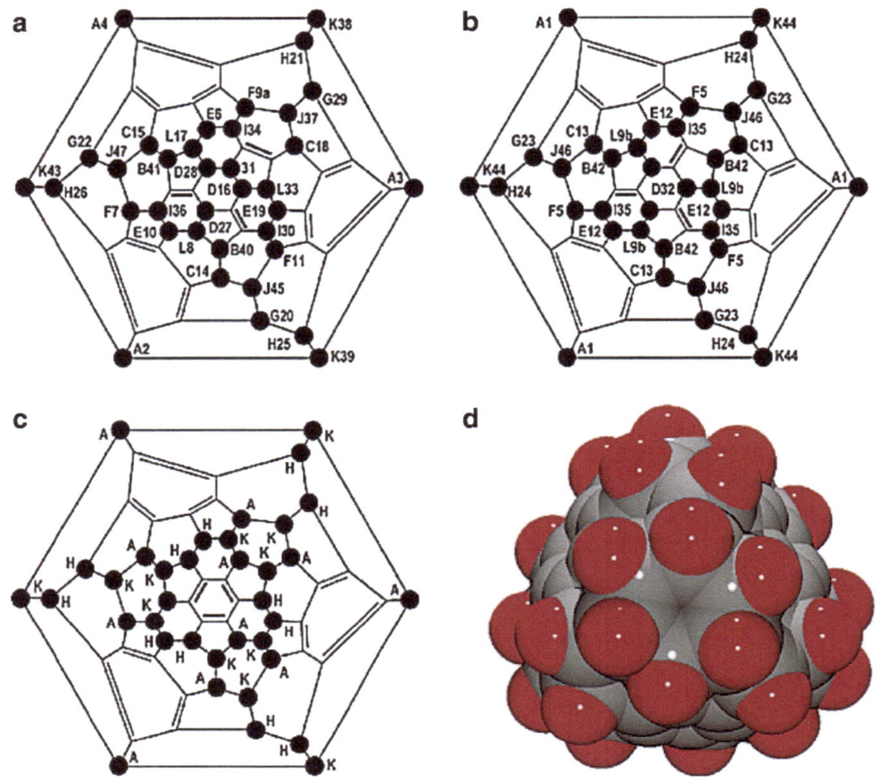

Fig. 6.11 Three isomers of $C_{60}H_{36}$ produced via transfer hydrogenation and identified using 2D NMR methods: Schlegel diagrams (**a–c**) and representative space filling model (**d**): (**a**) C_1 isomer, (**b**) a C_3 isomer and (**c**) a T isomer (Gakh et al. 2003) (Reprinted with permission from the American Chemical Society)

Yet another approach to prepare $C_{60}H_{36}$ isomers is the Zn-acid reduction which has been performed using both Brønsted (i.e., HCl) and Lewis acid (i.e., $ZnCl_2$) reagents. Several reports (Darwish et al. 1995; Cataldo 2003; Zhang et al. 2004a) describe these reactions and in each case, $C_{60}H_{36}$ isomers are the major products along with lesser amounts of more highly hydrogenated fulleranes. The reactions are simple, fast and reliable. In a typical reaction (Darwish et al. 1995), Zn was added to a solution of [60]fullerene in benzene or toluene that was maintained in the dark under an atmosphere of N_2. With rapid stirring, concentrated HCl was added over 5 minutes at room temperature. Within 1 hour, the solution color changed from violet to brown to colorless and the resulting crude product analyzed for multiple isomers of $C_{60}H_{36}$ as evidenced by ^{1}H NMR spectroscopy and mass spectrometry. Although HPLC separations provided partial resolution of crude product mixtures, a detailed characterization of individual isomers resulting from a Zn-acid reduction has yet to be reported.

6.6 [60]Fulleranes With More Than 36 Hydrogens: $C_{60}H_{>36}$

The formation of [60]fulleranes with more than 36 hydrogens is well documented using multiple methods including direct hydrogenation and catalytic hydrogenation methods but the products have generally escaped characterization by structure sensitive analytical techniques and no single isomer of formula $C_{60}H_{>36}$ has ever been isolated. Nonetheless, the possibility of forming a stabilized, highly hydrogenated [60]fullerane, perhaps even $C_{60}H_{60}$ with several hydrogen atoms located inside the cage, is very real and only limited by the creativity and imagination of the skilled synthetic chemist.

For many researchers, the lure of producing and studying highly hydrogenated fullerane species like, for example, $C_{60}H_{60}$, is intoxicating. Unfortunately, a $C_{60}H_{60}$ fullerane with all hydrogen atoms located outside the cage (i.e., *exo*) would suffer from severe angle and torsional strains. However, Saunders reported (Saunders 1991) MM3 calculations that predict stabilized $C_{60}H_{60}$ structures in which several of the hydrogen atoms are located inside the cage (i.e., *endo*). The most stable $C_{60}H_{60}$ derivative calculated by Saunders was a C_1 isomer with 10 hydrogen atoms *endo* and the remaining 50 hydrogens *exo*. Such a derivative could be prepared utilizing a chemistry that reversibly cleaves cage C–C cage bonds to form transient free radical methines that recombine following hydrogen inversion. While there are no experimental reports of fulleranes possessing covalently bound *endo* hydrogens, there are numerous suggestions in the literature that cage C–C bonds are weakened upon *exo* hydrogenation. These include the highly interesting reports of cage fragmentation and collapse during prolonged exposure to hydrogenation conditions. Thus, in an effort to prepare highly hydrogenated [60]fullerane, Talyzin and coworkers reported (Talyzin et al. 2004a, b, 2005a, b, 2006a, b) direct hydrogenations of [60]fullerene with H_2 at elevated pressures (50–120 bar) and temperatures (300–400 °C). These conditions led to complex mixtures of [60]fulleranes with a maximum of about 5.3 wt% hydrogen corresponding to an average structural formula of $C_{60}H_{40}$. High resolution atmospheric pressure photoionization Fourier transform ion cyclotron resonance mass spectrometry (HR API FT-ICR MS, of course) showed clear evidence for products containing 50–52 hydrogens (Talyzin et al. 2006b). Moreover, Talyzin and co-workers noted that fullerene sample weights initially increased upon hydrogen uptake but decreased after prolonged hydrogenation. HR API FT-ICR MS suggested partial cage collapse and fragmentation with the release of organic species including polycyclic aromatic hydrocarbons as well as fragmented fulleranes of formula $C_{54-59}H_x$ (Talyzin et al. 2006b). While isolation and careful characterization of fragmentation products has yet to be accomplished, the results tentatively suggest the possibility of selectively rupturing C–C bonds on highly hydrogenated fulleranes for synthetic gain.

Highly hydrogenated fulleranes of formula $C_{60}H_{>36}$ have been made by several other groups using a variety of methods. Thus, Darwish and co-workers performed a direct hydrogenation of [60]fullerene at high pressure (8630 psi, 595 bar) and 425 °C (Darwish et al. 2000). They reported formation of relatively stable [60]fulleranes that could be fractionated by HPLC into bands that were enriched in $C_{60}H_{42}$, $C_{60}H_{44}$, $C_{60}H_{48}$ and $C_{60}H_{52}$. Likewise, Talyzin and co-workers reported that a direct hydrogenation at even higher pressure (2 GPa, 20,000 bar) and temperatures (450–550°C) gave rise to $C_{60}H_{44-52}$ species as evidenced by Raman and IR spectroscopies, X-ray diffraction and HR API FT-ICR MS (Talyzin et al. 2006a). Interestingly, [60]fullerene is reported to polymerize under these high temperature, high pressure reaction conditions, only to depolymerize during hydrogenation. Much lower pressures can be utilized when catalysts are employed. Thus, Drelinkiewicz and co-workers reported catalytic hydrogenations using Pd, Ru and Rh catalysts at atmospheric pressure and 50–70 °C (Drelinkiewicz et al. 1996). While they claim formation of highly hydrogenated fulleranes with formulas $C_{60}H_{42-46}$, the analytical data (IR, ^{1}H NMR, UV–vis, elemental analysis) did not include mass spectrometry. Deshmukh and co-workers reported that a catalytic hydrogenation (Pd/C) of [60]fullerene under mild conditions (1 bar H_2, RT, 48 h) in a Lewis acidic ionic liquid gave $C_{60}H_{40}$ as evidenced by MALDI MS (Deshmukh et al. 2008). Gerst and co-workers demonstrated that a transfer hydrogenation of [60]fullerene in the presence of both 9,10-dihydroanthracene and [7H]benzanthrene (organic catalyst) gave a complex mixture of [60]fulleranes up to and including $C_{60}H_{44}$ (Gerst et al. 1993). Peera and co-workers utilized a Benkeser reduction of $C_{60}H_{36}$ to produce an equally complex mixture of [60]fulleranes with formulae $C_{60}H_{38-44}$ (Peera et al. 2003).

6.7 Conclusion

Since the first Birch reduction to prepare hydrogenated [60]fullerenes or [60]fulleranes (Haufler et al. 1990), many other synthetic methods have been developed and successfully demonstrated. The simplest of all [60]fullerane derivatives, 1,2-$C_{60}H_2$, is prepared in good isolated yield using either a $NaBH_4$ reduction (59% isolated; Wang et al. 2005) or a Zn(Cu) couple reduction (66% isolated; Bergosh et al. 1997). A few $C_{60}H_4$ (e.g., 1,2,3,4-$C_{60}H_4$; Henderson et al. 1994; Avent et al. 1994; Nossal et al. 2001a) and $C_{60}H_6$ (e.g., 1,2,33,41,42,50-$C_{60}H_6$; Meier et al. 1996) isomers have been formed and identified using structure sensitive analytical techniques like ^{1}H and ^{13}C NMR spectroscopy, but evidence for [60]fulleranes of formula $C_{60}H_{8-16}$ is based largely upon structure insensitive mass spectrometry data. Crown shaped C_{3v} $C_{60}H_{18}$ is the only [60]fullerane species that can be produced in high yield using multiple synthetic methods (Briggs et al. 2005; Wågberg et al. 2005; Kintigh et al. 2007; Athans et al. 2007). It is a stable, easily isolable species of limited solubility that persists for years. Of the synthetic methodologies available, the polyamine chemistries (conventional heating (Briggs et al. 2005) or microwave irradiation (Kintigh et al. 2007)) are best because the yields are excellent, the reactions are scalable, only standard glassware is required, the work-up is trivial, and no purification of C_{3v} $C_{60}H_{18}$ is required. Birch reductions (Haufler et al. 1990; Banks et al. 1993; Vasil'ev et al. 2000; Nossal et al. 2001b; Zhang et al. 2004b; Wang and Zhang 2006), transfer hydrogenations (Rüchardt et al. 1993) and Zn-acid conditions (Darwish et al. 1995;

Cataldo 2003; Zhang 2004a) all lead to mixtures of $C_{60}H_{36}$ isomers in good yield. Of these methods, the transfer hydrogenation reaction gives the cleanest product slate with formation of three isomers of C_1 symmetry (60–70%), C_3 symmetry (25–30%) and T symmetry (2–5%), respectively (Gakh 2003). However, no single $C_{60}H_{36}$ regioisomer has been isolated. The formation of [60]fulleranes with more than 36 hydrogens is well documented using multiple methods but the products have generally escaped characterization by structure sensitive analytical techniques. No single isomer of formula $C_{60}H_{>36}$ has ever been isolated.

Several high profile [60]fulleranes are noticeably missing from the portfolio of successfully synthesized species. These include $C_{60}H_{12}$ with six sets of adjacent (1,2) hydrogens placed about the [60]fullerane cage with octahedral geometry, T_h symmetric $C_{60}H_{24}$, and of course $C_{60}H_{60}$, perhaps with several hydrogen atoms located inside the cage. The formation of these and other exciting [60]fulleranes is only limited by the creativity and imagination of the skilled synthetic chemist.

6.8 Author Biographies

Jonathan Briggs earned his B.S. and M.S. degrees in the Department of Chemistry at the University of New Hampshire (Durham, NH) in 2000 and 2003, respectively, studying Diels–Alder reactions between [60]fullerene and various functionalized acenes. Since 2003, Jon has been employed as a Research Scientist at the University of New Hampshire. Glen Miller earned his B.S. and Ph.D. degrees from Clarkson University (Potsdam, NY) in 1987 and 1991, respectively, studying the formation and reactivity of carbodications in superacidic media. In 1990, he joined Exxon Research & Engineering Co. (Annandale, NJ) where he worked as Postdoctoral Fellow, Research Chemist and Senior Chemist before joining the University of New Hampshire in 1995. Currently a Professor of Chemistry and Materials Science, Glen Miller also directs the Materials Science Program and is Associate Director of the Nanoscale Science & Engineering Center for High-rate Nanomanufacturing at the University of New Hampshire.

References

Athans AJ, Briggs JB, Jia W, Miller GP (2007) J Mater Chem 17:2636–2641

Avent AG, Darwish AD, Heimbach DK, Kroto HW, Meidine MF, Parsons JP, Remars C, Roers R, Ohashi O, Taylor R, Walton DRM (1994) J Chem Soc Perkin Trans 2:15–22

Ballenweg S, Gleiter R, Krätschmer W (1993) Tetrahedron Lett 34:3737–3740

Banks MR, Dale MJ, Gosney I, Hodgson PKG, Jennings RCK, Jones AC, Lecoultre J, Langridge-Smith PRR, Maier JP, Scrivens JH, Smith MJC, Smyth CJ, Taylor AT, Thorburn P, Webster AS (1993) J Chem Soc Chem Commun 1149–1152

Becker L, Evans TP, Bada JL (1993) J Org Chem 58:7630–7631

Bergosh RG, Meier MS, Laske Cooke JA, Spielmann HP, Weedon BR (1997) J Org Chem 62:7667–7612

Billups WE, Luo W, Gonzalez A, Arguello D, Alemany LB, Marriott T (1997) Tetrahedron Lett 38:171–174
Briggs JB, Montgomery M, Silva LL, Miller GP (2005) Org Lett 7:5553–5555
Cataldo F (2003) Fullerenes Nanot Carbon Nanostruct 11:295–316
Darwish AD, Kroto HW, Taylor R, Walton DRM (1993) Fullerene Sci Technol 1:571–574
Darwish AD, Abdul-Sada AK, Langley GJ, Kroto HW, Taylor R, Walton DRM (1995) J Chem Soc Perkin Trans 2(12):2359–2365
Darwish AD, Avent AG, Taylor R, Walton DRM (1996) J Chem Soc Perkin Trans 2:2051–2054
Darwish AD, Taylor R, Loutfy R (2000) Fullerenes 2000-volume 9: functionalized fullerenes. Electrochem Soc Proc 11:179–185
Deshmukh RR, Lee JW, Shin US, Lee JY, Song CE (2008) Angew Chem Int Ed 47:8615–8617
Drelinkiewicz A, Byszewski P, Bielanski A (1996) React Kinet Catal Lett 59:19–27
Fagan PJ, Krusic PJ, Evans DH, Lerke SA, Johnston E (1992) J Am Chem Soc 114:9697–9699
Gakh AA, Romanovich AY, Bax A (2003) J Am Chem Soc 125:7902–7906
Gerst M, Beckhaus HD, Rüchardt C, Campbell EEB, Tellgmann R (1993) Tetrahedron Lett 34:7729–7732
Guarr TF, Meier MS, Vance VK, Clayton M (1993) J Am Chem Soc 115:9862–9863
Haufler RE, Conceicao J, Chibante LPF, Chai Y, Byrne NE, Flanagan S, Haley MM, O'Brien SC, Pan C, Xiao Z, Billups WE, Ciufolini MA, Hauge RH, Margrave JL, Wilson LJ, Curl RF, Smalley RE (1990) J Phys Chem 94:8634–8636
Henderson CC, Cahill PA (1993) Science 259:1885–1887
Henderson CC, Rohlfing CM, Assink RA, Cahill PA (1994) Angew Chem Int Ed Engl 33:786–788
Jin C, Hettich R, Compton R, Joyce D, Blencoe J, Burch T (1994) J Phys Chem 98:4215–4217
Kim E, Weck PA, Berber S, Tomanek D (2008) Phys Rev B. 78(1–4):113404
Kintigh J, Briggs JB, Letourneau K, Miller GP (2007) J Mater Chem 17:4647–4651
Mandrus D, Kele M, Hettich RL, Guiochon G, Sales BC, Boatner LA (1997) J Phys Chem B 101:123–128
Maroncelli M, Macinnis J, Fleming GR (1998) Science 243:1674–1681
Matsuzawa N, Fukunaga T, Dixon DA (1992) J Phys Chem 96:10747–10756
Meier MS, Corbin PS, Vance VK, Clayton M, Mollman M, Poplawska M (1994) Tetrahedron Lett 35:5789–5792
Meier MS, Weedon BR, Spielmann HP (1996) J Am Chem Soc 118:11682–11683
Meier MS, Spielmann HP, Bergosh RG, Haddon RC (2002) J Am Chem Soc 124:8090–8094
Miller GP (2006) C R Chimie 9:952–959
Niyazymbetov ME, Evans DH, Lerke SA, Cahill PA, Henderson CC (1994) J Phys Chem 98:13093–13098
Nossal J, Saini RK, Alemany LB, Meier M, Billups WE (2001a) Eur J Org Chem 4167–4180
Nossal J, Saini RK, Sadana AK, Bettinger HF, Alemany LB, Scuseria GE, Billups WE, Saunders M, Khong A, Weisemann R (2001b) J Am Chem Soc 123:8482–8495
Peera A, Saini RK, Alemany LB, Billups WE, Saunders M, Khong A, Syamala MS, Cross RJ (2003) Eur J Org Chem 21:4140–4145
Rüchardt C, Gerst M, Ebenhoch J, Beckhaus HD, Campbell EEB, Tellgmann R, Schwarz H, Weiske T, Pitter S (1993) Angew Chem Int Ed Engl 32:584–586
Saunders M (1991) Science 253:330–331
Sui Y, Qian J, Zhang J, Zhou X, Gu Z, Wu Y, Fu H, Wang J (1996) Fullerene Sci Technol 4:813–818
Talyzin AV, Shulga YM, Jacob A (2004a) Appl Phys A 78:1005–1010
Talyzin AV, Sundqvist B, Shulga YM, Peera AA, Imus P, Billups WE (2004b) Chem Phys Lett 400:112–116
Talyzin AV, Tsybin YO, Peera AA, Schaub TM, Marshall AG, Sundqvist B, Mauron P, Züttel A, Billups WE (2005a) J Phys Chem B 109:5403–5405
Talyzin AV, Tsybin YO, Schaub TM, Mauron P, Shulga YM, Züttel A, Sundqvist B, Marshall AG (2005b) J Phys Chem B 109:12742–12747

Talyzin AV, Dzwilewski A, Sundqvist B, Tsybin YO, Purcell JM, Marshall AG, Shulga YM, McCammon C, Dubrovinsky L (2006a) Chem Phys 325:445–451

Talyzin AV, Tsybin YO, Purcell JM, Schaub TM, Shulga YM, Noréus D, Sato T, Dzwilewski A, Sundzvist B, Marshall AG (2006b) J Phys Chem A 110:8528–8534

Vasil'ev Y, Wallis D, Nüchter M, Ondruschka B, Lobach A, Drewello T (2000) Chem Commun 1233–1234

Vieira SMC, Ahmed W, Birkett PR, Rego CA (2001) Chem Phys Lett 347:355–360

Vieira SMC, Ahmed W, Birkett PR, Rego CA, Kotsiris S, Drewello T (2004) Fullerenes Nanot Carbon Nanostruct 12:139–145

Wågberg T, Johnels D, Peera A, Hedenström M, Schulga YM, Tsybin YO, Purcell JM, Marshall AG, Noreus D, Sato T, Talyzin AV (2005) Org Lett 7:5557–5560

Wang NX, Zhang JP (2006) J Phys Chem A 110:6276–6278

Wang GW, Weedon BR, Meier MS, Saunders M, Cross RJ (2000) Org Lett 2:2241–2243

Wang GW, Li YJ, Li FB, Liu YC (2005) Lett Org Chem 2:595–598

Zhang JP, Wang NX, Xing YL, Yu AG, Yang YX, Wang WW, Sheng RL (2004a) J Chem Res 7:502–503

Zhang JP, Wang NX, Yang YX, Yu AG (2004b) Carbon 42:675–676

Chapter 7
Synthesis, Stability and Spectroscopy of Perdeuterofulleranes: $C_{60}D_{36}$ and $C_{70}D_{38}$ Evidences of Isotope Effects

Franco Cataldo[1,2], Susana Iglesias-Groth[3], and Arturo Manchado[3]

Abstract Perdeuterated fulleranes $C_{60}D_{36}$ and $C_{70}D_{38}$ were synthesized by the action of deuterium chloride on zinc dust. The atomic deuterium formed on the surface of the Zn granules was effective in deuterating C_{60} and C_{70} fullerenes in solution. The chemical structure of the perdeuterated fulleranes was studied in comparison to the corresponding hydrogenated derivatives $C_{60}H_{36}$ and $C_{70}H_{38}$ by electronic absorption spectroscopy and FT-IR spectroscopy. With the latter analytical technique the isotope effect on the infrared spectra was studied in detail. The thermal stability of the perdeuterated fullerenes shows a remarkable isotope effect in comparison to hydrogenated fullerenes as measured by thermogravimetry and differential thermal analysis. Even the oxidation resistance to air appeared improved in the case of the perdeuterated fullerenes.

7.1 Introduction

Fulleranes, the hydrogenated fullerenes, are an interesting and hot topic which may find application as reversible hydrogen storage and release medium (Withers et al. 1997; Peera et al. 2004; Schur et al. 2008), and for the astrochemical aspects (Iglesias-Groth 2004, 2006; Cataldo 2003a, b; Stoldt et al. 2001). In the latter case there is the need to have reference spectra for the search of fullerene derivatives in the interstellar medium.

At present numerous synthetic approaches have been developed in the synthesis of fulleranes. Wet methods can be distinguished from dry methods. Among the wet

[1]Istituto Nazionale di Astrofisica, Osservatorio Astrofisico di Catania, Via S. Sofia 78, 95123 Catania, Italy

[2]Actinium Chemical Research, Via Casilina 1626/A, 00133 Rome, Italy
e-mail: franco.cataldo@fastwebnet.it

[3]Instituto de Astrofisica de Canarias, Via Lactea s/n, E-38200 La Laguna, Tenerife, 2 CSIC, Spain
e-mail: sigroth@iac.es; amt@iac.es

F. Cataldo and S. Iglesias-Groth (eds.), *Fulleranes: The Hydrogenated Fullerenes*, Carbon Materials: Chemistry and Physics 2, DOI 10.1007/978-1-4020-9887-1_7,

routes we may include the Birch reduction, hydrazine, borane (Taylor 1999), diimide reduction (Avent et al. 1994) the reduction with Zn/HCl (Cataldo 2003a, b; Darwish et al. 1995, 1996a, b). Recently also a noteworthy fullerene hydrogenation with diethyltriamine has been reported (Kintigh et al. 2007). On the other hand dry methods range from the use of atomic hydrogen (Vieira et al. 2001, 2004; Eremtchenko et al. 2005; Petrie et al. 1995), to the catalytic hydrogenation (Talyzin et al. 2004), to the treatment of fullerenes under a very high H_2 pressure and high temperature in the absence of any catalyst (Talyzin et al. 2006a, b; Wågberg et al. 2008; Meletov et al. 2001). Hydrogen transfer reactions (Gerst et al. 1993; Ruchardt et al. 1993) have also been successfully employed in the fullerene hydrogenation, for instance by melting dihydroanthracene with fullerene in a sealed capillary.

When C_{60} is treated in solution with a mixture of Zn and HCl, depending on the conditions it is possible to isolate the highly hydrogenated $C_{60}H_{36}$. The success of such reaction is based on the formation of "nascent" hydrogen on the surface of the Zn granules treated with hydrochloric acid. "Nascent" hydrogen is simply atomic hydrogen chemisorbed on the metal surface and hence in a very reactive form ready to hydrogenate the substrate. In this chapter we report the details of the wet synthesis of hydrogenated and perdeuterated C_{60} and C_{70}. The perdeuterated fulleranes can be prepared by treating C_{60} with Zn and deuterium chloride instead of hydrogen chloride. The synthesis of the perdeuterofullerane $C_{60}D_{36}$ has been reported (Darwish et al. 1996a; Ruchardt et al. 1993) but it has never been isolated in macroscopic quantities and the FT-IR of the resulting perdeuterofullerane has never been published and analyzed. Furthermore, the thermal stability of $C_{60}D_{36}$ has never been analyzed in any detail. Similarly, the hydrogenated C_{70} fullerene prepared by any synthetic route has been fully characterized by a variety of analytical techniques as $C_{70}H_{38}$ (Darwish et al. 1995, 1996b; Kintigh et al. 2007; Wågberg et al. 2008), however only in one case the infrared spectrum of this molecule has been published (Darwish et al. 1995). The synthesis of perdeuterated C_{70} has been reported (Darwish et al. 1995) but its infrared spectrum never published and its properties never studied in any detail.

In the present chapter we report our results on these specific topics.

7.2 Experimental

7.2.1 Materials and Equipment

C_{60} and C_{70} fullerenes were obtained from Southern Chemical L.L.C. and were both 99.9+% pure grade. Toluene and n-hexane were purchased from Aldrich as well as DCl 37% in D_2O (deuterium enrichment 99%). Zn dust was from Aldrich as well and was characterized by a particle size <10 μm.

$C_{60}H_{36}$ was prepared as detailed in a previous work (Cataldo 2003a, b).

The FT-IR spectra were obtained on a Nicolet IR-300 spectrometer from Thermo-Electron Corp. The spectra were recorded either in transmittance or in reflectance mode using respectively KBr or Ge as windows.

The electronic absorption spectra were recorded on a Shimadzu UV160A spectrophotometer in n-hexane solutions.

The thermal stability was studied on a Linseis thermobalance model L81 + DTA under a nitrogen flow of 20 L/h and at a heating rate of 10°C/min.

7.2.2 Synthesis of $C_{60}H_{38}$ in n-Hexane

A saturated solution of C_{60} in n-hexane (about 13 mg C_{60} in 300 mL of n-hexane) was charged in a 1L conical flask together with 12 g of Zn dust. The flask was equipped with a rubber cap. Aqueous HCl (37% solution, 22 mL first addition and 15 mL second addition) was added to the mixture which was immediately closed with the cap hand-shaken. Periodically the cap was opened to permit the release of the excess of hydrogen produced from the reaction between HCl and Zn. The colour of the solution changed from violet to water-white. The complete reduction of C_{60} was checked by UV spectroscopy, by sampling the hexane solution and checking the spectrum. The reduction is complete when the UV spectrum is dominated by a maximum at about 217 nm which is due to $C_{60}H_{36}$ (Cataldo 2003a, b). No other features should be detectable. Knowing the initial concentration of C_{60}, based on the absorbance of the reduced solution it was possible to determine the molar extinction coefficient of $C_{60}H_{36}$: ε_{217} = 17140 L cm^{-1} mol^{-1}.

7.2.3 Synthesis of $C_{70}H_{38}$ in n-Hexane

Deuterated C_{60}: $C_{60}D_{36}$ was synthesized in n-hexane using the same procedure detailed above for $C_{60}H_{36}$ using Zn dust and 37% deuterium chloride (DCl) in heavy water (D_2O). After complete reduction also the electronic absorption spectrum of $C_{60}D_{36}$ is dominated by a maximum at 217 nm and the molar extinction coefficient was determined ε_{217} = 16480 L cm^{-1} mol^{-1}.

7.2.4 Synthesis of Perdeuterofullerane

C_{60} fullerene (96 mg) was dissolved in toluene (150 mL) in a conical flask with rubber stopper. Zn powder (24 g) was added and then 20 mL of DCl 37%. Immediately after the addition of DCl, the flask was closed with the rubber stopper and hand-shaken. Periodically the stopper was slightly raised to permit to discharge the pressure of D_2. After this treatment the solution changed its colour from violet to light orange. Then, other 10 mL of DCl were added and the mixture shaken again in the closed flask.

The organic layer was then separated from the aqueous and from the excess of Zn dust by decantation and with the aid of a separatory funnel. The toluene solution was distilled under the reduced pressure of a rotary pump in an oil bath kept at 130°C. After complete distillation of toluene a yellowish residue remained in the distillation flask. Nitrogen was admitted in the flask which was quickly transferred into a cold water bath. The reaction product immediately after the synthesis was collected from the flask and employed for recording of the FT-IR spectrum.

7.2.5 Synthesis of $C_{70}H_{38}$ in Toluene

A saturated toluene (200 mL) solution of C_{70} (280 mg) was transferred into a thick-walled 1 L conical flask equipped with a rubber cap. Zinc dust (18.0 g) was charged in the flask and after swirling, HCl 37% (26 mL) was added divided in two portions. After the addition of the first HCl portion (13 mL) the flask was closed with the cap and hand-shaken vigorously for few seconds, then the rubber cap was opened to release the excess of hydrogen pressure and closed immediately and shaken continuously. When the production of hydrogen faded, the flask was opened and the second HCl portion (13 mL) added. The flask was then closed and vigorously hand-shaken taking care time to time to open slightly the cap to release the excess of hydrogen under pressure. The solution after the hydrogen treatment becomes completely water-white while initially it was reddish in color. The C_{70} fullerene solution in toluene was separated from the zinc dust and the water layer by decantation, it was passed quickly through a paper filter, transferred in a round bottomed flask and distilled under reduced pressure in a water bath at 82°C. A white residue of hydrogenated C_{70} remained in the flask. As soon as produced it was sampled and used for recording the FT-IR spectrum minimizing any possible oxidation.

7.2.6 Synthesis of $C_{70}H_{38}$ in Benzene

A saturated solution of C_{70} (190 mg) in benzene (150 mL) was prepared and hydrogenated following the same procedure detailed in the previous Section 7.2.2. Immediately after the hydrogenation the solution was distilled under reduced pressure in a water bath at 50°C. The cream-white residue was collected for the FT-IR and thermogravimetric analysis.

7.2.7 Synthesis of $C_{70}D_{38}$ in Toluene

Perdeuterated [70] fullerene was prepared starting from a saturated toluene solution (200 mL) of C_{70} (280 mg) following the same procedure described in Section 7.2.2 for the synthesis of the $C_{70}H_{38}$. Of course the unique difference was the use of DCl 37% in D_2O instead of HCl 37% in H_2O.

The reaction product was isolated by distillation under reduced pressure as a white powder and employed immediately after the synthesis in the registration of the FT-IR spectrum and in the thermogravimetric analysis.

7.3 Results and Discussion

7.3.1 *Electronic Absorption Spectra of $C_{60}H_{36}$ and $C_{60}D_{36}$*

The hydrogenation of C_{60} in hydrocarbon solutions normally stops when 36 hydrogen or deuterium atoms have been added to the double bonds (Darwish et al. 1995; Taylor 1999, 2006). Thus, $C_{60}H_{36}$ and $C_{60}D_{36}$ represent the maximum achievable degree of fullerene hydrogenation. The addition of 36 hydrogen atoms leaves 12 residual double bonds in the fullerene cage out of the original 30 double bonds. The rather complex electronic absorption spectrum of C_{60} fullerene (with maxima at 213 nm ε_{213} = 135000, 257 nm ε_{257} = 175,000, 329 nm ε_{329} = 51,000, 404, 500, 540, 570, 600, 625) originating from the electronic transitions from the 30 weakly conjugated double bonds (Kroto et al. 1991), after hydrogenation is simplified into a single intense electronic transition band centred at 217 nm (Fig. 7.1). Among the numerous possible isomers of $C_{60}H_{36}$ calculations have shown that only four of them are stable and can exist. The isomers with T, S_6 and D_{3d} symmetry are characterized by the presence of four or two isolated benzenoid rings respectively in the hydrogenated fullerene cage. Instead, the other highly symmetric isomer with T_h symmetry, is characterized by 12 residual and isolated double bonds (Bensasson et al. 1997). In general the isolated benzenoid ring in partially hydrogenated polycyclic aromatic hydrocarbons such as octahydrophenanthrene, octahydroanthracene, dodecahydrochrysene or dodecahydrotriphenylene gives rise to the electronic transition at about 220 nm but it is always accompanied by strong absorption bands in the range of

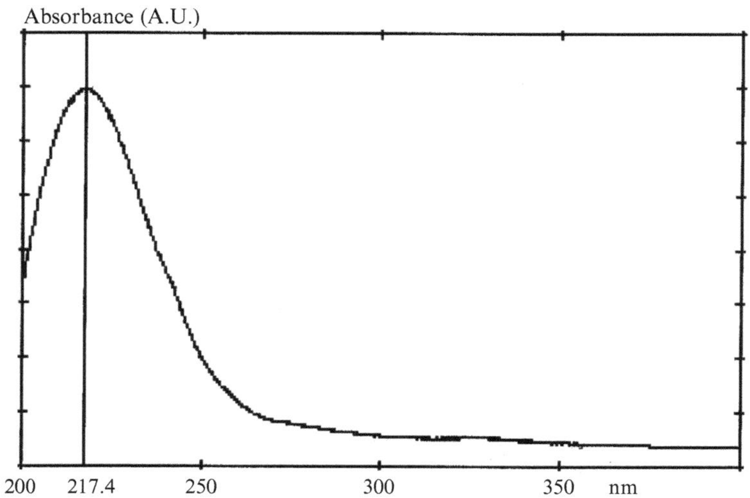

Fig. 7.1 Electronic absorption spectrum of $C_{60}H_{36}$ in n-hexane. The absorption maximum is indicated by the vertical line and is located at 217.4 nm with a ε_{217} = 17,140 L cm^{-1} mol^{-1}. A similar electronic absorption spectrum was recorded on $C_{60}D_{36}$ in n-hexane with a maximum of absorption located at 217 nm and ε_{217} = 16,480 L cm^{-1} mol^{-1}.

250–290 nm (Friedel & Orchin 1951) and such features at longer wavelengths are lacking in the spectrum of $C_{60}H_{36}$ (Fig. 7.1).

Another interesting aspect of $C_{60}H_{36}$ is its molar extinction coefficient ε. We have measured such coefficient since we started the hydrogenation from n-hexane solution of C_{60} where the solute concentration was determined spectrophotometrically by measuring the absorbance at 326 nm and using ε_{326} = 51,000 L cm^{-1} mol^{-1} (Kroto et al. 1991). After hydrogenation, by reading the absorbance at 217 nm and assuming the concentration of $C_{60}H_{36}$ as same as the starting C_{60} a value of ε_{217} = 17140 L cm^{-1} mol^{-1} has been calculated. The value of ε_{217} for $C_{60}H_{36}$ has the same order of magnitude as that reported by Bensasson et al. (1997) but it is 1.7 times lower. The difference is due to the fact that the $C_{60}H_{36}$ samples of Bensasson et al. (1997) were contaminated by significant amounts of anthracene impurities due to the synthetic procedure employed and their ε values were given with an uncertainty of >20%. Such contamination was completely absent in our samples which were synthesized by a different route. Additionally, the UV spectrum of $C_{60}H_{36}$ reported by Bensasson et al. (1997) was measured in CH_2Cl_2 which is not transparent to UV radiation as n-hexane so that the absorption peak was found at 230 nm and the higher ε may also be due to solvent absorption interference. However in transparent medium $C_{60}H_{36}$ has a unique absorption at 217 nm (Palit et al. 1998; Cataldo 2003a, b).

The electronic transition of $C_{60}H_{36}$ at 217 nm and the intensity of the transition (ε_{217} = 17140 L cm^{-1} mol^{-1}) are consistent with the presence of isolated double bonds in $C_{60}H_{36}$ and suggests that mainly the T_h isomer has been produced by the hydrogenation of C_{60} in solution. The T_h isomer is also considered the most stable among the possible isomers of $C_{60}H_{36}$ (Bensasson et al. 1997). For comparison, tetrasubsituted and isolated double bonds give rise to electronic transitions between 200 and 210 nm with a unique broad band similar to that shown in Fig. 7.1 and with $\varepsilon = 10^4$ L cm^{-1}mol^{-1} (Perkampus 1992). Simple conjugated dienes such as butadiene show a maximum of absorption just at 217 nm with ε = 21,000 L cm^{-1}mol^{-1} (Silverstein et al. 1981), again consistent with the spectrum of $C_{60}H_{36}$.

The replacement of hydrogen with deuterium in $C_{60}D_{36}$ does not alter the electronic transition of the molecular specie and indeed the electronic absorption spectrum of this molecule shown in Fig. 7.1 is very similar to that of the hydrogenated analogous with a maximum at 217 nm and ε_{217} = 16,480 L cm^{-1} mol^{-1}. Thus, also the intensity of the electronic transition of $C_{60}D_{36}$ is almost identical to that of $C_{60}H_{36}$.

7.3.2 The Electronic Absorption Spectra of Fulleranes $C_{70}H_{38}$ and $C_{70}D_{38}$

The maximum hydrogenation degree of C_{70} achievable by the action of Zn/HCl or Zn/DCl is represented by the two molecules $C_{70}H_{38}$ and $C_{70}D_{38}$ respectively (Darwish et al. 1995; Taylor 2006; Cataldo et al. 2009a, b; Wågberg et al. 2008), although in certain synthetic circumstances also $C_{70}H_{36}$ and $C_{70}H_{40}$ have been produced (Darwish et al. 1995; Taylor 2006). The electronic absorption spectrum of

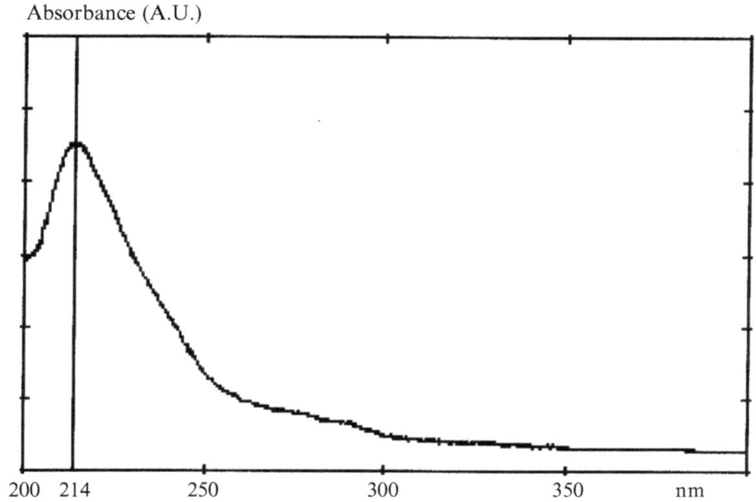

Fig. 7.2 Electronic absorption spectrum of $C_{70}H_{38}$ in n-hexane. The absorption maximum is indicated by the vertical line and is located at 214 nm with a ε_{214} = 6,300 L cm^{-1} mol^{-1}. A similar electronic absorption spectrum was recorded on $C_{70}D_{38}$ in n-hexane with a maximum of absorption located at 214 nm and ε_{217} = 5,800 L cm^{-1} mol^{-1}

$C_{70}H_{38}$ in n-hexane is shown in Fig. 7.2 and is characterized by a unique broad peak at 214 nm with additional weak features at 270–285 nm. The molar extinction coefficient was determined ε_{214} = 6,300 L cm^{-1} mol^{-1}. The molar extinction coefficient was determined in n-hexane starting from a ε C_{70} solution of known concentration as measured at 377 nm using ε_{C70} = 28,444 L cm^{-1} mol^{-1} (Wågberg et al. 2008). The C_{70} solution was subjected to hydrogenation with nascent hydrogen from Zn/HCl and the molar extinction coefficient of the resulting $C_{70}H_{38}$ was determined from its absorbance at 214 nm.

Although $C_{70}H_{38}$ shows an electronic spectrum similar to that of $C_{60}H_{36}$ (compare Fig. 7.1 with Fig. 7.2), the molar extinction coefficient of $C_{70}H_{38}$ is approximately 1/3 of that of $C_{60}H_{36}$. As already discussed previously, a weak electronic transition at 214 nm is not consistent with the presence of benzenoid rings in the hydrogenated fullerene derivative and this reasoning applies also in the case of $C_{70}H_{38}$ with an ε_{214} of only 6,300 L cm^{-1} mol^{-1}. In fact, theoretical calculations on $C_{70}H_{36}$ have shown that the most stable isomer does not posses any benzenoid ring and all the residual double bonds are isolated and distributed throughout the entire molecule (Fowler et al. 1996). Thus, as general rule, hydrogenation with nascent hydrogen from Zn/HCl seems to lead to fullerane isomers free from benzenoid rings. Of course, other synthetic methods such as the use of very high pressure and temperature can produce other isomers of fullerane with benzenoid rings as shown by Wågberg and colleagues 2008 and Fowler et al. (1996) for $C_{70}H_{38}$.

The isotopic substitution of hydrogen atoms with deuterium in fulleranes do not alter the electronic transitions. This has already been observed in other molecules of the fullerane family and can be observed in Fig. 7.2 for the present case of $C_{70}D_{38}$. This molecule displays an absorption maximum at 214 nm and shows also some weak features at 270–285 nm as in the case of its hydrogenated analogous. The molar extinction coefficient of $C_{70}D_{38}$ is slightly smaller than that of the hydrogenated analogous: a value $\varepsilon_{214} = 5{,}800$ L cm^{-1} mol^{-1} was measured.

7.3.3 *Aspects on Deuteration: the Isotope Effect*

The synthesis of $C_{60}H_{36}$ can be achieved in solution by treating C_{60} dissolved in a solvent with "nascent" hydrogen formed by the action of hydrochloric acid on zinc dust. Under appropriate conditions the reaction can be very smooth yielding almost exclusively $C_{60}H_{36}$ as shown by different analytical techniques by earlier investigators (Darwish et al. 1995, 1996b) and as confirmed later (Cataldo 2003a, b). Since the enthalpy of formation (at 298.15 K) of C_{60} is about 2,530 kJ/mol (Cataldo 1997; Karpushenkava et al. 2007) and ΔH_f° is only 1,206 kJ/mol for $C_{60}H_{36}$ (Karpushenkava et al. 2007), it is evident that the hydrogenation of C_{60} is largely a thermodynamically allowed reaction ($\Delta H_{react} = -1{,}324$ kJ/mol) and C_{60} acts as an hydrogen accumulator (Karpushenkava et al. 2007). Unfortunately, $C_{60}H_{36}$ suffers of certain instability (Cataldo 2003a, b) especially in presence of humidity and air probably due to allylic oxidation (Taylor 1999). Such instability appears typical of the fulleranes produced with the "wet" synthesis and instead is much less evident in fulleranes derived from dry, high pressure high temperature hydrogenation. Probably a role in the fulleranes instability prepared through the "wet" synthesis is played also by the fact that such synthesis is conducted in presence of conc. hydrochloric acid which may act as a catalyst in the above mentioned allylic oxidation. This observation is fundamental otherwise is not explainable why the fulleranes prepared through the dry synthesis are so stable.

An attempt to improve the stability of hydrogenated C_{60} may involve the replacement of hydrogen with deuterium. The C–D bond strength is 341.4 kJ/mol while the C–H bond strength involves an energy of 338.3 kJ/mol (Kerr 1987). Deuteration may imply also a kinetic isotope effect (Adler et al. 1976; Hibbert 1980). The zero point energy of C–H bond is 32.6 kJ/mol while for C–D bond is only 24.4 kJ/mol (Adler et al. 1976). Because the molecule containing a C–D bond has a lower zero point energy of 8.2 kJ/mol than the corresponding molecule with C–H bond, it has a higher activation energy to reach the transition state in the C–D bond breaking than the case of C–H. If the C–D or C–H bond breaking of a certain reaction is the rate determining step, then the rate of C–D bond scission will be slower than that of C–H scission and therefore an isotope effect, i.e. accumulation of deuterated molecules in the reactant, will be observed (Adler et al. 1976). The maximum calculated isotope effect in terms of kinetic rate constants has been reported to be $k_H/k_D = 7$ (Adler et al. 1976) or even higher (Hibbert 1980), which means that the reactions involving the C–D bonds breakage could be >7 times slower than those involving the C–H bond scission.

Curiously, it has been observed that the interstellar organic molecules show significant enrichment in deuterium with respect to the interstellar hydrogen (Shaw 2006; Tielens 2005). Of course kinetic isotope effects have been advocated to explain the improved deuterium content (Shaw 2006; Tielens 2005). Such enrichment has been detected also in organic molecules present in meteorites suggesting that their composition reflects the composition of the interstellar medium (Shaw 2006). Recently, the deuterium enrichment has been detected even in the insoluble organic matter of carbonaceous meteorites containing radicals and a polyaromatic structure (Gourier et al. 2008).

7.3.4 FT-IR Spectroscopy of $C_{60}D_{36}$ and $C_{60}H_{36}$

In the treatment of two atoms connected together, a simple harmonic oscillator model can be adopted involving the two masses connected with a spring having a force constant f_k. Thus, the vibrational frequency in wavenumbers Ω depends from the reduced mass μ, from f_k with c being the velocity of light.

$$\Omega = (1/2\pi c)[f_k/\mu]^{1/2} \tag{7.1}$$

In the case of the C–H and C–D bond, it can be easily demonstrated using Eq. 7.1 that the theoretical ratio $\Omega_H/\Omega_D = 0.964\sqrt{2} = 1.36$ (Colthup et al. 1990). Figure 7.3 shows the FT-IR spectra of $C_{60}D_{36}$ in comparison to $C_{60}H_{36}$. The two spectra are identical with the exclusion of the C–H stretching and bending bands (Cataldo et al. 2009a). More in detail the C–H stretching occurs at 2,905 and 2,841 cm^{-1} and is shifted to 2,157 and 2,092 cm^{-1} in the case of C–D stretching. Thus, the experimental $\Omega_H/\Omega_D = 1.35$ not far from the theoretical value of 1.36. A isotopic shift ratio of $\approx$1.32 has been reported previously in literature (Meletov et al. 2001) in the case for $C_{60}H_{36}$ and $C_{60}D_{36}$. Thus, in the present study the isotopic shift ratio Ω_H/Ω_D appears closer to the theoretical value of 1.36. It is interesting to note that if the ratio of the frequencies, following deuteration, is much less than the theoretical value, it can be assumed that the vibration is not simply a C–H stretching vibration but instead a mixed vibration involving interaction (coupling) with other vibrations. Since in our case this is not the case, the C–H vibration on fullerene cage is almost purely due to the stretching without coupling with other vibrations. Instead, coupling with other vibrations should be admitted in the case of the C–D stretching since the band appears broader and not split into two different bands as in the case of the C–H stretching (see Fig. 7.3). Furthermore, an additional contribution to the $\nu_{C\text{–}D}$ band broadening may derive from the larger Van der Waals dimensions of the deuterium atoms in comparison to the hydrogen atoms which may cause a closer vibrational interaction among adjacent atoms.

Concerning the C–D bending, Fig. 7.3 shows a peak at 966 cm^{-1} in $C_{60}D_{36}$ which appears shifted at about 1,274 cm^{-1} and overlapped to other features in the spectrum

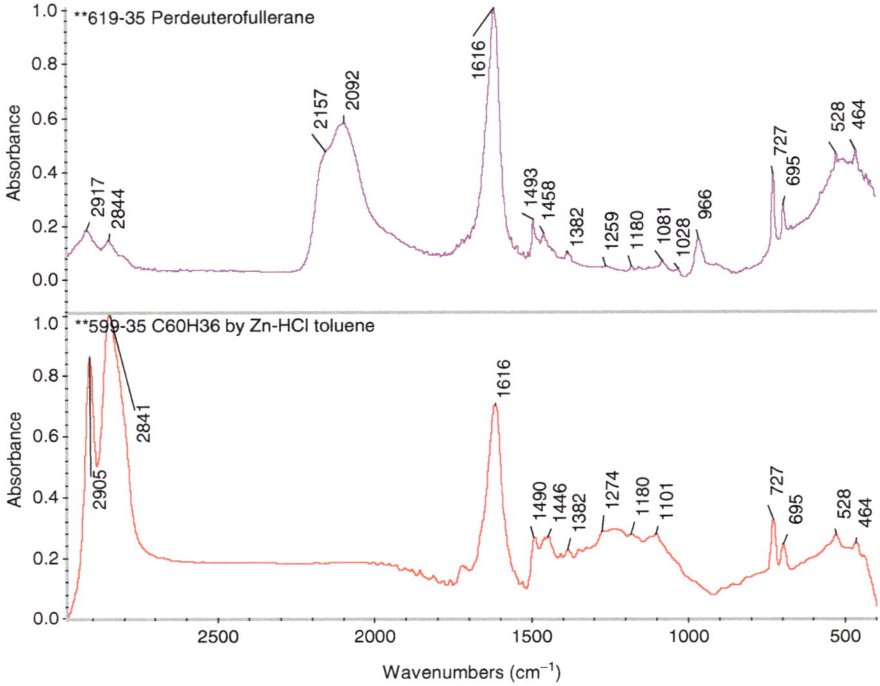

Fig. 7.3 FT-IR spectrum (in KBr) of $C_{60}D_{36}$ (*top*) in comparison to C_{60} H_{36} (*bottom*). Note the shift of the C–H stretching bands at 2,905 and 2,841 cm^{-1} which in the deuterated sample appear at 2,157 and 2,092 cm^{-1}

of $C_{60}H_{36}$ so that $\Omega_H/\Omega_D = 1.33$ in this case. Notoriously, the C–D bending is more intense than the C–H bending and deuteration studies are used to localize the position of this vibrational mode in the infrared spectra (Avram and Mateescu 1972).

It has been shown by mass spectrometry that the deuteration of C_{60} leads to products richer in deuterium content in comparison to the hydrogenation which instead leads almost exclusively to $C_{60}H_{36}$ (Taylor 1999; Darwish et al. 1995). However, the spectra in Fig. 7.3 show that $C_{60}H_{36}$ and its deuterated analogous display the same band pattern apart the discussed isotopic shift and hence it is reasonable to state that $C_{60}D_{36}$ has been obtained and studied. This observation has been made also by other authors who have used also Raman in addition to FT-IR spectroscopy (Meletov et al 2001).

Theoretical calculations have shown that $C_{60}H_{36}$ and consequently also its deuterated analogous are molecules with a number of possible isomers (Balasubramanian 2004). Symmetry considerations (Balasubramanian 2004) and thermodynamic calculations (Karpushenkava et al. 2007) restrict the number of stable isomers. Vibrational spectroscopic analysis combined with theoretical calculations have shown that the most stable isomers are those with symmetry S_6, T, T_h, D_{3d} (Meletov et al. 2001; Karpushenkava et al. 2007; Bini et al. 1998). The predominant

$C_{60}H_{36}$ isomer depends from the synthetic route. The synthesis under dry conditions under high H_2 pressure favours the D_{3d} isomer (Meletov et al. 2001; Bini et al. 1998) while the wet synthesis with Zn/HCl it was initially thought to favour the S_6 isomer. However, as discussed in the section of the electronic absorption spectra, considerations based on the absorption peak of both $C_{60}H_{36}$ and $C_{60}D_{36}$ and on their molar extinction coefficient have lead to assign the T_h symmetry to these molecules. T_h isomer is the sole isomer characterized by isolated double bonds and is free from benzenoid rings.

7.3.5 FT-IR Spectroscopy of $C_{70}D_{38}$ and $C_{70}H_{38}$

The FT-IR spectra of $C_{70}D_{38}$ and $C_{70}H_{38}$ are shown in Fig. 7.4. The C–H stretching bands are located at 2,903 and 2,829 cm^{-1} for the $C_{70}H_{38}$ sample synthesized in toluene and at 2,907 and 2,846 cm^{-1} for the sample prepared in benzene. The perdeuterated molecule $C_{70}D_{38}$ shows a broad band due to C–D stretching with a shoulder at 2,161 and a peak at 2,108 cm^{-1} so that the experimental value Ω_H/Ω_D = 2,903/2,

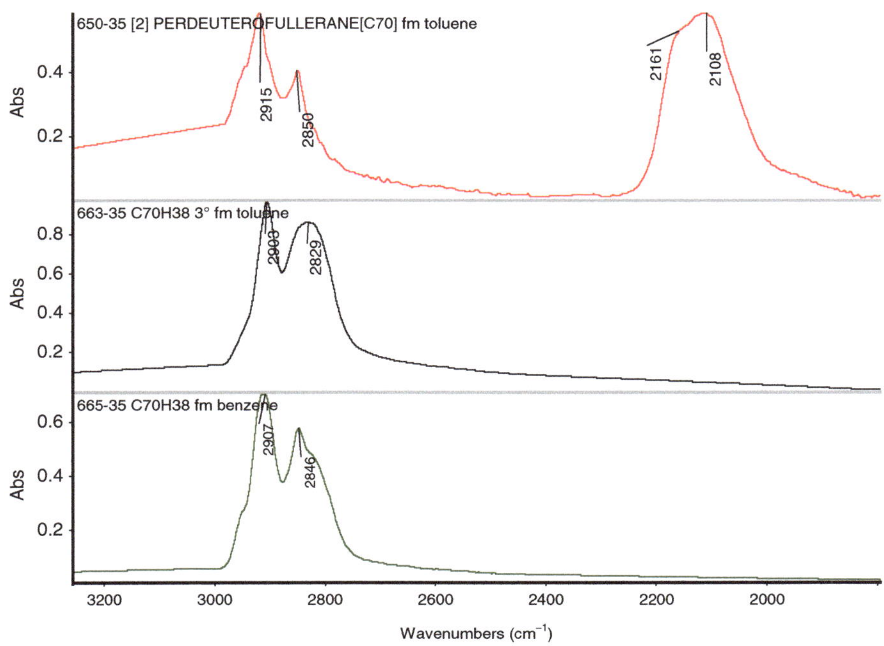

Fig. 7.4 FT-IR spectra (in KBr): (*top*) perdeuterated C_{70} fullerene $C_{70}D_{38}$ showing the C–D stretching bands at 2,161 and 2,108 cm^{-1}; the features at 2,915 and 2,850 cm^{-1} are due to residual C–H stretching. The spectra in the *middle* and at the *bottom* of the figure are due to $C_{70}H_{38}$ synthesized respectively in toluene and in benzene

161 ≈ 2,907/2,161 is 1.34 and the same value can be calculated also from the wavenumber ratios 2,829/2,108 and 2,846/2,108 for both the samples prepared in toluene and in benzene, in close agreement with the theoretical prediction. Thus, experimental stretching band shift after deuteration is so close to the theoretical value as already observed in the case of perdeuterated $C_{60}H_{36}$ that the C–H stretching frequencies are purely due to the stretching without coupling with other vibrations. Another feature of the hydrogenated C_{70} spectra reported in Fig. 7.4 regards the fact that the two resolved stretching bands observed in the hydrogenated molecule appear unresolved in a unique broad band after deuteration. This phenomenon has already been observed in the case of hydrogenated C_{60} and could be attributed to a certain degree of coupling of the C–D stretching vibrations which instead does not occur for the hydrogenated counterpart.

The FT-IR spectrum of $C_{70}D_{38}$ in the range between 1,800 and 400 cm^{-1} is shown in Fig. 7.5 in comparison to the spectra of $C_{70}D_{38}$ prepared in toluene and in benzene. There are no evidences of band shift due to deuteration with the obvious

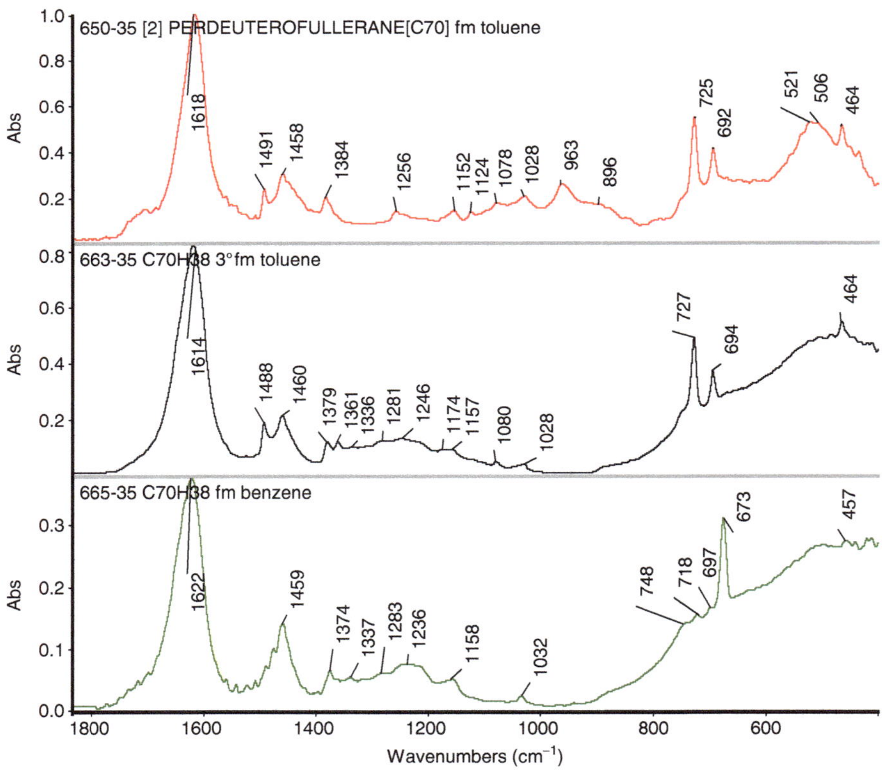

Fig. 7.5 FT-IR spectra (in KBr): (*top*) perdeuterated C_{70} fullerene $C_{70}D_{38}$ showing the C–D bending at 963 cm^{-1}. The spectra in the *middle* and at the *bottom* of the figure are due to $C_{70}H_{38}$ synthesized respectively in toluene and in benzene

exclusion of the C–D bending which appears at 963 cm^{-1}, almost at the same frequency of the C–D bending in perdeuterated fullerene (Cataldo et al 2009b) and therefore, the bending band of the hydrogenated counterpart should be localized at about 1,283 cm^{-1}. All the other vibrations in the spectral range considered in Fig. 7.3 are not affected by the isotopic substitution and hence are not connected to the C–H/C–D vibrations.

Fig. 7.5 shows also some small differences between the FT-IR spectrum of $C_{70}H_{38}$ prepared in toluene and that prepared in benzene. For example, the former shows two bands at 1,488 and 1,460 cm^{-1} as well as two sharp bands at 727 and 694 cm^{-1} while the latter sample shows only one band at 1,459 and at 673 cm^{-1} (although accompanied by three other bands at 748, 718 and 697 cm^{-1}). These spectral differences can be attributed to different hydrogenated isomeric composition of the two $C_{70}H_{38}$ samples. Hydrogenated $C_{70}H_{38}$ can be represented by a number of different isomeric structures as in the case of $C_{60}H_{36}$ (Taylor 1999). It is well known in organic chemistry that different solvents may affect the addition reactions leading to a different distribution of isomers (Reichardt 2003). Additionally, also different thermal processing during the work-up for the isolation of the $C_{70}H_{38}$ may have had an effect on isomeric composition of the products. In fact the sample synthesized in toluene was dried in vacuum at 82°C while that prepared in benzene was isolated at 50°C in vacuum.

7.3.6 Thermal Analysis (TGA, DTG and DTA) of $C_{60}D_{36}$ and $C_{60}H_{36}$

The thermal stability of fulleranes, their volatility (Dorozhko et al. 2001) and their ability to release hydrogen at high temperature have been the subject of numerous investigations. Particularly interesting is the ability of fulleranes to release molecular hydrogen at high temperatures regenerating the free fullerene. This fact has both technological implication for hydrogen storage and release (Withers et al. 1997; Peera et al. 2004; Karpushenkava et al. 2007; Schur et al. 2008) and astrochemical implications for instance in the explanation of the molecular hydrogen formation at low temperatures in the interstellar medium (Cataldo 2003b).

Previous studies (Peera et al. 2004; Schur et al. 2008; Cataldo 2003a; Avent et al. 1994; Talyzin et al. 2004, 2006a, b) have shown that $C_{60}H_{36}$ when heated in vacuum or under inert atmosphere is able to restore C_{60} fullerene, although the reaction is not fully reversible and also other decomposition products of fullerene and fullerane have been detected.

The thermogravimetric analysis (TGA) of $C_{60}H_{36}$ shows a weight loss larger than 7.7% at 510–520°C (Peera et al. 2004), a value which is larger than the theoretical content of hydrogen (4.76%) and which implies some partial decomposition of fullerane to other products. In particular by mass spectrometry it has been discovered that $C_{60}H_{36}$ decomposes between 480°C and 560°C liberating H_2 together with other hydrocarbons such as methane, ethane and benzene suggesting a partial cage breakdown (Talyzin et al. 2004, 2006b). Instead, under moderate heating $C_{60}H_{36}$ releases partially hydrogen producing $C_{60}H_{18}$ (Taylor 1999). C_{60} fulleranes with low degree of hydrogenation instead release the hydrogen content in stoichiometric amount between 480°C and 560°C.

In our TGA studies on $C_{60}H_{36}$ at a heating rate of 10°C/min and under nitrogen flow, we have found that the weight loss is almost double than the theoretical: –9.5% instead of the theoretical –4.76%, in line with the results of Talyzin and colleagues (2004). Even the perdeuterated $C_{60}D_{36}$ undergoes a weight loss which is double than that expected theoretically (found: –20%; theoretical: –9.1%). Thus it is confirmed that the hydrogen release from heavily hydrogenated fullerenes is not fully reversible since part of the substrate decomposes into various products (Talyzin et al. 2004, 2006b).

Concerning the decomposition temperature of hydrogenated and perdeuterated fullerenes, we have noticed an interesting difference which is evidenced in Fig. 7.6. The derivative thermogravimetry (DTG) of the two compounds $C_{60}H_{36}$ and $C_{60}D_{36}$ show two completely different peaks. In fact the maximum decomposition rate of $C_{60}H_{36}$ occurs at 501°C while that for $C_{60}D_{36}$ is shifted at 549°C. The value found for $C_{60}H_{36}$ is in line with that reported by other investigators (Peera et al. 2004). Instead the maximum decomposition rate of $C_{60}D_{36}$ at much higher temperature than its hydrogenated analogous can be attributed to the presence of C–D bond and hence to isotope effect. Heating implies the increase in the vibration of the C–H and C–D bonds until its rupture. The breakage of the C–H and C–D bonds appears to be a rate determining step in the thermal dehydrogenation of fullerane and

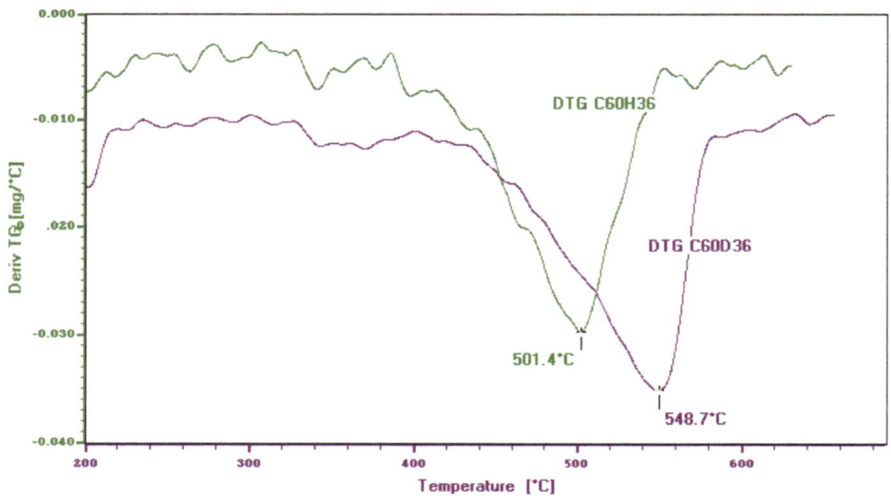

Fig. 7.6 Derivative thermogravimetry (DTG) under N_2 flow of $C_{60}H_{36}$ showing a maximum decomposition rate at 501°C. The perdeuterated $C_{60}D_{36}$ shows a decomposition at much higher temperature: 549°C. This temperature difference in the decomposition has been interpreted in terms of isotope effect due to deuteration

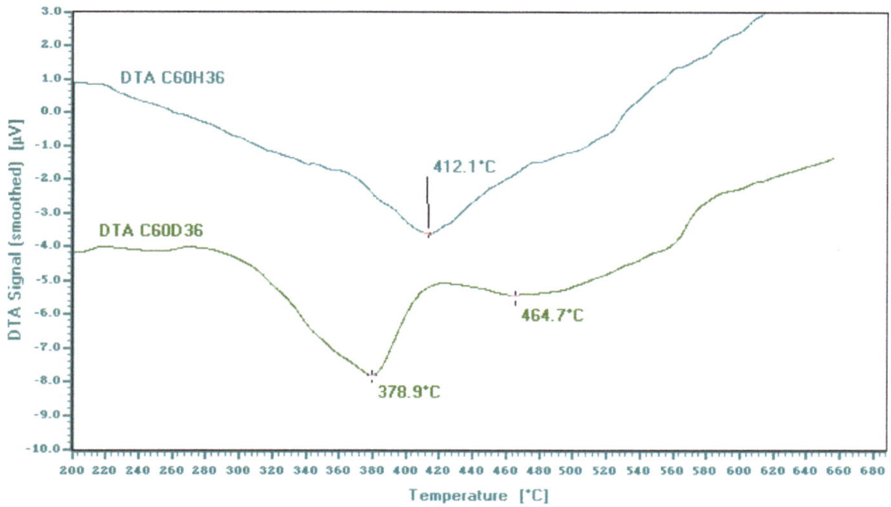

Fig. 7.7 Differential thermal analysis (DTA) in N_2 flow of $C_{60}H_{36}$ (*top trace*) showing a broad endothermic transition at 412°C, just at the onset of the decomposition.$C_{60}D_{36}$ shows two endothermic transitions at 379°C and 465°C. The second transition being the onset of the decomposition

perdeuterofullerane so that for the latter additional ≈25 kJ/mol energy should be administered to reach the maximum decomposition rate as justified by the higher temperature of the maximum decomposition rate. Although the dehydrogenation process of fulleranes is not a smooth process the difference in temperature of the maximum decomposition rate (and consequently the difference in energy) could be interpreted in terms of isotope effect.

The differential thermal analysis (DTA) of $C_{60}H_{36}$ and $C_{60}D_{36}$ also revealed a different thermal behaviour. As shown in Fig. 7.7 the former molecule shows an endothermic transition at 412°C which is just at the onset of the DTG large peak shown in Fig. 7.6. Conversely, $C_{60}D_{36}$ shows in Fig. 7.7 two endothermic transitions respectively at 379°C and 465°C. The latter being the onset of the decomposition DTG peak shown in Fig. 7.6. Thus, also the DTA measurement confirms a shift of +53°C between the onset decomposition of $C_{60}H_{36}$ and that of $C_{60}D_{36}$; this is in line with the shift of the deuterated compound of 48°C higher temperature in the peak of the maximum decomposition rate observed at the DTG.

7.3.7 *Thermal Stability of $C_{70}D_{38}$ in Comparison to $C_{70}H_{38}$*

As shown in the previous section, the thermogravimetric analysis of $C_{60}H_{36}$ and its deuterated counterpart does not match the theoretical content of hydrogen or deuterium. Talyzin and colleagues (2004, 2006b) have pointed out that the reason for this behaviour resides in the fact that fulleranes when heated do not release

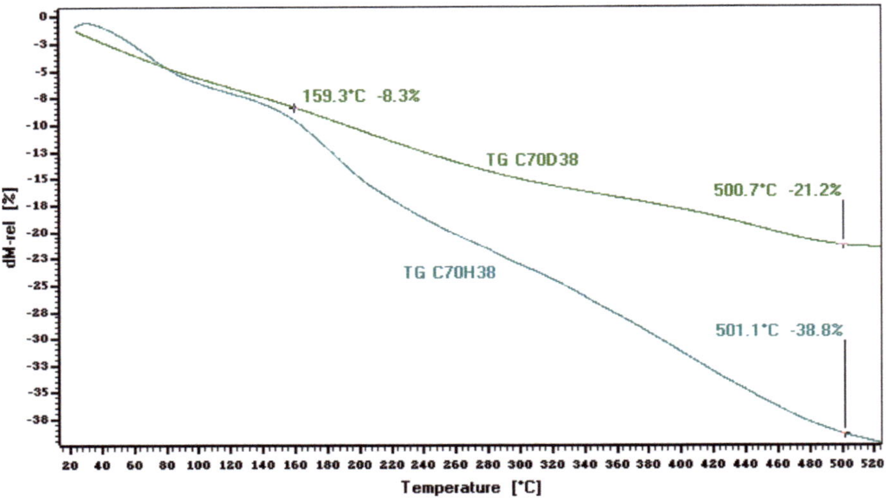

Fig. 7.8 Thermogravimetric analysis (TGA) under N_2 flow of $C_{70}D_{38}$ and $C_{70}H_{38}$. In both cases the weight loss exceeds by far the theoretical content of deuterium or hydrogen of the two molecules confirming that the action of heat involves the decomposition of the molecules with the release of a mixture of products leaving a carbonaceous residue of 78.8% in the case of $C_{70}D_{38}$ and 61.2% in the case of $C_{70}H_{38}$

smoothly only hydrogen but decompose into many products including methane and various aromatic hydrocarbons.

In this context also $C_{70}D_{38}$ and $C_{70}H_{38}$ does not represent an exception. In fact, the theoretical deuterium content of the former molecule is 8.3% but the weight loss measured with thermogravimetric analysis (TGA) at 500°C is more than double this value: 21.2% (see Fig. 7.8). Even more impressive is the case of $C_{70}H_{38}$ whose theoretical hydrogen content is 4.3% but at 500°C the weight loss is 38.8% (Fig. 7.8). Based on these data, it appears that fulleranes have a limited tendency to release reversibly the hydrogen or deuterium chemically linked to the cage but also a limited ability to evaporate in the gaseous phase. The predominant reaction is the thermal decomposition into a mixture of products (Talyzin et al. 2004, 2006b) leaving a carbonaceous residue in the crucible consisting of the 78.8% of the original weight of $C_{70}D_{38}$ and 61.2% of the original weight of $C_{70}H_{38}$.

The differential thermal analysis (DTA) of $C_{70}D_{38}$ and $C_{70}H_{38}$ shown in Fig. 7.9 reveals that at low temperature an endothermic transition occurs at about 70°C on both samples and the deuterated sample shows also an additional transition at about 170°C which was not detected in the hydrogenated counterpart. At present we cannot offer a clear explanation of these thermal transitions. Instead, the broad endothermal transition occurring at 416°C for $C_{70}D_{38}$ and at 344°C for $C_{70}H_{38}$ are clearly due to the complete decomposition of the samples. As observed in the case of $C_{60}D_{36}$ and $C_{60}H_{36}$, the decomposition of the deuterated molecule occurs at much higher temperature than its hydrogenated analogous. More precisely in the present case the

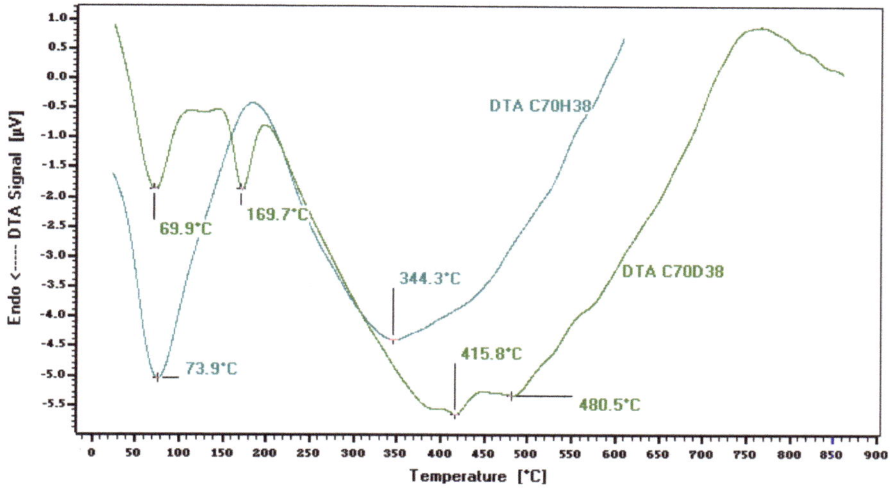

Fig. 7.9 Differential thermal analysis (DTA) under N_2 flow of $C_{70}D_{38}$ and $C_{70}H_{38}$. The former molecule shows an endothermal decomposition peak at 416°C while the latter reaches the decomposition peak at much lower temperature: 344°C. Isotope effect can be advocated for the explanation of the better stability of the perdeuterated C_{70} in comparison to its hydrogenated analogous

temperature shift in the decomposition peak of the deuterated molecule was at +72°C higher than the hydrogenated, a value comparable to +53°C shift observed in the case of decomposition of $C_{60}D_{36}$ versus $C_{60}H_{36}$. The explanation of such shift in the decomposition temperature resides in the isotope effect. Although the dehydrogenation process of fulleranes is not a smooth process, since it is accompanied by a number of side reactions as discussed above, the shift in the decomposition temperature implies that the breakage of the C–H and C–D bonds appears to be a rate determining step in the thermal dehydrogenation of fullerane and perdeuterofullerane so that for the deuterated molecule additional ≈25 kJ/mol energy should be administered to reach the decomposition peak.

7.3.8 Stability of $C_{60}D_{36}$ Exposed to Air at Room Temperature

$C_{60}H_{36}$ is not particularly stable to air and humidity especially if it has been produced by a wet synthesis (Darwish et al. 1995 1996a, b; Cataldo 2003a, b). Improved stability has been observed instead on $C_{60}H_{36}$ prepared by dry synthetic routes (Peera et al. 2004; Talyzin et al. 2004, 2006a). Evidently, traces of solvents used in the synthesis, water, reagents remain trapped in the solid and exert some kind of catalytic effect in the decomposition. The decomposition mechanism has been postulated to involve allylic oxidation (Taylor 1999) which is completely reasonable and involves free radicals. Hydrogen atoms in α position to the double bonds are abstracted by the action of adventitious free radicals or by heat and light. Thus, a free radical stabilized

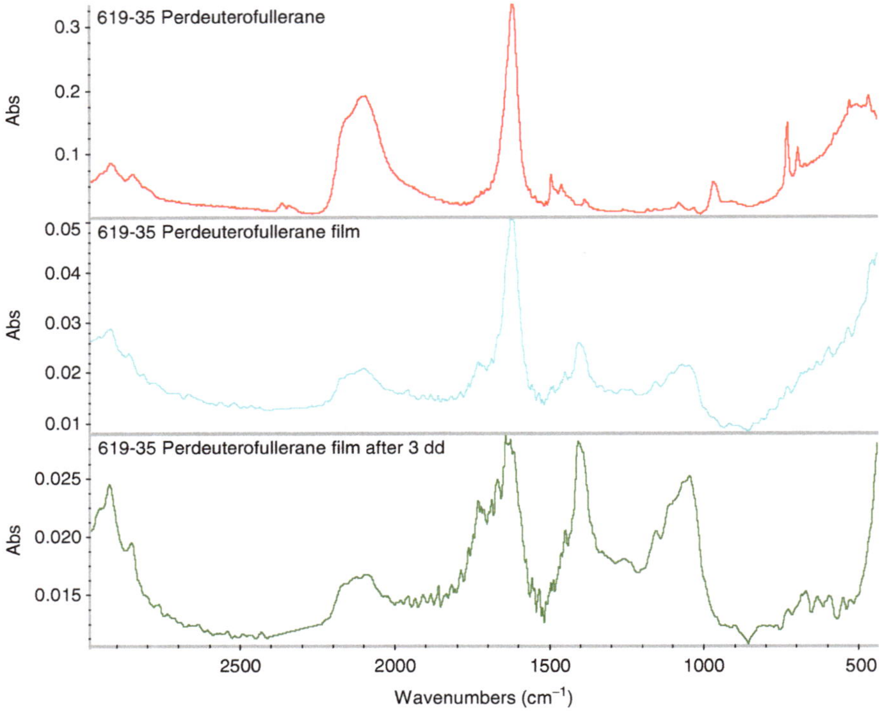

Fig. 7.10 FT-IR of perdeuterofullerane crystalline film on KBr plate. From *top* to *bottom* are shown the spectra of $C_{60}D_{36}$ immediately after synthesis, after 1 day and 3 days exposure to air respectively at room temperature

by conjugation with double bond remains in place of the hydrogen atom. Usually a hydroperoxide or a peroxide derived from oxygen from air saturates the free radical. The fate of the hydroperoxide is its conversion to ketonic groups.

Figure 7.10 (top) shows the FT-IR spectrum of freshly prepared $C_{60}D_{36}$. The exposure to air after 1 day causes alterations in the spectrum (Fig. 7.10, middle). In particular it can be noticed the reduction of the intensity of the C–D stretching band at 2,092 cm^{-1} and the complete disappearance of the C–D bending at 966 cm^{-1}. Evidences of oxidation can be inferred by the C=O stretching band at 1,710 cm^{-1} and by the C–OH and C–OOH bending at about 1,040 cm^{-1} supporting the allylic oxidation mechanism. After 3 days exposure to air an increase in the relative intensity of the ketone, hydroxyl and hydroperoxide bands can be observed (Fig. 7.10, bottom).

The same oxidation bands were observed, much more intense, also on $C_{60}H_{36}$ but after the exposure to air for less than 1 h (Cataldo 2003a, b). $C_{60}H_{36}$ turns it colour from white-yellow to orange in 1 h or less; this phenomenon takes much longer time for $C_{60}D_{36}$ synthesized in the same way, with Zn/DCl. Thus, perdeuterofullerene $C_{60}D_{36}$ appears to be much more stable to air than $C_{60}H_{36}$ and this different behaviour can be explained once again by the isotope effect.

7.3.9 Oxidation Stability of $C_{70}D_{38}$ and $C_{70}H_{38}$

Fulleranes are decomposed by the action of oxygen and it has been reasonably suggested that the mechanism of such degradation is free-radical mechanism involving allylic oxidation (Taylor 1999). Such oxidation mechanism is also responsible for rubber and plastics degradation and involves always as a first step the C–H bond breakage in α position to a double bond (Brydson 1978; Jessop et al. 1994). Since $C_{60}H_{36}$ and $C_{70}H_{38}$ are not fully saturated molecules such mechanism is fully operative.

As detailed discussed in a previous section and elsewhere (Kohen and Limbach 2006), when the C–H bond breakage is a rate determining step, then the hydrogen substitution with deuterium leads to a measurable primary isotope effect with improved stability toward oxidation of the deuterated molecule or macromolecule in comparison to its hydrogenated counterpart (Jessop et al. 1994; Kohen and Limbach 2006). Figure 7.11 shows qualitatively this phenomenon when compared with Fig. 7.12 (first two spectra on top of the figure). The FT-IR spectrum of $C_{70}D_{38}$ after

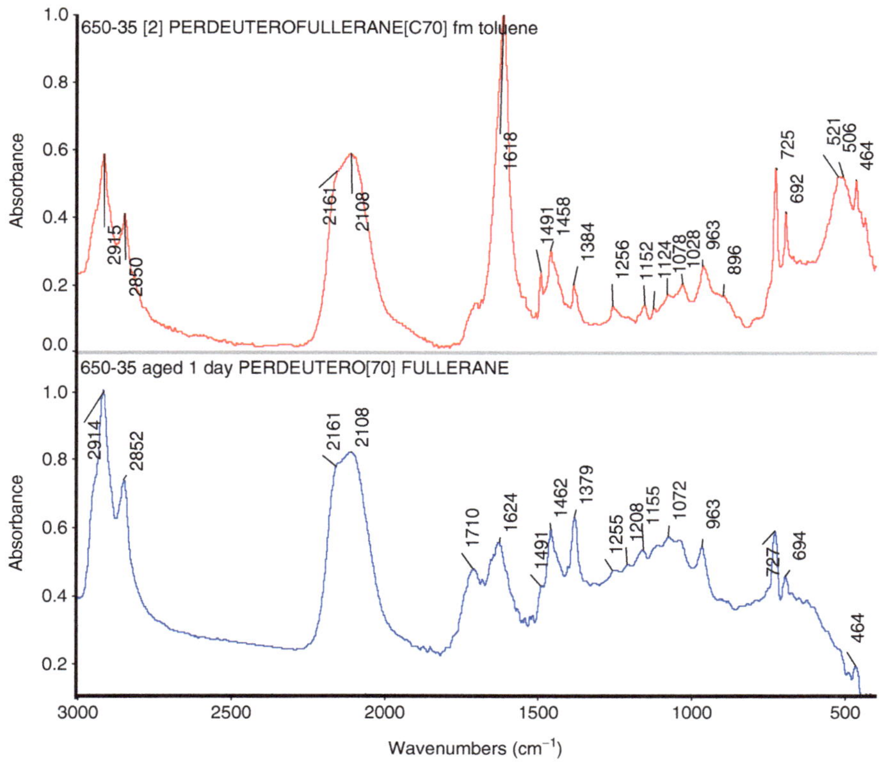

Fig. 7.11 FT-IR spectra (in KBr): perdeuterated [70] fullerene $C_{70}D_{38}$ as produced (*top*) and after 1 day exposure to air (*bottom*). Note that the oxidation is testified by the ketone band at 1,710 cm^{-1} and by the reduction of the intensity of the band at about 1,620 cm^{-1}

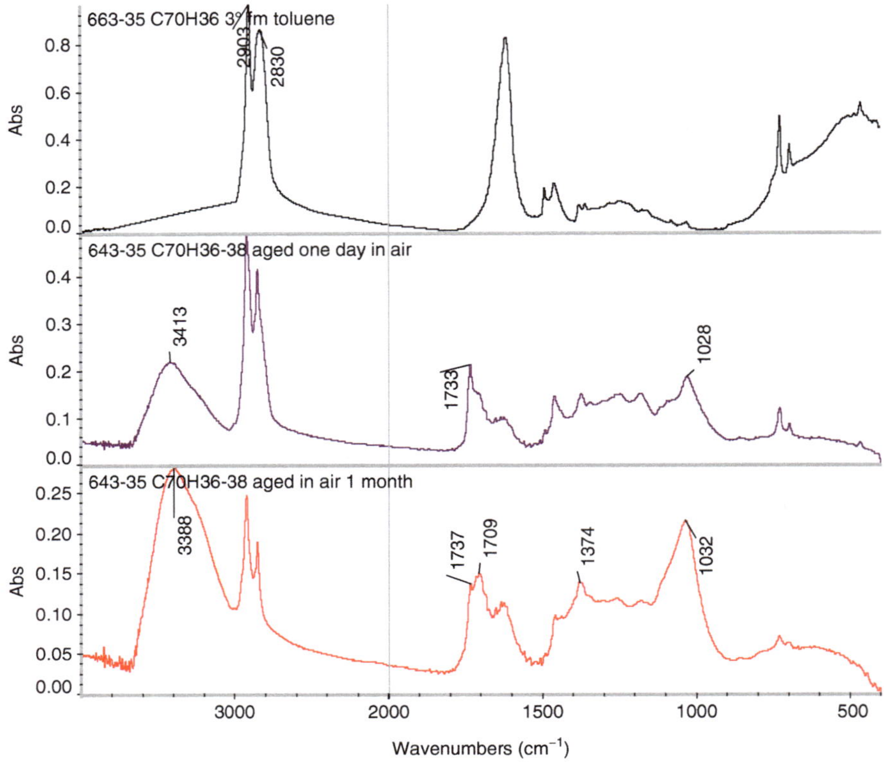

Fig. 7.12 FT-IR spectra (in KBr): $C_{70}H_{38}$ (*top*) when exposed to air for 1 day (*second from top*) shows the development of a sharp band ketone band at 1,737 cm^{-1} accompanied by the OH stretching band at 3413 cm^{-1} and the OH bending band at 1,028 cm^{-1}. Exposure of $C_{70}H_{38}$ for 1 month to air (*third spectrum from top*) causes a much more intense ketone band split at 1,733 and 1,709 cm^{-1} combined with intense OH stretching and bending respectively at 3,384 and 1,032 cm^{-1}

1 day exposure to air shows signs of oxidation first of all suggested by the new ketone band at 1,710 cm^{-1} (Fig. 7.11 bottom) but also by the new band at 1,072 due to the C–OH bending mode, all accompanied by the reduction in intensity of the band at about 1,620 cm^{-1}. Instead, the C–D stretching and bending respectively at 2,161, 2,108 and 963 cm^{-1} appear not altered by the oxidation. The formation of ketone band implies necessarily the fullerene cage opening but localized at the site where allylic oxidation has taken place while the rest of the cage remains unchanged and deuterated. For comparison, the behaviour of $C_{70}H_{38}$ when exposed to air for 1 day can be followed in the first two FT-IR spectra at the top of Fig. 7.12. One day exposure to air of the hydrogenated molecule shows the development of a sharp band ketone band at 1,737 cm^{-1} accompanied by the OH stretching band at 3413 cm^{-1} and the OH bending band at 1,028 cm^{-1} with the simultaneous reduction in intensity of the band located at 1,631 cm^{-1}. The spectral changes appears more dramatic in the case of $C_{70}H_{38}$ where for instance the absorbance ratio $A_{1737}/A_{1631} > 1$ while in the case of $C_{70}D_{38}$ the ratio

$A_{1710}/A_{1624} < 1$, implying a higher oxidation degree for the hydrogenated sample in comparison to the deuterated sample and hence confirming, although qualitatively, the stabilizing primary isotope effect in the deuterated molecule toward oxidation.

Exposure of $C_{70}H_{38}$ for 1 month to air (Fig. 7.12, third spectrum from top) show a much more intense ketone band, this time split at 1,733 and 1,709 cm^{-1} combined with intense OH stretching and bending respectively at 3,384 and 1,032 cm^{-1}. The oxidation in such sample can be considered extensive.

Acknowledgements The present research work has been supported by grant AYA2007-64748 from the Spanish Ministerio de Ciencia e Innovacion.

References

Adler RW, Baker R, Brown JM (1976) Mechanism in organic chemistry. Wiley, New York, Chapter 1
Avent AG, Darwish AD, Heimbach DK, Kroto HW, Meidine MF, Parsons JP, Remars C, Roers R, Ohashi O, Taylor R, Walton DRM (1994) J Chem Soc Perkin Trans 2:15
Avram M, Mateescu GhD (1972) Infrared spectroscopy applications in organic chemistry. Wiley-Interscience, New York
Balasubramanian K (2004) Chem Phys Lett 400:78
Bensasson RV, Hill TJ, Land EJ, Leach S, McGarvey DJ, Truscott TG, Ebenhoch J, Gerst M, Ruchardt C (1997) Chem Phys 215:111
Bini R, Ebenhoch J, Fanti M, Fowler PW, Leach S, Orlandi G, Rüchardt Ch, Sandall JPB, Zerbetto F (1998) Chem Phys 232:75
Brydson JA (1978) Rubber chemistry. Applied Science, London
Cataldo F (1997) Fullerene Sci Technol 5:1615
Cataldo F (2003a) Fullerenes Nanot Carbon Nanostruct 11:295
Cataldo F (2003b) Fullerenes Nanot Carbon Nanostruct 11:317
Cataldo F, Iglesias-Groth S, Manchado A (2009a) Fullerenes Nanot Carbon Nanostruct 17:378
Cataldo F, Iglesias Groth S, Manchado Torres A (2009b) Fullerenes Nanot Carbon Nanostruct 17:401
Colthup NB, Daly LH, Wiberley SE (1990) Introduction to infrared and Raman Spectroscopy, 3rd edn. Academic, San Diego, CA, pp 152–155
Darwish AD, Abdul-Sada AK, Langley GJ, Kroto HW, Taylor R, Walton DRM (1995) J Chem Soc. Perkin Trans 2:2359
Darwish AD, Kroto HW, Taylor R, Walton DRM (1996a) Chem Soc Perkin Trans 2:1415
Darwish AD, Abdul-Sada AK, Langley GJ, Kroto HW, Taylor R, Walton DRM (1996b) Synth Met 77:303
Dorozhko PA, Lobach AS, Popov AA, Senyavin VM, Korobov MV (2001) Chem Phys Lett 336:39
Eremtchenko M, Ottking R, Krischok S, Doring S, Temirov R, Schafer JA (2005) Fullerenes Nanot Carbon Nanostruct 13:131
Fowler PW, Sandall JPB, Austin SJ, Manolopoulos DE, Lawrenson PDM, Smallwood JM (1996) Synth Met 77:97
Friedel RA, Orchin M (1951) Ultraviolet spectra of aromatic compounds. Wiley, New York, Spectra 13
Gerst M, Beckhaus HD, Rüchardt Ch, Campbell EEB, Tellgmann R (1993) Tetrah Lett 34:7729
Gourier DRF, Delpoux O, Binet L, Vezin H, Moissette A, Derenne S (2008) Geochim Cosmochim Acta 72:1914

Hibbert F (1980) Isotopes and organic reaction mechanisms. In: Elvidge JA, Jones JR (eds) Isotopes: essential chemistry & applications. Chemical Society of Chemistry, London, Chapter 9
Iglesias-Groth S (2004) Astrophys J 608:L37
Iglesias-Groth S (2006) Month Not Roy Astron Soc 368:1925
Jessop PG, Morris RH, Azizian H (1994) Polymer 35:1952
Karpushenkava LS, Kabo GJ, Diky VV (2007) Fullerenes Nanot Carbon Nanostruct 15:227
Kerr JA (1987) Bond strength in diatomic molecules. In: Weast RC (ed) CRC handbook of chemistry and physics, 68th edn. CRC Press, Boca Raton, FL, p F170
Kintigh J, Briggs JB, Letourneau K, Miller GP (2007) J Mater Chem 2007:4647
Kohen A, Limbach HH (2006) Isotope effects in chemistry and biology. CRC Press/Taylor & Francis, Boca Raton, FL
Kroto HW, Allaf AW, Balm SP (1991) Chem Rev 91:1213
Meletov KP, Assimopoulos S, Tsilika I, Bashkin IO, Kulakov VI, Khasanov SS, Kourouklis GA (2001) Chem Phys 263:379
Palit DK, Mohan H, Mittal JP (1998) J Phys Chem 102:4456
Peera AA, Alemany LB, Billups WE (2004) Appl Phys A78:995
Perkampus HH (1992) UV–VIS atlas of organic compounds. VCH, Weinheim
Petrie S, Becker H, Baranov VI, Bohme DK (1995) Int J Mass Spectrom Ion Process 145:79
Reichardt C (2003) Solvents and solvent effects in organic chemistry. Weinheim, Wiley-VCH
Ruchardt C, Gerst M, Ebenhoch J, Beckhaus HD, Campbell EEB, Tellgmann R, Schwartz H, Weiske T, Pitter S (1993) Angew Chem Int Ed 32:584
Schur DV, Zaginaichenko S, Veziroglu TN (2008) Int J Hydrogen Energy 33:3330
Shaw AM (2006) Astrochemistry. Wiley, Chichester
Silverstein RM, Bassler GC, Morrill TC (1981) Spectrometric identification of organic compounds, 4th edn. Wiley, New York, p 313
Stoldt CR, Maboudian R, Carraro C (2001) Astrophys J 548:L225
Talyzin AV, Shulga YM, Jacob A (2004) Appl Phys A 78:1005
Talyzin AV, Dzwilewski A, Sundqvist B, Tsybin YO, Purcell JM, Marshall AG, Shulga YM, McCammon C, Dubrovinsky L (2006a) Chem Phys 325:445
Talyzin AV, Tsybin YO, Purcell JM, Schaub TM, Shulga YM, Noreus D, Sato T, Dzwilewski A, Sundqvist B, Marshall AG (2006b) J Phys Chem A 110:8528
Taylor R (1999) Lecture notes on fullerene chemistry. A handbook for chemists. Imperial College Press, London
Taylor R (2006) CR Chimie 9:982
Tielens AGGM (2005) The physics and chemistry of the interstellar medium. Cambridge University Press, Cambridge
Vieira SMC, Ahmed W, Birkett PR, Rego CA (2001) Chem Phys Lett 347:355
Vieira SMC, Ahmed W, Birkett PR, Rego CA, Kotsiris S, Drewello Th (2004) Fullerenes Nanot Carbon Nanostruct 12:139
Wågberg Th, Hedenström M, Talyzin AV, Sethson I, Purcell YO, Marshall JM, Noréus AG, Johnels D (2008) Angew Chem Int Ed 47:2796
Withers J, Loufty R, Lowe T (1997) Fullerene Sci Technol 5:1

Chapter 8
Isotope Effect in the UV Photolysis of Hydrogenated and Perdeuterated Fulleranes

Franco Cataldo[1], Susana Iglesias-Groth[2], and Arturo Manchado[2]

Abstract Fulleranes, the hydrogenated fullerenes $C_{60}H_{36}$ and $C_{60}D_{36}$ have been synthesized in n-hexane where they show an absorption maximum at 217 nm in their electronic absorption spectra. Similarly $C_{70}H_{38}$ and $C_{70}D_{38}$ show an absorption maximum in n-hexane at 214 nm.

The interstellar light extinction curve shows a "bump" just at 217 nm which is attributed to hydrogenated interstellar carbon dust. Remarkably also fulleranes which can be considered an analogous of hydrogenated carbon dust show a maximum of absorption just at 217 nm. The width of this absorption appears to be consistent with the observed widths of the UV bump. The photolysis of fulleranes with monochromatic light at 254 nm causes a shift of the absorption maximum from 217 nm to longer wavelengths. This band shift of the absorption maximum has already been observed in the photolysis or in the thermal processing of hydrogenated carbon dust. The band shift is due to the release of molecular hydrogen, a process which causes the aromatization of the carbon dust. $C_{60}H_{36}$ and $C_{60}D_{36}$ are photolyzed at a rate $k_{C60H36} = 2.45 \times 10^{-3}\ s^{-1}$ while $k_{C60D36} = 2.27 \times 10^{-3}\ s^{-1}$. This implies an isotope effect so that $C_{60}H_{36}$ is photolyzed 1.08 times faster than $C_{60}D_{36}$. In the interstellar medium this implies a deuterium enrichment of the hydrogenated interstellar carbon dust. The presence of a measurable isotope effect suggests that the photolysis of $C_{60}H_{36}$ and $C_{60}D_{36}$ involves the rupture of the C–H and C–D bond with release of H_2.

The hydrogenated C_{70} fullerenes $C_{70}H_{38}$ and $C_{70}D_{38}$ have been photolyzed at 254 nm under Ar or He flow. The photolysis rate constant have been measured: $k = 1.54 \times 10^{-3}\ s^{-1}$ for $C_{70}H_{38}$ while the photolysis of $C_{70}D_{38}$ occurs at a rate of $1.17 \times 10^{-3}\ s^{-1}$. A remarkable isotope effect in the photolysis of the two molecules has been

[1]Istituto Nazionale di Astrofisica, Osservatorio Astrofisico di Catania, Via S. Sofia 78, 95123, Catania, Italy
and
Actinium Chemical Research srl, Via Casilina 1626/A, 00133 Rome, Italy
e-mail: franco.cataldo@fastwebnet.it

[2]Instituto de Astrofisica de Canarias, Via Lactea s/n, E-38200, La Laguna, Tenerife, 2 CSIC, Spain
e-mail: sigroth@iac.es; amt@iac.es

F. Cataldo and S. Iglesias-Groth (eds.), *Fulleranes: The Hydrogenated Fullerenes*, Carbon Materials: Chemistry and Physics 2, DOI 10.1007/978-1-4020-9887-1_8,

determined $k_H/k_D = 1.32$. The photolysis mechanism and products of $C_{70}H_{38}$ and $C_{70}D_{38}$ have been discussed and based on the isotope effect, the rate determining step involves, as for other fulleranes, the activation of the C–H and C–D bond.

Fullerane $C_{60}H_{18}$ and its deuterated analogous $C_{60}D_{18}$ were synthesized in n-hexane solution by a reduction reaction of C_{60} under the action of HCl or DCl on Zn dust. The resulting solutions were subjected to UV irradiation at 254 nm from a low pressure mercury lamp under He. It was found that at 212 nm the photolysis rate constant of $C_{60}H_{18}$ molecule ${}^{H}k_{212} = 8.68 \times 10^{-4}\ s^{-1}$ was significantly higher than that of its deuterated analogous $C_{60}D_{18}$:${}^{D}k_{212} = 5.93 \times 10^{-4}\ s^{-1}$. Similarly, at 256 nm it was confirmed the result that $C_{60}D_{18}$ was photolyzed more slowly than $C_{60}H_{18}$. In fact, also in this case ${}^{H}k_{256} = 6.83 \times 10^{-4}\ s^{-1}$ is significantly higher than that of its deuterated analogous $C_{60}D_{18}$:${}^{D}k_{212} = 3.74 \times 10^{-4}\ s^{-1}$. Kinetic isotope effect involving the C–H and C–D bond activation has been advocated to explain the differences in photodecomposition speed of $C_{60}H_{18}$ in comparison to $C_{60}D_{18}$.

8.1 Introduction

Fullerenes have been predicted to be present in space (Hare and Kroto 1992). The main sources of carbon dust and molecules in the interstellar medium are the late-type carbon-rich stars, but there is a class of stars which lies in the transition between the asymptotic giant branch (AGB) to the planetary nebula stage which is very promising as sources of fullerenes. The stars in such class are rather rare, are helium rich and extremely depleted in hydrogen content in their gaseous shell so that the carbon vapour ejected from the star could be cooled in appropriate and ideal environment for the fullerene formation (Kroto 2006). In fact, the presence of hydrogen is known to have negative effects in the formation of the fullerene cage and to favour the production of polycyclic aromatic hydrocarbons (PAHs) and other products (Goeres and Sedlmayr 1993). The prototype of such stars where fullerene may be present is *R Coronae Borealis* and the corresponding class of RCrB stars (Unsold and Baschek 2002). In the interstellar medium C_{60} fullerene should be present as neutral molecule or should undergo ionization to C_{60}^{+}. Such cation may be responsible for some spectral features in the diffuse interstellar bands (DIBs), absorption bands detected in the spectra of our galaxy and beyond (Foing and Ehrenfreund 1994, 1997). Fullerene cations undergo multiple addition of atomic hydrogen forming hydrogenated derivatives (Petrie et al. 1995) but also neutral fullerenes add easily atomic hydrogen at very low temperatures (Howard 1993), it is reasonable to think that fulleranes, the hydrogenated fullerenes derivatives should be present in the interstellar medium (Petrie and Bohme 2000).

Calculated spectra for hydrogenated fullerenes have been published in comparison with the unidentified infrared emission bands (Webster 1991; Stoldt et al. 2001). The infrared spectrum of $C_{60}H_{36}$ has been compared also with the infrared features of other astrophysical objects like the proto-planetary nebulae (Cataldo 2003a, b). An inventory about fullerenes and hydrogenated derivatives in the interstellar medium and their theoretical spectral properties can be found in the recent works of Iglesias-Groth (2004, 2005, 2006).

Fulleranes at different degree of hydrogenation can be prepared through various synthetic approaches (Taylor 1999, 2006; Nossal et al. 2001; Cataldo 2003a, b). Deuterated derivatives of both C_{60} and C_{70} have been synthesized and their respective infrared spectra studied (Cataldo et al. 2009a, b). The hydrogenation with "nascent" hydrogen in solution normally stops at the $C_{60}H_{36}$ stage and it is not possible to add more hydrogen to C_{60}. The fullerene hydrogenation in solution can also produce $C_{60}H_{18}$ under opportune reaction conditions (Darwish et al. 1995); alternatively, a thermal treatment of $C_{60}H_{36}$ can convert it into $C_{60}H_{18}$ (Taylor 1999, 2006). Recently, a convenient and selective synthesis of $C_{60}H_{18}$ has been reported by using H_2 at a pressure of 100 bar at 500°C for 10 h (Wagberg et al. 2005).

An important property of molecules of astrochemical interest regards their stability to UV photons. A plethora of organic molecules are today known in different space environments (Ehrenfreund and Charnley 2000), some of them known to be the precursors of life. Photochemical processing may lead to the production in space of other interesting molecules from common precursors or may lock certain molecules into dust or dust-forming nanoparticles (Glassgold 1996; Van Dishoeck and Blake 1998). In this context, the photophysical and photochemical properties of fullerenes and their hydrogenated counterpart, the fulleranes, are of interest either for the detection of these species and to estimate their survival and their fate in the harsh space environment. The photophysical properties of fullerenes are well known (Guldi and Kamat 2000), the photochemistry and the radiation chemistry of these molecules have been partially explored as well (Cataldo et al. 2009c). Instead almost nothing is known about the photochemistry of the hydrogenated fullerenes although a couple of papers on their photophysical properties have been published (Bensasson et al. 1997; Palit et al. 1998).

Here, we summarize a series of studies dedicated to the exploration of the photochemistry of fulleranes (Cataldo et al. 2009d) studying the photolysis of fullerenes ad different degrees of hydrogenation or deuteration in high dilution, using hydrocarbon solvents which are inert and transparent to UV radiation and a monochromatic radiation source emitting at 245 nm. All the photolysis experiments were conducted under inert atmosphere to avoid any interference from oxygen from air. In other words, the same approach we have used for polyynes to check their photostability (Cataldo et al. 2008) was applied in the present case to the fulleranes $C_{60}H_{36}$ and $C_{60}D_{36}$.

8.2 Experimental

8.2.1 Reagents and Equipment

C_{60} and C_{70} fullerenes were obtained from MTR and was 99.95% pure grade. n-Hexane (HPLC grade) tetradecane (special grade granted without olefins) and HCl 37% were obtained from Aldrich as well as the deuterium chloride (DCl) 37%

in fully deuterated water (D_2O, deuterium enrichment 99%). Zn dust was from Aldrich and was characterized by a particle size <10 μm. The inert gases employed in the photolysis experiments (helium or argon) were 99.99% pure and supplied from Fluka.

The C_{3v} isomer of $C_{60}H_{18}$ was kindly donated by Dr. Alexandr Talyzin from Umeå University, Sweden. It was obtained by high pressure hydrogenation of C_{60} (Wagberg et al. 2005).

Electronic absorption spectra were recorded on a Shimadzu UV160A spectrophotometer using standard quartz cuvettes.

The FT-IR spectra of fullerane and C_{60} were recorded in KBr matrix at room temperature. For the thermal treatment of the $C_{60}H_{36}$ samples use was made of a thermogravimetric apparatus from Linseis, model L81+DTA working under N_2 flow (18 L/h).

For the photolysis experiments, a low pressure mercury lamp (12 W) from Helios Italquartz (Milan, Italy) was employed. The lamp emits monochromatic photons at about 245 nm. The UV irradiations were performed at room temperature in a quartz reactor fitted with inlet and outlet gas valves with the lamp inserted inside the reactor (Ninomiya and Naito 1989). All irradiations were performed under inert atmosphere with rigorous exclusion of air. These conditions were achieved by bubbling continuously Ar or He through the solutions.

8.2.2 Photolysis of $C_{60}H_{36}$

A saturated solution of C_{60} in n-hexane (about 13 mg C_{60} in 300 mL of n-hexane) was charged in a 1L conical flask together with 12 g of Zn dust. The flask was equipped with a rubber cap. Aqueous HCl (37% solution, 22 mL first addition and 15 mL second addition) was added to the mixture. The flask was immediately closed with the cap and hand-shaken. Periodically the cap was opened to permit the release of the excess of hydrogen produced from the reaction between HCl and Zn. The colour of the solution changed from violet to water-white. The complete reduction of C_{60} was checked by UV spectroscopy, by sampling the hexane solution and checking the spectrum. The reduction is complete when the UV spectrum is dominated by a maximum at about 217 nm which is due to $C_{60}H_{36}$ (Cataldo 2003a, b). No other features should be detectable. Knowing the initial concentration of C_{60}, based on the absorbance of the reduced solution it was possible to determine the molar extinction coefficient ε of $C_{60}H_{36}$:ε_{217} = 17,140 L cm^{-1} mol^{-1}.

The hexane solution of $C_{60}H_{36}$ was decanted from the aqueous phase, quickly filtered through paper filter, dried 5 min over anhydrous $CaCl_2$ and charged in the UV reactor. The irradiation was started after 10 min Ar or He bubbling and bubbling was continued also during irradiation. During photolysis the gradual formation of a whitish precipitate was noticed which causes light scattering in the electronic absorption spectra. Periodically samples were taken from the lamp to study the photolysis kinetics from the electronic absorption spectra.

The photolysis of $C_{60}H_{36}$ was repeated five times with good repeatability.

An attempt to collect and characterize the white precipitate was made after photolysis by distilling off in a water bath under reduced pressure n-hexane leaving a concentrated dispersion of the white insoluble photoproduct in 5 mL of residual solvent. When exposed to air the photoproduct turns yellow and then orange. It is even more sensitive to air than the starting molecule $C_{60}H_{36}$. In fact the FT-IR spectrum shows only extensive oxidation: strong ketone band and OH groups stretching and bending. Presumably the photolysis involves the dimerization or oligomerization of $C_{60}H_{36}$.

8.2.3 Photolysis of $C_{60}D_{36}$

Deuterated C_{60}:$C_{60}D_{36}$ was synthesized in n-hexane using the same procedure detailed above for $C_{60}H_{36}$ using Zn dust and 37% deuterium chloride (DCl) in heavy water (D_2O). After complete reduction also the electronic absorption spectrum of $C_{60}D_{36}$ is dominated by a maximum at 217 nm and the molar extinction coefficient was determined $\varepsilon_{217} = 16480$ L cm^{-1} mol^{-1}.

The photolysis of hexane solution of $C_{60}D_{36}$ was made as detailed above for $C_{60}H_{36}$. The photolysis experiment on $C_{60}D_{36}$ was repeated twice with good repeatability.

8.2.4 Photolysis of $C_{70}H_{38}$

The hydrogenation of C_{70} was achieved using the same general procedure described above for the synthesis of $C_{60}H_{36}$. A solution of 10 mg of C_{70} in 300 mL of n-hexane was shaken in a conical flask with cap with 17 g Zn dust and 20 mL aqueous HCl (37%) added in two portions. The cap was periodically opened to discharge the excess of H_2 accumulated in the flask. The solution changed its colour from wine-red to water-white. After reduction the n-hexane solution was separated from the aqueous phase and Zn dust, quickly filtered through a paper filter, dried over anhydrous $CaCl_2$ and charged into the photochemical reactor. After complete reduction $C_{70}H_{38}$ displayed an absorption maximum at 214 nm in the UV spectrum; $\varepsilon_{214} = 6{,}300$ L cm^{-1} mol^{-1}.

The photolysis was performed under He bubbling and periodically samples were taken from the solution to follow the kinetics of the reaction. The photolysis experiment on C_{70} was repeated twice with good repeatability.

8.2.5 Photolysis of $C_{70}D_{38}$

$C_{70}D_{38}$ was synthesized in n-hexane using the same procedure detailed above for $C_{70}H_{38}$ using Zn dust and 37% deuterium chloride (DCl) in heavy water (D_2O).

After complete reduction also the electronic absorption spectrum of $C_{70}D_{38}$ was dominated by a maximum at 214 nm and the molar extinction coefficient was determined $\varepsilon_{214} = 5800$ L cm^{-1} mol^{-1}.

The photolysis of hexane solution of $C_{70}D_{38}$ was made as detailed above for $C_{70}H_{38}$. The photolysis experiment on $C_{70}D_{38}$ was repeated three times with good repeatability.

8.2.6 Preparation of $C_{60}H_{18}$ in n-Hexane and Subsequent Photolysis Under He

A stock solution of 13 mg of C_{60} dissolved in 300 mL of n-hexane is charged in a conical flask (1 L volume) fitted with a cork cap. The conical flask was already charged with 10.0 g of zinc dust. The hydrogenation reaction is started by the adding to the mixture (hexane solution + Zn) 15 mL of HCl 37% in one shot. Immediately after the addition of hydrochloric acid the flask is closed with the cap and hand shaken. After appropriate intervals of time the cork is opened slightly to permit the release of the excess of hydrogen. The original violet colour of the C_{60} solution is lost. The solution is settled under H_2 atmosphere. The organic layer is easily collected by decantation, quickly filtered through a paper filter, and the resulting water-white solution charged into the quartz reactor and treated immediately with a slow stream of helium. The electronic absorption spectrum of $C_{60}H_{18}$ is different from that of $C_{60}H_{36}$ and characterized by an absorption band at about 330 nm which is completely absent in the spectrum of $C_{60}H_{36}$ (Darwish et al. 1995; Palit et al. 1998). Thus, by stopping the hydrogenation at the early stages and be checking the resulting electronic absorption spectrum it is possible to confirm that a stock solution of $C_{60}H_{18}$ has been obtained rather than a $C_{60}H_{36}$ solution.

The UV irradiation was then started under continuous He bubbling through the solution to exclude rigorously any interference from oxygen. Periodically samples of the solution were taken from the reactor and the progress of the reaction was made by monitoring the changes in the ultraviolet absorption bands of the $C_{60}H_{18}$ in a UV spectrophotometer.

8.2.7 Preparation of $C_{60}D_{18}$ in n-Hexane and Subsequent Photolysis Under He

Exactly the same procedure as described in Section 8.2.2 was repeated with the variant to replace 15 mL of HCl with 15 mL of DCl (37%). In such way the deuterated derivative $C_{60}D_{18}$ was obtained in n-hexane solution and subsequently irradiated by UV photons under He.

8.2.8 *Photolysis of C_{3v} Isomer of $C_{60}H_{18}$ in Tetradecane Under Ar*

C_{3v}-$C_{60}H_{18}$ (15 mg) was stirred in 150 mL of tetradecane at 40°C for 4 h. Although the dissolution of C_{3v}-$C_{60}H_{18}$ was not complete, the resulting yellow solution was decanted from the indissolved fullerane and charged into a quartz reactor. The reactor was fitted with inlet and outlet valves for argon gas and with a low pressure mercury lamp of 12 W. The photolysis was conducted under continuous Ar bubbling through the solution to exclude rigorously any interference from oxygen. Periodically samples of the solution were taken and the progress of the reaction was made by monitoring the changes in the ultraviolet absorption bands of the $C_{60}H_{18}$ in a UV spectrophotometer.

8.3 Results and Discussion

8.3.1 *Kinetics of the Fulleranes $C_{60}H_{36}$ and $C_{60}D_{36}$ Photolysis*

Figure 8.1 shows the electronic absorption spectrum of $C_{60}H_{36}$ in n-hexane. The UV photolysis causes evident changes in the absorption spectrum of $C_{60}H_{36}$ as shown in Fig. 8.1. First of all the maximum of absorption is shifted from 217.4 nm of the

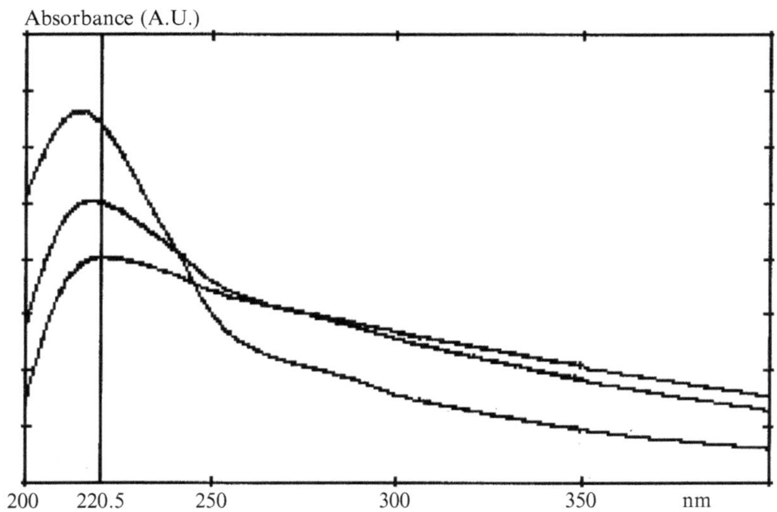

Fig. 8.1 Electronic absorption spectrum of $C_{60}H_{36}$ in n-hexane after photolysis under He. The spectra from top to bottom are taken after photolysis of 120, 240 and 360 s respectively. Note both the gradual shift of the absorption maximum from 217.4 nm of pristine $C_{60}H_{36}$ to 220.5 nm after 360 s photolysis. Note also the broadening of the peak and the increase in the absorption tale due to light scattering caused by the formation of an insoluble whitish suspension of photoproducts

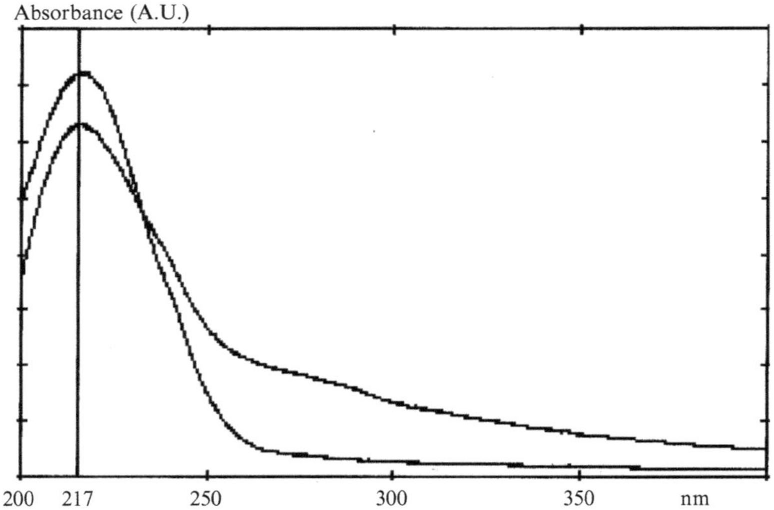

Fig. 8.2 Electronic absorption spectrum of $C_{60}D_{36}$ in n-hexane. The absorption maximum is indicated by the vertical line and is located at 217 nm with a ε_{217} = 16,480 L cm^{-1} mol^{-1}. After 120 s photolysis the intensity of the peak at 217 nm is reduced and the baseline with $\lambda > 260$ nm shows a strong increase in optical density due to light scattering caused by the formation of a white insoluble photoproduct

pristine sample to 220.5 nm after 360 s of UV irradiation. The shift is accompanied by a reduction in intensity of the absorption peak and by an increase of the base line (β) absorption and tailing absorption at longer wavelengths due to the formation of insoluble photoproduct in the solution which causes light scattering. In fact, during UV irradiation the solution becomes whitish and opalescent. A similar behaviour was observed also in the photolysis of the deuterated molecule $C_{60}D_{36}$. In fact, in Fig. 8.2 it is shown the electronic absorption spectrum of $C_{60}D_{36}$ before photolysis and after 120 s of UV irradiation. The photolysis causes a reduction of the absorption peak at 217 nm and an increase in the tailing absorption above 250 nm with an increase in the baseline absorption, behaviour completely comparable to that shown by $C_{60}H_{36}$ during UV irradiation, due to the formation of insoluble photoproduct which causes light scattering in the solution.

Thus, in the calculation of the photolysis rate constant, the increase in the baseline absorption has been taken into account and the kinetic rate constant has been determined using the absorbance read at 217–220 nm.

The photolysis kinetics has been determined according to the following equation which conforms to the pseudofirst order kinetic law with v being the photolysis velocity (Yeremin 1979):

$$\upsilon = -dc/dt = kc \tag{8.1}$$

Since the concentration *c* corresponds to absorbance A thanks to the Lambert–Beer law:

$$A = \varepsilon bc \tag{8.2}$$

where ε is the molar extinction coefficient and b is the path length of the measuring cell, after substitution Eq. 8.2 in Eq. 8.1 and integrating we get:

$$\upsilon = Ln(A_t / A_0) = -kt \tag{8.3}$$

Since we have taken into account the increase in absorption of the baseline (β) during photolysis, the Eq. 8.3 has been modified as follows:

$$\upsilon = Ln[(A_t - / \beta_t)/(A_0 - / \beta_0) = -kt \tag{8.4}$$

Thus, k is the slope obtained by plotting Ln $[(A_t - \beta_t)/(A_0 - \beta_0)]$ against the photolysis time t and represents the pseudofirst order rate constant of the reaction.

In Fig. 8.3 are reported the results of our measurements made on the photolysis of both $C_{60}H_{36}$ and $C_{60}D_{36}$; the results are linear and follow well Eq. 8.4 with very high correlation coefficients (R^2) as shown in Fig. 8.3. The photolysis rate constant of $C_{60}H_{36}$ appears slightly higher than that of $C_{60}D_{36}$; in fact, $k_{C60H36} = k_H = 2.45 \times 10^{-3}\ s^{-1}$ while $k_{C60D36} = k_D = 2.27 \times 10^{-3}\ s^{-1}$ so that $k_H/k_D = 1.08$.

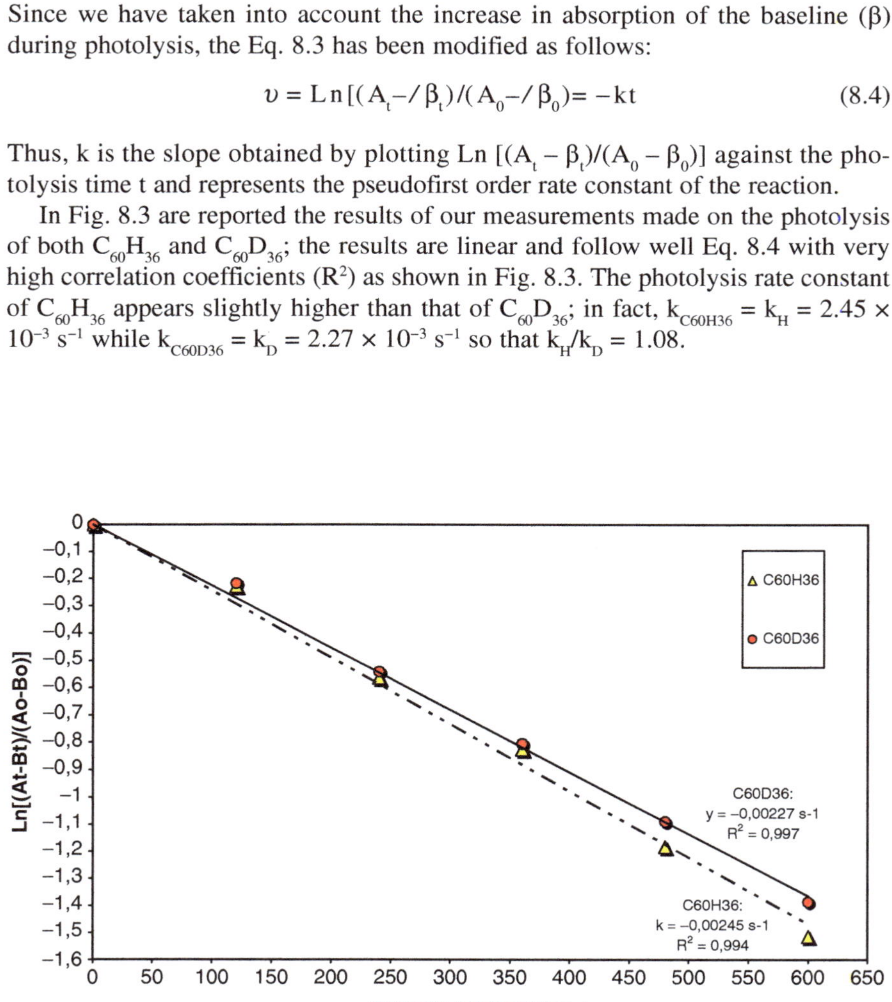

Fig. 8.3 Determination of the pseudofirst rate constant for the photolysis of $C_{60}H_{36}$ and $C_{60}D_{36}$ in n-hexane. The difference in photolysis rate constant between the hydrogenated and deuterated fullerene is small: $k_{C60H36} = 2.45 \times 10^{-3}\ s^{-1}$ and $k_{C60D36} = 2.27 \times 10^{-3}\ s^{-1}$. The ratio $k_H/k_D = 1.08$

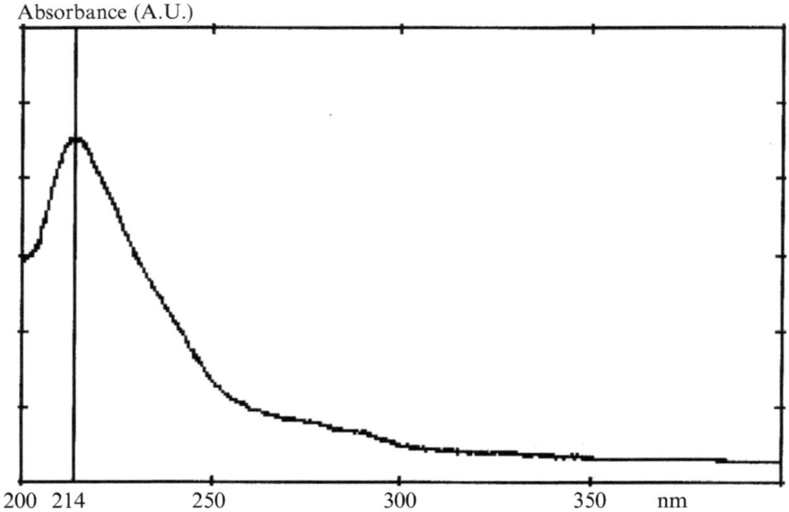

Fig. 8.4 Electronic absorption spectrum of $C_{70}H_{38}$ in n-hexane. The absorption maximum is indicated by the vertical line and is located at 214 nm with a ε_{214} = 6,300 L cm^{-1} mol^{-1}

8.3.2 Photolysis of $C_{70}H_{38}$ and $C_{70}D_{38}$

$C_{70}H_{38}$ and $C_{70}D_{38}$ solutions in n-hexane appear very sensitive to oxygen. The exposure to air for few minutes causes changes in the electronic absorption spectrum. The sensitivity to air is much more evident for $C_{70}H_{38}$ and $C_{70}D_{38}$ rather than for $C_{60}H_{36}$ and $C_{60}D_{36}$. As in another case (Cataldo et al. 2008), the UV irradiation was conducted rigorously under inert atmosphere using He or Ar bubbled continuously into the hexane solution.

As shown in Figs. 8.4 and 8.5, the UV irradiation causes a reduction in intensity of the main electronic transition of pristine $C_{70}H_{38}$ located at 214 nm and a shift of the absorption maximum toward longer wavelengths was noticed. This trend during photolysis is similar to that recorded in the case of $C_{60}H_{36}$ and $C_{60}D_{36}$ and interpreted as hydrogen release from the fullerane cage. The spectral changes in the photolysis of $C_{70}D_{38}$ are identical to those observed for its C_{60} hydrogenated analogous.

A common phenomenon in the UV photolysis of fulleranes is the production of an insoluble photoproduct which remains suspended in the organic solvent causing opalescence and light scattering in the electronic absorption spectrum. The effects of the light scattering caused by the photoproduct can be appreciated in Fig. 8.5 from the strong tailing absorption after the peak, the peak broadening and from the rise of the spectrum baseline. In the measurement of the photolysis kinetics the effect of baseline increase has been taken into consideration for a correct calculation as already reported in the previous section.

Another feature of the photolysis of all fulleranes studied is the fact that the kinetics can be described by pseudofirst order law. Figure 8.6 shows the results of

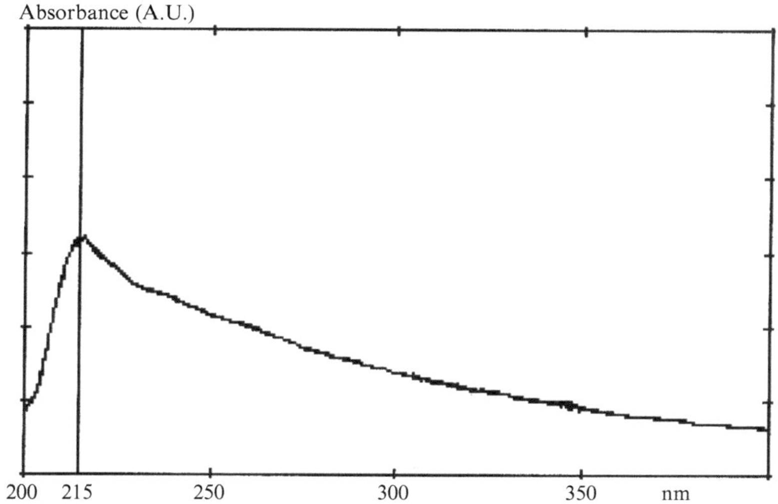

Fig. 8.5 Electronic absorption spectrum of $C_{70}H_{38}$ in n-hexane after 240 s photolysis under He. Note the strong reduction in the absorbance of the peak now located at 215 nm (*indicated by the vertical line*) and the strong increase in the baseline due to light scattering

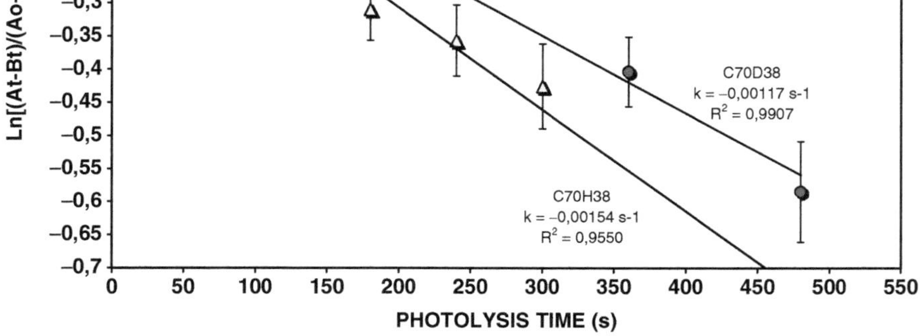

Fig. 8.6 The kinetics of photolysis of both $C_{70}H_{38}$ and $C_{70}D_{38}$ can be described by pseudofirst order law. The photolysis rate constant of $C_{70}H_{38}$ is 1.54×10^{-3} s^{-1} while the photolysis of $C_{70}D_{38}$ occurs at a rate of 1.17×10^{-3} s^{-1}. The differences in the two rate constant between the hydrogenated and deuterated molecule can be ascribed to primary isotope effects

Table 8.1 Summary photolysis rate constants

Molecule	k (s^{-1})	Reference
$C_{60}H_{36}$	2.45×10^{-3}	This work
$C_{60}D_{36}$	2.27×10^{-3}	This work
$C_{60}H_{18}$	8.64×10^{-4}	Cataldo et al. 2009d
$C_{60}D_{18}$	5.93×10^{-4}	Cataldo et al. 2009d
Polyynes	$2.4–5.2 \times 10^{-3}$	Cataldo et al. 2008

our measurements during UV irradiation of $C_{70}H_{38}$ and $C_{70}D_{38}$. The absorbance changes at about 214–220 nm have been plotted against the photolysis time. The photolysis rate constant of $C_{70}H_{38}$ was found $1.54 \times 10^{-3}\ s^{-1}$ while the photolysis of $C_{70}D_{38}$ occurred at a rate of $1.17 \times 10^{-3}\ s^{-1}$. The differences in the two rate constant between the hydrogenated and deuterated molecule can be ascribed to primary isotope effects. Further details in the kinetic isotope effect and further details can be found in monographic text (Kohen and Limbach 2006).

All fulleranes have in common the activation of the C–H bond under the action of the UV light with release of molecular hydrogen. In fact, fulleranes show a remarkable photolytic isotope effect has been observed (see Table 8.1 for a summary of data). For $C_{70}H_{38}$ and its deuterated analogous $k_H/k_D = 1.32$ a value close to that found for $C_{60}H_{18}$.

8.3.3 *Photolysis of $C_{60}H_{18}$ and $C_{60}D_{18}$ in n-Hexane Under Helium: Determination of the Kinetic Isotope Effect*

The solution of $C_{60}H_{18}$ has been irradiated inside a quartz reactor with a low pressure mercury lamp having a monochromatic emission at 245 nm. The solution was kept under continuous He blanket to avoid any interference from air. Periodically samples from the irradiated solution were taken to measure the electronic absorption spectrum.

As shown in Fig. 8.7, the freshly prepared solution of $C_{60}H_{18}$ in n-hexane is characterized by three distinct maxima at 212, 260 and 340 nm (Palit et al. 1998). A similar spectrum is displayed by the deuterated molecule $C_{60}D_{18}$ in Fig. 8.8. Since the replacement of hydrogen with deuterium implies only a variation of mass of the substituents but not the electronic properties of the molecule. Thus, the two electronic absorption spectra are in Figs. 8.7 and 8.8 respectively due to $C_{60}H_{18}$ and its deuterated analogous are necessarily identical.

The irradiation with UV photons of 245 nm (about 5.07 eV) causes a slow reduction in intensity of the absorption bands at 212, 260 and 340 nm. This can be visualized both in Figs. 8.7 and 8.8 where also the absorption spectra after about 1,300 s irradiation are reported. In both cases the three-pattern bands is still detectable after the photolysis although the bands appear weaker and broadened. Furthermore, the baseline absorption shows an increase as the photolysis proceeds. This can be observed by looking at the spectra at $\lambda = 400$ nm and is due to the

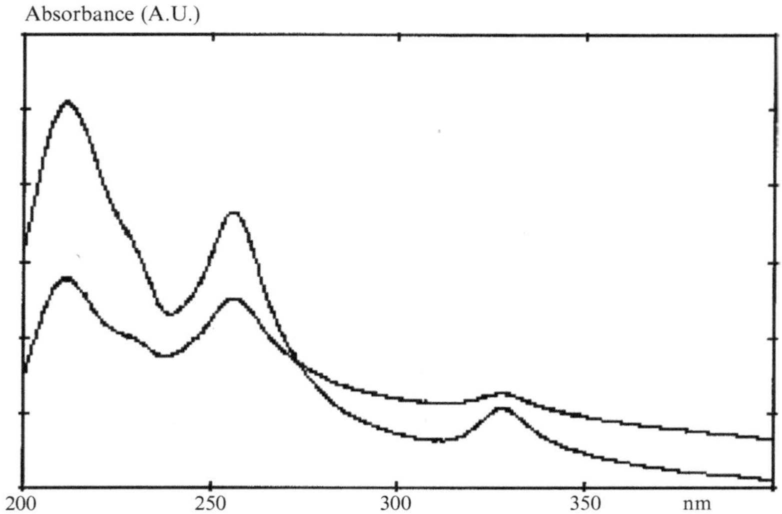

Fig. 8.7 Electronic absorption spectra in n-hexane. The upper trace is due to pristine $C_{60}H_{18}$ and the lower trace with low intensity peaks is due to $C_{60}H_{18}$ after 1,230 s photolysis under He. Note the increase of baseline at 400 nm due to light scattering due to the formation of insoluble products

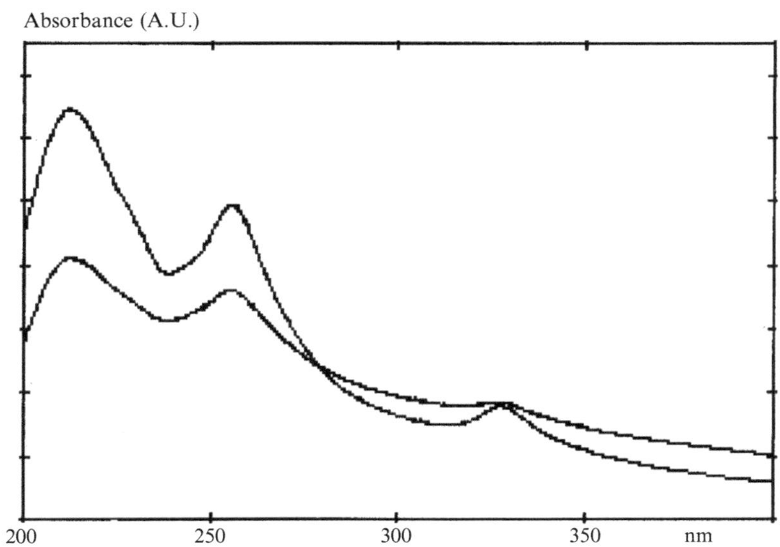

Fig. 8.8 Electronic absorption spectra in n-hexane. The upper trace is due to pristine $C_{50}D_{18}$ and the lower trace with low intensity peaks is due to $C_{60}D_{18}$ after 1,363 s photolysis under He. Note the increase of baseline at 400 nm due to light scattering due to the formation of insoluble products

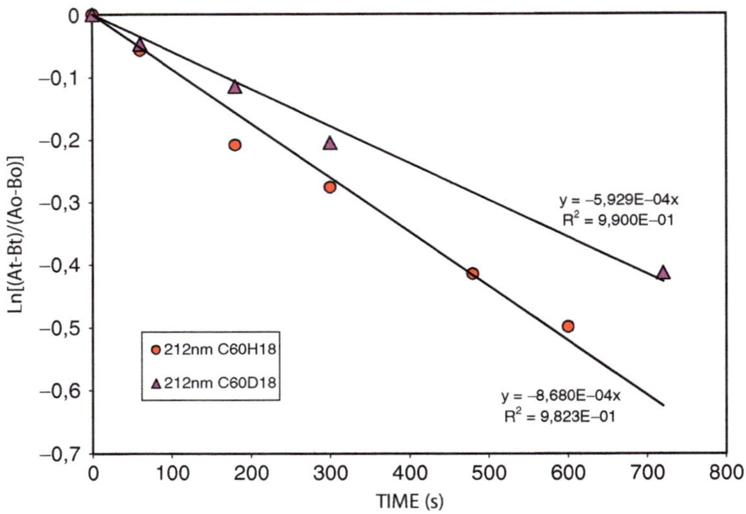

Fig. 8.9 Photolysis of $C_{60}H_{18}$ and $C_{60}D_{18}$ in n-hexane under He. The absorbance changes measured at 212 nm have been treated according to Eq. 8.4. The photolysis rate constant of $C_{60}H_{18}$ is ${}^{H}k_{212} = 8.68 \times 10^{-4}\ s^{-1}$ whereas for $C_{60}D_{18}$ ${}^{D}k_{212} = 5.93 \times 10^{-4}\ s^{-1}$

formation of an insoluble product which causes some opalescence and hence some light scattering.

In both cases the kinetic rate constant has been determined by using the absorbance read at 212 and 256 nm.

The photolysis kinetics has been determined according to Eqs. 8.1–8.4 as detailed in a previous section and conforming to the pseudofirst order kinetic law.

The results of the two photolysis experiments of $C_{60}H_{18}$ versus $C_{60}D_{18}$ recorded at 212 nm are reported in Fig. 8.9. The experimental data can be best fitted with the linear equation (Eq. 8.4) and the resulting rate constant k can be derived from the slope of the curves. The correlation factors (R^2) found in both cases are very high.

The photolysis rate constant of $C_{60}H_{18}$ molecule ${}^{H}k_{212} = 8.68 \times 10^{-4}\ s^{-1}$ is significantly higher than that of its deuterated analogous $C_{60}D_{18}$:${}^{D}k_{212} = 5.93 \times 10^{-4}\ s^{-1}$. A similar treatment of the absorbance data measured at 256 nm and reported in Fig. 8.10 confirms the result that $C_{60}D_{18}$ is photolyzed more slowly than $C_{60}H_{18}$. In fact, also in this case ${}^{H}k_{256} = 6.83 \times 10^{-4}\ s^{-1}$ is significantly higher than that of its deuterated analogous $C_{60}D_{18}$:${}^{D}k_{212} = 3.74 \times 10^{-4}\ s^{-1}$.

To explain this differences in the photolysis rate constant we should advocate once again the kinetic isotope effect and more precisely, most probably the primary isotope effect (Kohen and Limbach 2006). Because of the C–D bonds, the deuterated fullerane have lower zero-point energy than the hydrogenated fullerane (at same degree of hydrogenation). Assuming that the photolysis implies the activation of the C–H and C–D bond, then the activation energy necessary to reach the transition state is higher for the deuterated molecule in comparison to its hydrogenated analogous. Thus, if the rupture of the C–H and C–D bond occurs at the transition

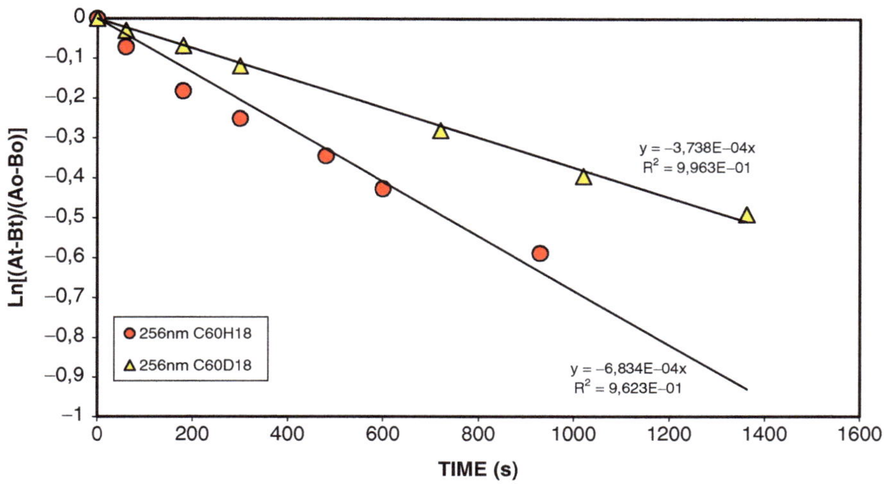

Fig. 8.10 Photolysis of $C_{60}H_{18}$ and $C_{60}D_{18}$ in n-hexane under He. The absorbance changes measured at 256 nm have been treated according to Eq. 8.4. The photolysis rate constant of $C_{60}H_{18}$ is ${}^{H}k_{212} = 6.83 \times 10^{-4}\ s^{-1}$ whereas for $C_{60}D_{18}$ ${}^{D}k_{212} = 3.74 \times 10^{-4}\ s^{-1}$

state, the rate of photolysis of the hydrogenated molecule should be higher than that of the deuterated molecule. Our data, taken from Figs. 8.9 and 8.10 show that the rate constant ratios are:

$${}^{H}k_{212}/{}^{D}k_{212} = 1.46$$

and

$${}^{H}k_{212}/{}^{D}k_{212} = 1.83$$

suggesting that the photolysis of $C_{60}H_{18}$ is from 1.5 to 1.8 times faster than the photolysis of $C_{60}D_{18}$. The isotope effect found is quite high suggesting more a primary kinetic isotope effect rather than secondary isotope effect (Kohen and Limbach 2006).

Based on our results it can be expected that if fulleranes are present in the interstellar medium they should show significant enrichment in deuterium with respect to the interstellar hydrogen, as it happens for other hydrogenated organic molecules present in the interstellar medium. Such enrichment has been detected also in organic molecules present in meteorites suggesting that their composition reflects the composition of the interstellar medium. Recently, the deuterium enrichment has been detected even in the insoluble organic matter of carbonaceous meteorites containing radicals and a polyaromatic structure.

8.3.4 Photolysis of C_{3v}-$C_{60}H_{18}$ in Tetradecane Under Ar

An authentic and fully characterized sample of C_{3v}-$C_{60}H_{18}$ prepared by high pressure hydrogenation of C_{60} (Wagberg et al. 2005) appears somewhat different in its

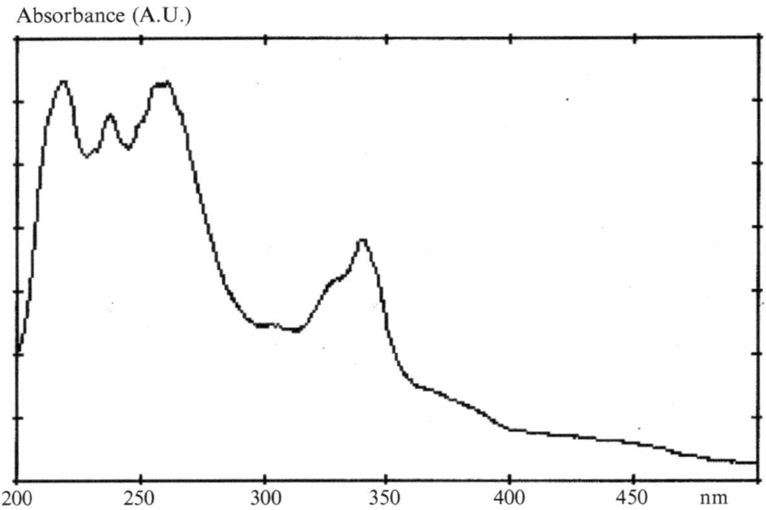

Fig. 8.11 Electronic absorption spectrum of pure C_{3v}-$C_{60}H_{18}$ in tetradecane. The spectrum is quite more complex than that shown in Fig. 8.1 of $C_{60}H_{18}$ generated in situ in n-hexane from Zn/HCl

properties than the $C_{60}H_{18}$ prepared in situ by hydrogenation of C_{60} dissolved in n-hexane (see previous section). First of all C_{3v}-$C_{60}H_{18}$ appears almost insoluble in n-hexane. Even after prolonged stirring the dissolution of C_{3v}-$C_{60}H_{18}$ as checked spectrophotometrically is negligible. This poses some doubt about the C_{3v} symmetry of the $C_{60}H_{18}$ prepared with the Zn/HCl or Zn/DCl synthetic route. Pure C_{3v}-$C_{60}H_{18}$ shows instead a certain solubility in a relatively long chain paraffin such as tetradecane (see also the experimental section). Such solubility was expected on the basis of the fullerene solubility in long chain fatty acids and their esters (Cataldo and Braun 2007; Cataldo 2008). The electronic absorption spectrum of C_{3v}-$C_{60}H_{18}$ in tetradecane is shown in Fig. 8.11 and appears much more complex than that of $C_{60}H_{18}$ prepared from Zn/HCl or Zn/DCl synthetic route (compare with Fig. 8.11 with Fig. 8.7). More in detail the electronic spectrum of C_{3v}-$C_{60}H_{18}$ is characterized by a series of absorption maxima at 218, 237, 249, 255 and 261 nm. Additional features are found at 303, 326 and 340 nm (Fig. 8.11). Because of these important spectral differences, it can be anticipated that the $C_{60}H_{18}$ obtained the Zn/HCl synthesis may have a different symmetry than that the C_{3v} which instead it has been ascertained (Wagberg et al. 2005) for the sample synthesized in the solid state.

The photolysis of C_{3v}-$C_{60}H_{18}$ in tetradecane has been performed with the low pressure mercury lamp as UV source under Ar in the same general conditions employed for the other $C_{60}H_{18}$ samples. Also in this case the decomposition and the disappearance of C_{3v}-$C_{60}H_{18}$ has been followed by spectrophotometric measurements of the irradiated solution. After 1560 s irradiation, the spectrum of the irradiated C_{3v}-$C_{60}H_{18}$ (Fig. 8.12) appears completely different from the spectrum of the pristine

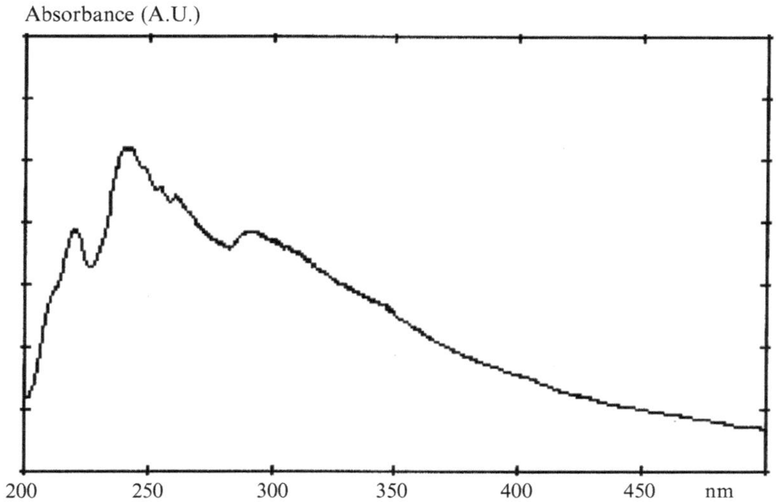

Fig. 8.12 Electronic absorption spectrum of C_{3v}-$C_{60}H_{18}$ in tetradecane after 1,560 s photolysis under Ar

sample (Fig. 8.11) and is characterized by three absorption maxima at 217, 237 and 292 nm. The absorption maximum originally located at 340 nm appears completely bleached. Thus, the spectrum of Fig. 8.12 is completely different from the spectra of $C_{60}H_{18}$ or $C_{60}D_{18}$ prepared and irradiated for ≈1,300 s in n-hexane (Fig. 8.8), confirming once again that the samples prepared and irradiated in n-hexane have certainly a different symmetry than that of the authentic C_{3v}-$C_{60}H_{18}$. The different spectral evolution of C_{3v}-$C_{60}H_{18}$ under UV irradiation has not hindered the possibility to measure the decomposition kinetics. Figure 8.13 shows the treatment of the absorbance data read at 217 and 255 nm. Such data have been processed according to Eq. 8.4 and the linear response has been achieved. Thus for C_{3v}-$C_{60}H_{18}$ the pseudofirst rate constant is $k_{217} = 4.04 \times 10^{-4}\ s^{-1}$ and $k_{255} = 2.25 \times 10^{-4}\ s^{-1}$. These values are significant different from those measured on $C_{60}H_{18}$ in n-hexane. However, this does not come as a surprise considering that C_{3v}-$C_{60}H_{18}$ in tetradecane has a different absorption spectrum than that of $C_{60}H_{18}$ and furthermore, the spectral evolution of C_{3v}-$C_{60}H_{18}$ under UV irradiation is completely different than that observed in the case of $C_{60}H_{18}$ in n-hexane.

8.3.5 Discussion of the Photolysis Data in an Astrochemical Context

Primary isotope effects are defined as concerning reactions in which, at some stage – normally the rate determining step – a bond to the isotopically labelled compound

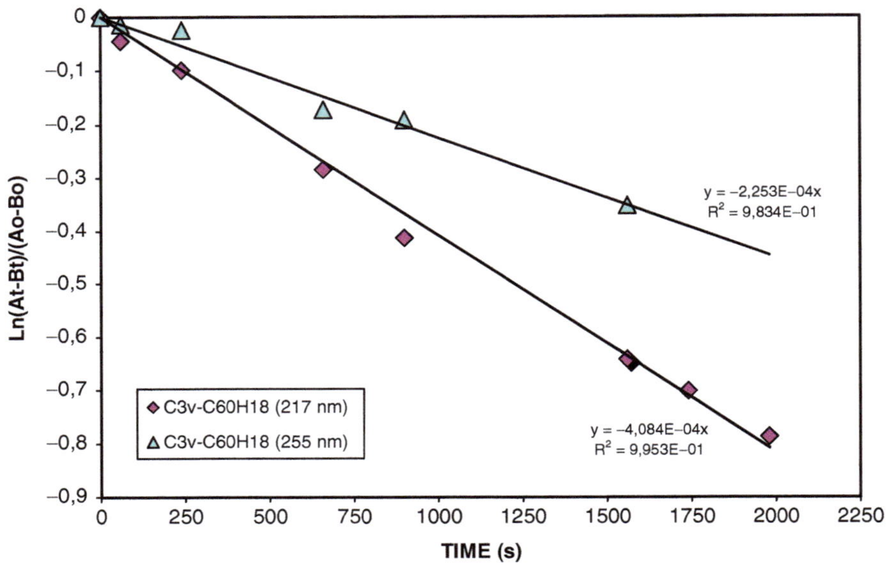

Fig. 8.13 Photolysis of C_{3v}-$C_{60}H_{18}$ in tetradecane under Ar. The absorbance changes measured at 217 and at 255 nm have been treated according to Eq. 8.4. The photolysis rate constants of C_{3v}-$C_{60}H_{18}$ are $k_{217} = 4.04 \times 10^{-4}\ s^{-1}$ whereas $k_{255} = 2.25 \times 10^{-4}\ s^{-1}$

is formed or broken (Bamford and Tipper 1983; Kohen and Limbach 2006). The measurable difference in photolysis rate between $C_{60}H_{36}$ and $C_{60}D_{36}$ can be explained in terms of isotope effect so that the photolysis of the former molecule is 1.08 times faster than the photolysis of the deuterated analogous. Because of the presence of C–D bonds, the deuterated fullerane have lower zero-point energy than the hydrogenated fullerene; assuming that the photolysis implies the rupture of the C–H and C–D bond as the rate determining step, then the activation energy necessary to reach the transition state is higher for the deuterated molecule in comparison to its hydrogenated analogous. Thus, if the rupture of the C–H and C–D bond occurs at the transition state, the rate of photolysis of the hydrogenated molecule is higher than that of the deuterated molecule.

It is not the first time that we observe an isotope effect in hydrogenated fullerenes. Elsewhere (Cataldo et al. 2009d) we have reported about the photolysis of $C_{60}H_{18}$ and $C_{60}D_{18}$ in n-hexane in the same photolysis conditions adopted in the present work. The isotope effect was much more relevant for the fullerenes at lower hydrogenation degree since the photolysis of $C_{60}H_{18}$ was found from 1.5 to 1.8 times faster than the photolysis of $C_{60}D_{18}$. The photolysis rate constants of $C_{60}H_{18}$ and $C_{60}D_{18}$ was found about one order of magnitude lower than that observed for the couple $C_{60}H_{36}$ and $C_{60}D_{36}$ (Cataldo et al. 2009d). Probably the reason for this different photochemical behaviour resides in the cage strain (being more strained the $C_{60}H_{36}$ cage rather than the $C_{60}H_{18}$) and in the different photophysical properties of $C_{60}H_{36}$ and $C_{60}D_{36}$. In particular Palit et al. (1998) have studied the photophysical properties of $C_{60}H_{18}$ and $C_{60}H_{36}$ synthesized with the same procedure adopted by us in cyclohexane solution and reported remarkable differences in the fluorescence quantum yield, in the singlet and triplet lifetime of the two fulleranes $C_{60}H_{18}$ and $C_{60}H_{36}$.

The fact that we are measuring an appreciable isotope effect in the photolysis of $C_{60}H_{18}$ and $C_{60}H_{36}$ implies that the C–H and C–D bonds are involved in the photolytic process. Indeed the UV photons of 254 nm have an energy of 112.7 kcal/mol, sufficient to break the C–H bonds of fulleranes (bond energy ≈100 kcal/mol). Combining the activation of the C–H/C–D bond with the formation of an insoluble photoproduct during the fullerane photolysis, it can be deduced that the fulleranes are loosing hydrogen (or deuterium) and then are forming fullerane dimers or oligomers by coupling reactions among fullerane cages. This causes the consumption of the fullerane in solution during photolysis and its accumulation as insoluble photoproduct. Unfortunately, attempts to isolate the photoproduct and to study its chemical structure were unsuccessful (see experimental section) for the reason that the photoproducts are very sensitive to air and undergoes severe oxidation during the isolation procedure.

This study suggests that in space conditions fulleranes may loose hydrogen or deuterium under the action of UV radiation and it is reasonable to think that an equilibrium could be formed between the hydrogen added to the cage by simple addition reaction for instance of atomic hydrogen and the hydrogen lost due to photolytic degradation. In this process a significant enrichment of the fullerene cage in deuterium could occur because of the measured isotope effects. Such deuterium enrichment has already been detected in other organic molecules present in meteorites suggesting that their composition reflects the composition of the interstellar medium (Tielens 2005; Shaw 2006). The deuterium enrichment has been detected indeed in the insoluble organic matter of carbonaceous meteorites containing radicals and a polyaromatic structure (Gourier et al. 2008).

Experimental studies have shown that hydrogenated carbon dust grains undergo various processing phenomena in the interstellar medium (Colangeli et al. 1999; Mennella 2001). In other words, carbon dust is hydrogenated by atomic hydrogen but can loose molecular hydrogen by various mechanisms which may be thermal or may include the actions of cosmic rays or UV photons. The thermal or radiation processing causes the loss of molecular hydrogen and the aromatization of the carbon dust. Thus, the hydrogenated carbon dust displays a feature at about 217 nm which is shifted gradually toward longer wavelengths as the thermal or radiation processing takes place, the shift is due to the fact that the sp^3 hybridized carbon atoms are transformed into aromatic sp^2 carbon atoms due to the release of hydrogen. The interstellar light extinction curve shows a "bump" centred at 217.5 nm and is thought to be due to hydrogenated carbon dust (Ehrenfreund and Charnley 2000). As noticed by one of us (Cataldo 2003a, b), it is really curious that the hydrogenated fullerene $C_{60}H_{36}$ displays an absorption spectrum with a maximum just at 217.4 nm (Fig. 8.1). Even the half width of this band (40 nm) coincides with that measured on the interstellar light extinction curve (Unsold and Baschek 2002).

It has been proposed that fullerane $C_{60}H_{36}$ can be taken as a model of the interstellar hydrogenated carbon dust (Cataldo 2003a, b), but it is also possible that this molecule is a contributor to the interstellar light extinction curve together with other dust.

In Fig. 8.2 we have shown that the UV photolysis of $C_{60}H_{36}$ causes a gradual shift of the absorption maximum from 217.5 nm to longer wavelengths going beyond 220 nm, exactly the same phenomenon observed previously (Colangeli et al. 1999) in the UV irradiation of hydrogenated carbon dust.

8.3.6 *Photolysis Stability of $C_{60}H_{36}$: A Comparison with Other Fulleranes and with Polyynes*

Polyynes are carbon chains having general formula $H–(C{\equiv}C)_n–H$ which are of remarkable importance in astrochemistry and astrophysics (Cataldo 2004, 2006a, b). In fact they have been detected by radioastronomy in different astronomical enviroments like late-type carbon-rich stars, asymptotic giant branch (AGB) stars, dark molecular clouds, hot molecular cores and even in comets and in the atmospheres of planets and satellites of the Solar System (Millar 2004). As summarized in Table 1, under identical experimental conditions used in the present work for the photolysis of $C_{60}H_{36}$, the polyynes C_8H_2, $C_{10}H_2$, $C_{12}H_2$ and $C_{14}H_2$ exhibit a photolysis rate constant comprised between 2.4×10^{-3} and 5.2×10^{-3} s^{-1} (Cataldo et al. 2008). Thus, the carbon chains which are rather diffuse in the Universe display comparable to faster UV degradation than the fullerane $C_{60}H_{36}$. Instead, fulleranes with lower degree of hydrogenation such as $C_{60}H_{18}$ show a much better photostability (Cataldo et al. 2009d). In fact in Table 8.1 the photoysis rate constant of $C_{60}H_{18}$ 3 times lower than that of polyynes and fullerane $C_{60}H_{36}$. Thus, if polyynes are present in so many different astrophysical environment, then also fulleranes which are more stable in general and to UV photons in particular should be present in space.

8.4 Conclusions

Fulleranes $C_{60}H_{36}$ and its deuterated analogous $C_{60}D_{36}$ have been synthesized in an UV-transparent medium: n-hexane. In this solvent both molecules show an absorption maximum at about 217 nm and the molar extinction coefficient has been determined $\varepsilon_{217} = 17{,}140$ L cm^{-1} mol^{-1} and $\varepsilon_{217} = 16{,}480$ L cm^{-1} mol^{-1} respectively. The wavelength of the peak and the width of this absorption appears to reproduce the main characteristics of the UV bump in the interstellar extinction curve. The UV irradiation of $C_{60}H_{36}$ and $C_{60}D_{36}$ with a monochromatic light at 245 nm causes the photolysis of these molecules which was monitored by electronic absorption spectroscopy. The photolysis causes a gradual shift of the absorption maximum from 217 nm to longer wavelengths and a reduction in the absorption intensity. The shift in the absorption maximum from 217 nm to longer wavelengths has been reported also in the photolysis of hydrogenated carbon dust in the laboratory simulation of the UV and thermal processing of the interstellar carbon dust analogous.

Measurable isotope effect has been detected in the photolysis of the two molecules $C_{60}H_{36}$ and $C_{60}D_{36}$ with the hydrogenated molecule being photolyzed at 1.08 times faster than the deuterated analogous. This implies that the photolysis activates the C–H bonds and the rate determining step in the fullerane photolysis should be due to the C–H bond rupture. Thus, in the interstellar medium fulleranes may loose molecular hydrogen forming fulleranes at lower hydrogenation degree or forming back $C_{60.}$ By reacting with atomic hydrogen C_{60} forms again fulleranes at various degree of hydrogenation. Therefore a dynamic equilibrium could be foreseen where atomic hydrogen is converted into molecular hydrogen through the catalysis of the fullerene surface. Since fullerenes or fullerene-like structures are present in carbon dust, then the process leading to molecular hydrogen from atomic hydrogen should occur widely in the Universe. Additionally, the isotope effect should lead to a deuterium enrichment of the surface of the interstellar carbon dust as well as the surface of interstellar fulleranes.

Thermal processing of $C_{60}H_{36}$ at 630°C under inert atmosphere causes the regeneration of C_{60} and the release of H_2. Under determined conditions this process is fully reversible: C_{60} can be hydrogenated by atomic hydrogen and then by thermal treatment can release back all hydrogen as H_2 producing C_{60}.

The photolysis rate constants of $C_{60}H_{36}$ is comparable or slightly lower than that of polyynes while it is three times faster than the photolysis rate constant of $C_{60}H_{18}$, a fullerane with a lower degree of hydrogenation than $C_{60}H_{36}$.

Acknowledgements The present research work has been supported by grant AYA2007-64748 of the Spanish Ministerio de Ciencia e Innovacion.

References

Bamford CH, Tipper CFH (1983) Modern kinetics methods. Elsevier, Amsterdam, pp 120–132
Bensasson RV, Hill TJ, Land EJ, Leach S, McGarvey DJ, Truscott TG, Ebenhoch J, Gerst M, Ruchardt C (1997) Chem Phys 215:111–123
Cataldo F (2003a) Fullerenes Nanot Carbon Nanostruct 11:295–316
Cataldo F (2003b) Fullerenes Nanot Carbon Nanostruct 11:317–331
Cataldo F (2004) Int J Astrobiol 3:237–246
Cataldo F (2006a) Int J Astrobiol 5:37–45
Cataldo F (2006b) Origin Life Evol Biosph 36:467–475
Cataldo F (2008) Solubility of fullerenes in fatty acids esters: a new way to deliver *in vivo* fullerenes. Theoretical calculations and experimental results. In: Cataldo F, Da Ros T (eds) Medicinal chemistry and pharmacological potential of fullerenes and carbon nanotubes. Springer, Berlin, pp 317–335
Cataldo F, Braun T (2007) Fullerenes Nanot Carbon Nanostruct 15:331–339
Cataldo F, Strazzulla G, Iglesias-Groth S (2008) Int J Astrobiol 7:107–116
Cataldo F, Iglesias-Groth S, Manchado A (2009a) Fullerenes Nanot Carbon Nanostruct 17:378
Cataldo F, Iglesias-Groth S, Manchado A (2009b) Fullerenes Nanot Carbon Nanostruct 17:401
Cataldo F, Strazzulla A, Iglesias-Groth S (2009c) Month Not Roy Astrophys Soc 394:615

Cataldo F, Iglesias-Groth S, Manchado A (2009d) Fullerenes Nanot Carbon Nanostruct 17:414
Colangeli L, Mennella V, Bussoletti E, Palumbo P, Rotundi A (1999) Laboratory analogues for interstellar carbon dust. In: d'Hendecourt L, Joblin C, Jones A (eds) Solid interstellar matter the ISO revolution. Springer, Heidelberg, Lecture 9
Darwish AD, Abdul-Sada AK, Langley GJ, Kroto HW, Taylor R, Walton DRM (1995) J Chem Soc. Perkin Trans 2:2359–2365
Ehrenfreund P, Charnley SB (2000) Annu Rev Astronom Astrophys 38:427–483
Foing BH, Ehrenfreund P (1994) Nature 369:296–298
Foing BH, Ehrenfreund P (1997) Astron Astrophys Lett 317:L59–L62
Glassgold AE (1996) Annu Rev Astronom Astrophys 34:241–277
Goeres A, Sedlmayr E (1993) Fullerene Sci Technol 1:563–570
Gourier DRF, Delpoux O, Binet L, Vezin H, Moissette A, Derenne S (2008) Geochim Cosmochim Acta 72:1914–1923
Guldi DM, Kamat P (2000). Photophysical properties of pristine fullerenes, functionalized fullerenes, and fullerene-containing donor-bridge acceptor systems. In: Kadish KM, Ruoff RS (eds) Chapter 5 in Fullerenes: chemistry, physics and technology. Wiley-Interscience, New York
Hare JP, Kroto HW (1992) Acc Chem Res 25:106–112
Howard JA (1993) Chem Phys Lett 203:540–544
Iglesias-Groth S (2004) Astrophys J 608:L37–L40
Iglesias-Groth S (2005) Astrophys J 632:L25–L28
Iglesias-Groth S (2006) Month Not Roy Astron Soc 368:1925–1930
Kohen A, Limbach HH (2006) Isotope effects in chemistry and biology. CRC Press/Taylor & Francis, Boca Raton, FL
Kroto HW (2006) Introduction: Space-Pandora's Box. In: Rietmeijer FJH (ed) Natural fullerenes and related structures of elemental carbon. Springer, Dordrecht, pp 1–5
Mennella V (2001) Astron Astrophys 367:355–361
Millar T (2004) Organic molecules in the interstellar medium. In: Ehrenfreund P (ed) Astrobiology: future perspectives. Kluwer, Dordrecht, pp 17–31
Ninomiya I, Naito T (1989) Photochemical synthesis. Academic, London, p 221
Nossal J, Saini RK, Alemany LB, Meier M, Billups WE (2001) Eur J Org Chem 2001:4167–4180
Palit DK, Mohan H, Mittal JP (1998) J Phys Chem 102:4456–4461
Petrie S, Bohme DK (2000) Astrophys J 540:869–885
Petrie S, Becker H, Baranov VI, Bohme DK (1995) Int J Mass Spectrom 145:79–88
Shaw AM (2006) Astrochemistry. Wiley, Chichester
Stoldt CR, Maboudian R, Carraro C (2001) Astrophys J 548:L225–L228
Taylor R (1999) Lecture notes on fullerene chemistry. A handbook for chemists. Imperial College Press, London, pp 56–70
Taylor R (2006) CR Chimie 9:982–1000
Tielens AGGM (2005) The physics and chemistry of the interstellar medium. Cambridge University Press, Cambridge
Unsold A, Baschek B (2002) The new cosmos: an introduction to astronomy and astrophysics, 5th edn. Springer, Berlin, p 247
Van Dishoeck EF, Blake GA (1998) Annu Rev Astronom Astrophys 36:317–368
Wagberg T, Johnels D, Peera A, Hedenstrom M, Schulga YM, Tsybin YO, Purcell JM, Marshall AG, Noreus D, Sato T, Talyzin AV (2005) Org Lett 7:5557–5560
Webster A (1991) Nature 352:412–414
Yeremin EN (1979) The foundation of chemical kinetics. Mir, Moscow, pp 21–26

Chapter 9
Characterization of Hydrogenated Fullerenes by NMR Spectroscopy

Mattias Hedenström[1], Thomas Wågberg[2], and Dan Johnels[3]

Abstract NMR spectroscopy is so far the only analytical technique that has been used to get a detailed structural characterization of hydrogenated fullerenes. A substantial amount of information derived from different NMR experiments can thus be found in the literature for a number of fullerenes hydrogenated to various degrees. These studies have benefitted from the fact that chemical shifts of ^{1}H and ^{13}C and in some cases also ^{3}He can be used to obtain structural information of these compounds. Such results, together with discussions about different NMR experiments and general considerations regarding sample preparations, are summarized in this chapter. The unique information, both structural and physicochemical, that can be derived from different NMR experiments ensures that this technique will continue to be of central importance in characterization of hydrogenated fullerenes.

9.1 Introduction

This chapter will give an overview of the use of NMR spectroscopy in characterization of hydrogenated fullerenes. This first section serves as a general introduction of the importance of NMR as a characterization method in this field. Thereafter, different NMR techniques and methods that have been implemented to derive different chemical and structural properties of hydrogenated fullerenes are reviewed. To conclude this chapter, a more detailed description of the different hydrogenated

[1]Department of Chemistry, Computational life Science (CLIC), Umeå University, SE-90187, Umeå, Sweden
e-mail: mattias.hedenstrom@chem.umu.se

[2]Department of Physics, Umeå University, SE-90187, Umeå, Sweden
e-mail: thomas.wagberg@physics.umu.se

[3]Department of Chemistry, Umeå University, SE-90187, Umeå, Sweden
e-mail: Dan.Johnels@chem.umu.se

F. Cataldo and S. Iglesias-Groth (eds.), *Fulleranes: The Hydrogenated Fullerenes*, Carbon Materials: Chemistry and Physics 2, DOI 10.1007/978-1-4020-9887-1_9,

fullerenes studied with NMR will be given. The majority of NMR studies on hydrogenated fullerenes have concentrated on determining the structure, i.e. the hydrogenation pattern, whereas only a few studies have focused on determining the physical and chemical properties of these systems. Consequently, this chapter will focus on presenting the current knowledge regarding structure and symmetry as studied by NMR spectroscopy. It should be noted that some good reviews on hydrogenated fullerenes already have been published (Hirsch 1994; Nossal et al. 2001a).

NMR spectroscopy has proven to be an invaluable tool in structure determination of hydrogenated fullerenes ever since the first synthesis of $C_{60}H_{36}$ (Haufler et al. 1990) and the first structure determination by NMR was reported for $C_{60}H_2$ by Henderson and Cahill in 1993 (Henderson and Cahill 1993). Today, a multitude of both one- and two-dimensional NMR experiments are available for this purpose. Recent developments in NMR technology with higher magnetic field strengths and cryoprobes have dramatically increased the sensitivity and improved the usefulness of NMR in this field even further.

Different ^{1}H and ^{13}C NMR experiments are evidently the main tools for studying hydrogenated fullerenes with NMR spectroscopy but also the chemical shift of ^{3}He encapsulated within the fullerene cage can be used to determine structure and isomer distribution of hydrogenated fullerenes. Determination of the hydrogen pattern in these molecules can be a daunting task considering the, in many cases, immense number of possible isomers (Balasubramanian 1991; Dunlap et al. 1994). So, what kind of structural information can be gained from NMR spectroscopy that could help us solve this problem? The two NMR properties mainly used for structure determination are chemical shifts and J-couplings. The chemical shifts, i.e. the peak positions in the spectrum reflect the chemical environment of the observed nucleus and is thus important in structure determination of all kinds of organic molecules. Although the chemical shifts per se contain useful structural information, they are usually not sufficient for unambiguous structure determination of organic molecules. This limitation is even more pronounced for hydrogenated fullerenes considering that the hydrogens experience a relatively similar chemical environment and all resonances therefore appear in a narrow chemical shift range. In addition to peak position, the number of ^{1}H and ^{13}C resonances in the spectra also yields important information as it is related to molecular symmetry. Once the symmetry of the molecule has been established, the number of possible structures can often be reduced drastically. Structure determination is most often highly dependent on information about how different hydrogens and carbons are positioned in relation to each other in the molecule. In NMR spectroscopy, this information is revealed by J-couplings that correlate hydrogens and/or carbons close to each other in the structure. The most commonly observed J-couplings are $^{3}J_{HH}$-couplings between ^{1}H on adjacent carbon atoms and $^{1}J_{CH}$-couplings between a ^{1}H directly bonded to a ^{13}C atom. $^{1}J_{CC}$-couplings can also be observed using special techniques. The size of the J-coupling is measured in hertz and is dependent on the type of nuclei, the hybridization state and the angle between the interacting nuclei. $^{3}J_{HH}$-couplings can be observed in a ^{1}H spectrum or a two-dimensional (2D) COSY spectrum while observation of proton-carbon and carbon–carbon coupling usually relies on different 2D NMR experiments such as HSQC or INADEQUATE (see Section 9.5 for description of 2D NMR experiments).

NMR spectroscopy has also been used to study molecular dynamics and partial alignment of hydrogenated fullerenes in solution by relaxation measurements.

9.2 Synthesis and Sample Preparation

A very large number of synthetic techniques have been used to hydrogenate fullerenes. Most often these result in a complex mixture containing fullerene molecules with different hydrogenation degree as well as isomers with same number of attached hydrogens. Although some synthetic techniques can produce nearly pure isomeric material (Wågberg et al. 2005), studies on purified samples often relies upon separation with High Pressure Liquid Chromatography (HPLC) (Bucsi et al. 1997). In this review we do not focus on synthesis techniques and therefore only briefly mention some of the techniques that have been used. Among the most commonly used technique is the so called Birch reduction which has been reported in numerous studies (Banks et al. 1993; Haufler et al. 1990; Nossal et al. 2001b). Other techniques include hydrozirconation (Ballenweg et al. 1993), diimide reduction (Avent et al. 1994), rhodium-catalyzed hydrogenation (Avent et al. 1994), hydrazine reduction (Billups et al. 1997a), dissolving metal reduction (Bergosh et al. 1997) and polyamine mediated hydrogenation (Briggs et al. 2005; Kintigh et al. 2007). Hydrogenation under high pressure at elevated temperature has also been used successfully to synthesize different hydrogenated fullerenes (Wågberg et al. 2005, 2008).

Special precautions should be taken when preparing samples of hydrogenated fullerenes for NMR analysis as most of them are unstable and decompose readily in solution. Exposure to light and air should be kept to a minimum as light-catalyzed oxidation, especially of the higher hydrides, proceeds rapidly. This has been observed in many studies of both hydrogenated C_{60} and C_{70} (Avent et al. 1994; Becker et al. 1993; Bergosh et al. 1997; Billups et al. 1997b; Darwish et al. 1995, 1996a; Drelinkiewicz et al. 1996; Lobach et al. 1998). Complete degradation of $C_{60}H_{36}$ into fullerenols was, for example, observed when $C_{60}H_{36}$ dissolved in toluene was subjected to air and light for 48 h (Darwish et al. 1995). The use of deoxygenized solvents is highly recommended but does not always suffice to completely avoid decomposition as oxygen is readily trapped within the fullerene crystal lattice (Darwish et al. 1996a). In the absence of light, hydrogenated fullerenes can be stored in solution for a limited period of time. As an example, Darwish and co-workers found that $C_{60}H_{36}$ is reasonably stable for up to 2 weeks in solution (Darwish et al. 1995). To avoid excessive decomposition, our recommendation is to prepare the NMR sample under inert atmosphere using freeze-pump-thawed deoxygenated solvents, sealing the tube properly directly after dissolving the material and keep it protected from light. By this method Wågberg et al. prepared $C_{60}H_{18}$ and $C_{70}H_{38}$ solutions that were stable for at least several months (Wågberg et al. 2005, 2008).

Hydrogenated fullerenes are significantly less soluble than their parent fullerenes which limit the range of solvents that can be used for NMR analysis. Common solvent for hydrogenated fullerenes include CS_2 (Avent et al. 1994; Bergosh et al. 1997; Meier et al. 1996; Spielmann et al. 1998), toluene-d_8 (Darwish et al. 1995; Henderson and Cahill 1993; Henderson et al. 1994a), C_6D_6 (Becker et al. 1993; Wågberg et al. 2005). Mixtures between 1-methylnaphthalene and CD_2Cl_2 and C_6D_6 and CS_2 have also been used (Alemany et al. 1997a, b; Billups et al. 1997a; Cross et al. 1996). More recently, $CDCl_3$ (Gakh et al. 2003; Kintigh et al. 2007; Wågberg et al. 2008) and 1,2-dichlorobenzene-d_4 (*o*-$C_6D_4Cl_2$) (Nossal et al. 2001b; Peera et al. 2003; Wågberg et al. 2008) have been employed in studies of higher hydrogenated fullerenes. Wågberg et al. found significantly increased solubility of $C_{70}H_{38}$ in *o*-$C_6D_4Cl_2$ or $CDCl_3$ compared to C_6D_6 (Wågberg et al. 2008) and less decomposition in *o*-$C_6D_4Cl_2$ compared to CS_2 has also been reported for $C_{60}H_{36}$ (Nossal et al. 2001b). Wågberg et al. proposed that trace amounts of acid present in $CDCl_3$ could have a severe affect on compound stability (unpublished results). A similar observation has been reported earlier (Rüchardt et al. 1993).

9.3 ^{1}H NMR Spectroscopy

1D ^{1}H NMR is probably the most widely used NMR technique in the studies of hydrogenated fullerenes. The high sensitivity of this nucleus makes it attractive as experimental times can be kept rather short even for dilute solutions, especially important considering the inherent instability of these compounds. Proton chemical shifts ranging from 2.63 to 7.23 ppm have been observed in hydrogenated fullerenes. Table 9.1 summarizes the ^{1}H NMR data reported in the literature for hydrogenated fullerenes derived from C_{60} and C_{70}. The ^{1}H chemical shifts are to a varying degree dependent on the solvent and the reported chemical shifts are consistently lower in toluene than in CS_2 or C_6D_6. Compared to ^{13}C the spectral range is rather narrow and peak overlap can pose a serious problem, especially in mixtures of higher hydrogenated fullerenes. As mentioned in Section 9.1, J-coupling connectivity are perhaps the most important information derived from ^{1}H spectra when it comes to structure analysis because the splitting pattern can give information about both the number and the chemical shifts of neighboring ^{1}H. $^3J_{HH}$-coupling constants ranging between 9 and 17 Hz are usually observed in hydrogenated fullerenes. It has been noted that $^3J_{HH}$-coupling constants for 6,6-bonds (13.9–16.3 Hz) differ significantly from those for 5,6-bonds (9.3–9.8 Hz) in both $C_{60}H_n$ and $C_{70}H_n$ (Avent et al. 1994; Bergosh et al. 1997; Henderson et al. 1994b; Meier et al. 1996; Spielmann et al. 2000). Long-range couplings are commonly observed in organic compounds containing C–C double bonds. $^5J_{HH}$- and even $^7J_{HH}$-couplings are therefore often observed in hydrogenated fullerenes which can result in quite complex splitting patterns. Long range couplings are much smaller than $^3J_{HH}$-couplings (usually 1–5 Hz). Another complication is that J-coupled protons with similar chemical shifts generate second-order spectra that

Table 9.1 Observed ^{1}H chemical shifts for hydrogenated C_{60} and C_{70}

Entry[a]	Structure	Reference	Symmetry	Solvent	[1]H chemical shifts[b]
1 (1a)	1,2,-$C_{60}H_2$	1	C_s	Toluene	5.93(2H,s)
2 (1a)	1,2-$C_{60}H_2$	2	C_s	C_6D_6/CS_2 (1:1)	6.14(2H,s)
3 (1a)	1,2-$C_{60}H_2$	3	C_s	CS_2	6.97(2H,s),
4 (1a)	1,2-$C_{60}H_2$	4	C_s	CS_2	7.17(2H,s)
5 (1a)	1,2-$C_{60}H_2$	3	C_s	C_6D_6	5.89(2H,s)
6 (1b)	1,4-$C_{60}H_2$	3	C_s	CS_2	7.23(2H,s)
7 (1b)	1,4-$C_{60}H_2$	3	C_s	C_6D_6	6.13(2H,s)
8 (2a)	1,2,3,4-$C_{60}H_4$	3	C_s	CS_2	6.02(4H,AA′BB′,m)
9 (2a)	1,2,3,4-$C_{60}H_4$	3	C_s	C_6D_6	4.97(4H,AA′BB′,m)
10 (2a)	1,2,3,4-$C_{60}H_4$	4	C_s	Toluene	5.03(4H,AA′BB′,m)
11 (2b)	1,2,18,36-$C_{60}H_4$	3	C_s	CS_2	6.38(2H,s)[c]
12 (2b)	1,2,18,36-$C_{60}H_4$	5	C_s	CS_2	6.42(2H,s), 6.11(2H,AB,d)
13 (2b)	1,2,18,36-$C_{60}H_4$	4	C_s	Toluene	5.36(2H,s), 5.05 (2H,AB,d)
14 (2b)	1,2,18,36-$C_{60}H_4$	3	C_s	C_6D_6	5.49(2H,s)[c]
15 (2b)	1,2,18,36-$C_{60}H_4$	2,6	C_s	CS_2/C_6D_6	5.49(2H,s), 5.18(2H,AB,d)
16 (2c)	1,2,33,50-$C_{60}H_4$[d]	5	C_2	CS_2	6.66(4H,AB,d)
17 (2d)	1,2,55,60-$C_{60}H_4$	4	D_{2h}	Toluene	6.33(4H,s)
17 (2d)	1,2,57,58-$C_{60}H_4$	3	C_s	CS_2	6.27(4H,s)
19 (2d)	1,2,57,58-$C_{60}H_4$	3	C_s	C_6D_6	5.34(4H,s)
20 (2d)	1,2,57,58-$C_{60}H_4$	6	C_s	C_6D_6/CS_2	6.45(4H,s)
21 (3a)	1,2,18,22,23,36-$C_{60}H_6$[d]	5	C_3	CS_2	5.25(6H,AB,d)
22 (3b)	1,2,33,41,42,50-$C_{60}H_6$	5,7	D_3	CS_2	6.0(6H,s)
23 (4)	$C_{60}H_{18}$	8	C_{3v}	CS_2	4.49(3H,d), 4.15(6H,m), 3.95(3H,dt), 3.45 (6H,m)
24 (4)	$C_{60}H_{18}$	9	C_{3v}	o-$C_6D_4Cl_2$	4.3(3H,d), 3.95(6H,m), 3.7(3H,dt), 3.23(6H,m)
25 (4)	$C_{60}H_{18}$	10	C_{3v}	C_6D_6	4.10 (3H,d), 3.65(6H,m), 3.26(3H,dt), 2.92(6H,m)
26 (5a)	$C_{60}H_{36}$	11	C_1	$CDCl_3$	3.69(2H), 3.46(2H), 3.41(2H), 3.34(2H), 3.32(2H), 3.30(4H), 3.29 (8H), 3.28(2H), 3.18(2H), 3.16(2H), 3.06(2H), 2.92(2H), 2.88(2H), 2.63(2H)

(continued)

Table 9.1 (continued)

Entry[a]	Structure	Reference	Symmetry	Solvent	^{1}H chemical shifts[b]
27 (5b)	$C_{60}H_{36}$	11	C_3	$CDCl_3$	3.47(6H), 3.32(6H), 3.27(6H), 3.22(6H), 3.17(6H), 2.93(6H)
28 (5c)	$C_{60}H_{36}$	11	T	$CDCl_3$	3.86(12H), 3.67(12H), 3.19(12H)
29 (6a)	1,2-$C_{70}H_2$	12	C_1	Toluene	4.00(2H,AB,d)
30 (6a)	1,2-$C_{70}H_2$	3	C_1	CS_2	5.08(2H,AB,d)
31 (6a)	1,2-$C_{70}H_2$	13	C_1	CS_2	5.24(2H,AB,d)
32 (6a)	1,2-$C_{70}H_2$	3	C_1	C_6D_6	3.97(2H,AB,d)
33 (6b)	5,6-$C_{70}H_2$	12	C_s	Toluene	3.91(2H,s)
34 (6b)	5,6-$C_{70}H_2$	3	C_s	CS_2	4.96(2H,s)
35 (6b)	5,6-$C_{70}H_2$	3	C_s	C_6D_6	3.87(2H,s)
36 (7a)	1,2,3,4-$C_{70}H_2$	3	C_s	CS_2	4.30(4H,AA′BB′,m)
37 (7b)	1,2,5,6-$C_{70}H_4$	3	C_1	CS_2	5.41(1H), 4.62(1H), 4.42(1H), 4.28(1H)
38	1,2,13,30-$C_{70}H_4$[d]	3	C_s	CS_2	6.38(2H,s), 5.09(2H,AB,d)
39 (7c)	1,2,56,57-$C_{70}H_4$	13	C_2	CS_2	5.22(4H,AB,d)
40 (7d)	1,2,67,68-$C_{70}H_4$	13	C_2	CS_2	5.62(4H,AB,d)
41 (8)	7,19,23,27,33,37,44,53-$C_{70}H_8$	13	C_s	CS_2	5.50(4H,m), 5.24(4H,m)
42 (9)	7,8,19,26,33,37,45,49,53,63-$C_{70}H_{10}$	13	C_s	CS_2	5.45(2H,m), 5.38(2H,m), 5.33–5.29(4H,m), 5.27–5.25(2H,m)
43 (10)	$C_{70}H_{38}$	14	C_2	o-$C_6D_4Cl_2$	4.08(2H), 4.07(2H), 4.06(2H), 3.96(2H), 3.89(2H), 3.87(4H), 3.78(2H), 3.75(2H), 3.68(2H), 3.55(2H), 3.52(2H), 3.41(2H), 3.34(4H), 3.18(2H), 3.14(2H), 2.96(2H), 2.76(2H)

[a]Number in parenthesis refer to the structures in Section 9.7

[b]Only the center of the spin-system is given for AB and AA′BB′ spin-systems. The numbers of protons and splitting pattern is given in parenthesis (d = doublet, q = quartet, s = singlet, m = multiplet, dt = doublet of a triplet)

[c]Associated with unassigned AB quartet

[d]Tentative assignment

1 – (Henderson and Cahill 1993), 2 – (Alemany et al. 1997a), 3 – (Avent et al. 1994), 4 – (Henderson et al. 1994b), 5 – (Bergosh et al. 1997), 6 – (Billups et al. 1997a), 7 – (Meier et al. 1996), 8 – (Darwish et al. 1996a), 9 – (Briggs et al. 2005), 10 – (Wågberg et al. 2005), 11 – (Gakh et al. 2003), 12 – (Henderson et al. 1994a), 13 – (Spielmann et al. 1998), 14 – (Wågberg et al. 2008)

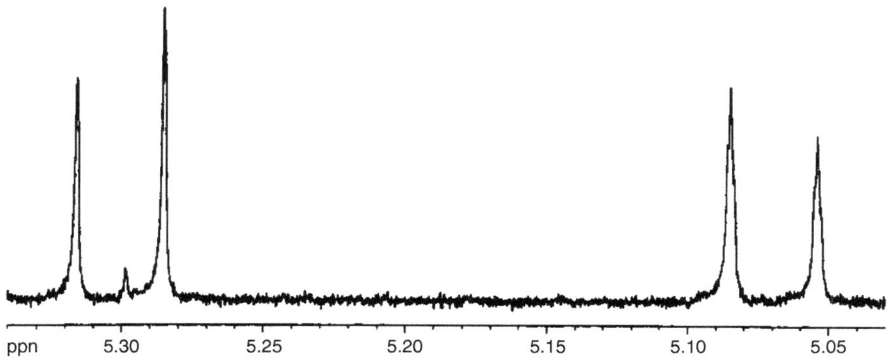

Fig. 9.1 Part of the ^{1}H NMR spectrum of $C_{60}H_4$ at 500MHz in a 2:1 solution of C_6D_6/CS_2 showing a typical AB spin-system (Alemany et al. 1997a)

need to be simulated in order to deduce the composition of the spin-system and the different J-couplings (Alemany et al. 1997a; Avent et al. 1994; Henderson et al. 1994b; Spielmann et al. 1998, 2000).

In studies of hydrogenated fullerenes, the concept of spin-systems is widely used to describe the hydrogen pattern on the fullerene surface. Spin-systems contain ^{1}H connected via J-couplings and the splitting patterns of the ^{1}H resonances can thus be used to determine the types of spin-system in the molecule. A simple spin-system containing only two protons where the $\delta_A - \delta_B \gg J_{AB}$ (i.e. where the difference in chemical shift is much larger than the couplings constant) is denoted AX where A is the most down-field member of the spin-system. Additional members of the spin-system are assigned letters between A and X, roughly correlated with their positions in the spectra in relation to A and X. In case of strong coupling where deviation from the 1:1 ratio of the doublets in a two-spin systems is observed, the spins-system is denoted AB instead of AX (see Fig. 9.1). Chemically but not magnetically equivalent protons are notated with a prime, e.g. AA′X and when several magnetically equivalent protons are present then number is given in subscript, e.g. AB_3. It should be noted, however, that these conventions are not strictly followed in all spin-system descriptions.

Although the vast majority of ^{1}H NMR studies on hydrogenated fullerenes have focused on structure determination, it can also be used to study molecular dynamics and partial alignment of the molecule in solution (see Section 9.7.1).

9.4 ^{13}C NMR Spectroscopy

The much wider chemical shift dispersion of ^{13}C results in a significantly higher spectral resolution than in the corresponding ^{1}H spectra. However, the sensitivity is much lower compared to ^{1}H NMR as a consequence of the low natural abundance (approximately 1%) and low gyromagnetic ratio of ^{13}C. These limiting factors, together with the fact that ^{1}H NMR in many cases yield sufficient information for structure determination, have resulted in fewer reports of ^{13}C chemical shifts for hydrogenated fullerenes (Table 9.2). In general, ^{13}C spectra of hydrogenated fullerenes contain a larger number of signals than the corresponding ^{1}H spectrum. Although ^{13}C chemical shift values have been reported for a number of different hydrogenated fullerenes, complete assignment of all ^{13}C resonances is very challenging and has only been performed in a few cases (Meier et al. 2002a, 2003; Wågberg et al. 2008). The reason is that information about J_{CC}-couplings is often required. This information can be gained with the INADEQUATE experiment (Bax et al. 1980). As a result of the low natural abundance of ^{13}C the likelihood of two ^{13}C atoms being adjacent to each other in the structure is very low. This experiment is therefore very insensitive and in general feasible only when using ^{13}C-enriched material (Meier et al. 2002a, b, 2003; Wågberg et al. 2008). Partial assignment of carbon resonances can be performed with various other 2D NMR experiments (discussed below) but Weedon and co-workers have shown that also one-dimensional ^{1}H-coupled ^{13}C spectra yield a wealth of information that can be used both to determine hydrogenation pattern and to assign part of the ^{13}C resonances (Bergosh et al. 1997; Meier et al. 1996; Spielmann et al. 1998). From the number of ^{13}C resonances the symmetry of the system could be deduced and by simulating the appearance of the ^{13}C multiplets, the magnitude of different $^{1}J_{CH}$- and $^{2}J_{CH}$-coupling constants could be determined. For hydrogenated fullerenes derived from C_{60} and C_{70} $^{1}J_{CH}$-couplings are similar and in the range of 135–145 Hz, while the $^{2}J_{CH}$-couplings on the other hand differ significantly between 1,2-$C_{60}H_2$ (6.7 Hz) and 1,2-$C_{70}H_2$ (approximately 5Hz) although the geometry of the sp^2-carbons adjacent to the hydrogenated sp^3-carbons is very similar in both molecules (Spielmann et al. 1998).

It is clear that hydrogen addition has a major impact on the chemical environment of the carbons in the fullerene cage, manifested by a chemical shift dispersion of over 16 ppm for the remaining sp^2-carbons (Table 9.2, entries 1 and 2) by an addition of only two hydrogens. Bergosh et al. noted, somewhat surprisingly, that ^{13}C immediately adjacent to the site of addition are found most downfield of all sp^2-carbons in $C_{60}H_2$ and $C_{60}H_6$ (Bergosh et al. 1997). These sp^2-carbons are directly bonded to sp^3-carbons and are also the least pyramidalized and are thus expected to be shifted upfield rather than downfield. Meier et al. studied this phenomenon in more detail and noted that pyramidalization angles alone are not sufficient to correctly predict ^{13}C chemical shifts in hydrogenated fullerenes without considering changes in ring currents (Meier et al. 2002a, 2003) (see also Section 9.7.1). The ^{13}C chemical shifts correlate well with values calculated using HF-GIAO theory (Jameson 1996). $^{1}J_{CC}$ coupling constants increases with shorter bond lengths and increasing π-bond order (Berger 1984; Unkefer et al. 1983) and this has also been observed in hydrogenated fullerenes where $^{1}J_{CC}$ along 6,6-bonds are larger (66–71 Hz) than for 5,6-bonds (52–58 Hz) (Meier et al. 2002a).

Table 9.2 Observed ^{13}C chemical shifts for hydrogenated C_{60} and C_{70}

Entry[a]	Structure	Reference	Symmetry	Solvent	^{13}C chemical shifts[b]
1 (1a)	1,2-$C_{60}H_2$	1	C_s	CS_2	153.28, 148.38, 147.90, 146.93, 146.87, 146.61, 146.10, 145.99, 145.28, 143.91, 143.66, 143.14, 142.57, 142.52, 142.20, 141.05, 136.92, 54.10
2 (1a)	1,2-$C_{60}H_2$	2	C_s	*o*-$C_6D_4Cl_2$	152.69(2), 148(2), 147.61(1), 146.61(2), 146.54(2), 146.33(1), 145.72(2), 145.65(2), 144.94(2), 143.53(1), 142.76(2), 142.19(2), 142.17(2), 141.81(2), 140.59(2), 136.56(2), 53(1)
3 (3b)	1,2,33,41,42,50-$C_{60}H_6$	1	D_3	CS_2	158.05, 153.97, 153.66, 151.65, 151.45, 145.71, 144.18, 142.60, 141.55, 52.28
4 (3b)	1,2,33,41,42,50-$C_{60}H_6$	2	D_3	*o*-$C_6D_4Cl_2$	157.07, 153.15, 152.97, 151, 150.70, 145.06, 143.47, 142.04, 140.93, 51.47
5 (4)	$C_{60}H_{18}$	3	C_{3v}	*o*-$C_6D_4Cl_2$	151.30, 149.30, 146.10, 141.30, 133.60, 43.50, 43.20, 37.70, 37.00
6 (4)	$C_{60}H_{18}$	4	C_{3v}	C_6D_6	149.50, 141.50, 133.90, 43.10, 42.70, 37.50, 36.60[c]
7 (5a)	$C_{60}H_{36}$	5	C_1	$CDCl_3$	47.99, 47.79, 44.10, 44.00, 42.30 , 39.67, 39.36, 38.01, 37.94, 37.47, 36.84, 36.81, 36.74, 36.00, 35.98, 31.52, 31.37, 31.34[d]
8 (5b)	$C_{60}H_{36}$	5	C_3	$CDCl_3$	48.02, 44.97, 37.56, 37.33, 35.92, 31.31[d]
9 (5c)	$C_{60}H_{36}$	5	T	$CDCl_3$	49.25, 43.61, 31.37[d]
10 (6a)	1,2-$C_{70}H_2$	6	C_s	CS_2	157.48(2), 156.34(2), 152.36(2), 151.90(1), 151.86(2), 151.81(2), 151.41(2), 151.18(2), 150.49(2), 150.37(2), 150.36(2), 150.19(2), 149.83(2), 149.57(2), 149.20(2), 148.55(2), 147.95(2), 147.79(2), 147.47(1), 147.34(2), 146.96(2), 146.55(2), 143.70(2), 143.46(2), 143.43(2), 143.40(2), 142.93(2), 141.80(2), 140.85(2), 138.68(2), 135(2), 134.50(2), 131.95(2), 131.72(4), 46.37(1), 44.06(1)
11 (8b)	1,2,56,57-$C_{70}H_4$	6	C_2	CS_2	159.50, 158.43, 156.99, 156.83, 155.24, 153.41, 152.69, 152.27, 152.07, 151.02, 150.74, 150.61, 149.23, 148.70, 148.11, 147.85, 146.14, 144.99, 144.55, 144.41, 143.88, 143.10, 141.34, 141.16, 140.63, 140.58, 139.05, 138.86, 138.35, 135.08, 134.60, 133.10, 132.32, 46.86, 44.70
12 (8c)	1,2,67,68-$C_{70}H_4$	6	C_2	CS_2	158.44, 158.12, 153.13, 152.42, 152.36, 151.68, 151.66, 151.01, 150.49, 150.42, 150.27, 149.20, 148.97, 147.64, 147.51, 147.37, 147.30, 144.88, 144.56, 143.82, 143.20, 143.19, 143.11, 142.84, 142.43, 140.29, 137.48, 136.29, 135.44, 134.71, 134.04, 134.01, 131.78, 46.20, 44.58

(continued)

Table 9.2 (continued)

Entry[a]	Structure	Reference	Symmetry	Solvent	^{13}C chemical shifts[b]
13 (9)	7,19,23,27,33,37,44,53-$C_{70}H_8$	6	C_s	CS_2	158.22(2), 155.51(2), 154.40(1), 153.21(1), 153.06(2), 153.03(2), 151.79(2), 151.67(2), 151.53(2), 151.52(2), 151.27(2), 151.19(2), 151.14(2), 151.03(2), 150.83(2), 150.27(2), 149.96(2), 148.59(2), 148.21(2), 147.91(2), 147.66(2), 147.58(2), 146.99(2), 146.28(2), 145.56(2), 142.91(2), 142.68(1), 139.62(2), 136.62(2), 132.88(2), 131.32(2), 130.08(2), 129.60(2), 48.37(2), 48.27(2), 48.10(2), 47.97(2)
14 (10)	7,8,19,26,33,37,45,49,53,63-$C_{70}H_{10}$	7	C_s	CS_2	154.43(2), 153.94(1), 153.23(2), 153.07(2), 152.55(1), 152.30(2), 152.24(2), 151.85(2), 151.54(2), 151.08(2), 151.05(2), 151.03(2), 150.89(2), 150.66(2), 150.32(2), 150.07(2), 149.70(2), 149.50(1), 149.20(2), 148.95(2), 148.44(2), 148.72(2), 148.31(2), 146.98(2), 144.86(2), 143.42(2), 141.09(1), 132.16(2), 131.56(2), 130.99(2), 129.39(2), 129.13(2), 48.09(2), 47.58(2), 47.55(2), 47.37(2), 45.64(2)
15 (11)	$C_{70}H_{38}$	8	C_2	o-$C_6D_4Cl_2$	144.08, 142.28, 140.69, 139.86, 137.64, 137.22, 136.54, 136.44, 136.16[e], 135.69, 131.94, 131.91, 131.29, 125.28, 123.36, 49.15, 47.49, 45.78, 45.68, 45.41, 45.05, 44.85, 44.41, 42.85, 42.78, 42.63, 42.08, 41.82, 41.46, 35.99, 32.67, 31.57, 31.32, 29.76

[a]Number in parenthesis refer to the structures in Section 9.7
[b]Number in parenthesis refers to the relative peak intensity (where reported)
[c]Partial assignment
[d]Only sp^3-carbons
[e]Overlapping signals
1 – (Bergosh et al. 1997), 2 – (Meier et al. 2002a), 3 – (Briggs et al. 2005), 4 – (Wågberg et al. 2005), 5 – (Gakh et al. 2003), 6 – (Spielmann et al. 1998), 7 – (Spielmann et al. 2000), 8 – (Wågberg et al. 2008)

9.5 Two-Dimensional NMR Techniques

Many different 2D NMR experiments are today available and have become routine experiments in structure determination of all types of organic molecules. Structure determinations of many of the higher hydrogenated fullerenes have relied heavily upon information derived from different 2D NMR experiments. The 2D NMR experiments described here are all based upon correlation between resonances via J-couplings and can be divided in two groups, homonuclear or heteronuclear experiments. In homonuclear experiments, J-couplings between the same types of nuclei are observed and give rise to off-diagonal cross-peaks in the spectra. Peaks in heteronuclear experiments, on the other hand, originate from correlations between different types of nuclei, e.g. ^{1}H and ^{13}C. The most commonly used homonuclear 2D NMR experiment is the COSY (Marion and Wüthrich 1983) (Correlation Spectroscopy) or double-quantum filtered COSY (DQF-COSY) (Rance et al. 1983) experiment where cross-peaks are observed between J-coupled protons. This information can be used to determine the types of spin-systems occurring in the molecule and has aided the structure determination of both $C_{60}H_n$ and $C_{70}H_n$ (Avent et al. 1994; Briggs et al. 2005; Nossal et al. 2001b; Spielmann et al. 1998; Wågberg et al. 2005). Cross-peak intensity in a COSY spectrum is related to the size of the coupling constant and can thus be used to discriminate between $^{3}J_{HH}$ couplings and long-range couplings (Briggs et al. 2005). Assignment of the ^{13}C resonances are usually performed with different 2D experiments that utilize either the $^{1}J_{CH}$ couplings, such as the HSQC (Heteronuclear Single Quantum Coherence) (Kay et al. 1992) experiment or long-range couplings between ^{1}H and ^{13}C such as the HMBC (Heteronuclear Multiple Bond Correlation) (Bax et al. 1986) experiment. These experiments generate spectra with cross-peaks between ^{1}H and the J-coupled ^{13}C neighbor(s) and directly give the assignment of the ^{13}C resonances close to the protons once the proton resonances have been assigned. As the correlation to the resonance of the bound ^{13}C in the second dimension improves the resolution dramatically, these spectra are also helpful in determining the number of ^{1}H resonances. This is illustrated in Fig. 9.2 where a ^{1}H spectrum and

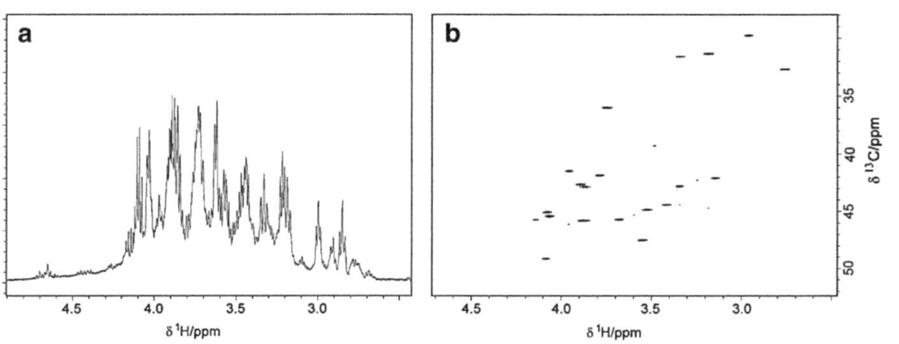

Fig. 9.2 (**a**) ^{1}H spectrum from $C_{70}H_{38}$. (**b**) The corresponding ^{1}H-^{13}C HSQC spectrum

a ^{1}H-^{13}C HSQC spectrum for $C_{70}H_{38}$ are shown. The narrow chemical shift range and complex structure of this molecule result in severe overlap of the ^{1}H resonances in the ^{1}H spectrum. However, all ^{1}H resonances are resolved in the HSQC spectrum because the ^{13}C atoms covalently bound to the hydrogens have different chemical shifts.

In order to get a complete assignment of all ^{13}C resonances, J_{CC}-couplings need to be studied. A suitable experiment to obtain this information is INADEQUATE (Incredible Natural Abundance Double Quantum Transfer Experiment) (Bax et al. 1980) which has the same information content as a COSY experiment but for ^{13}C instead of ^{1}H and cross-peaks are thus observed between J-coupled ^{13}C nuclei. As mentioned earlier, this experiment is very insensitive as two ^{13}C nuclei have to be vicinal which is very unlikely as a result of the low natural abundance of ^{13}C. This problem can be partially circumvented using ^{13}C-enriched material but even so the INADEQUATE experiment requires long experimental times. Another 2D experiment that has proven useful in determining the hydrogenation pattern is the HSQC-COSY (Lerner and Bax 1987) experiment which is an extension of the HSQC experiment where in addition to signals between directly bound ^{1}H and ^{13}C resonances, signals also appear for J-coupled protons. A closely related experiment is HSQC-TOCSY (Lerner and Bax 1987) where cross-peaks from the whole or part of the proton spin-system can be observed for each protonated ^{13}C.

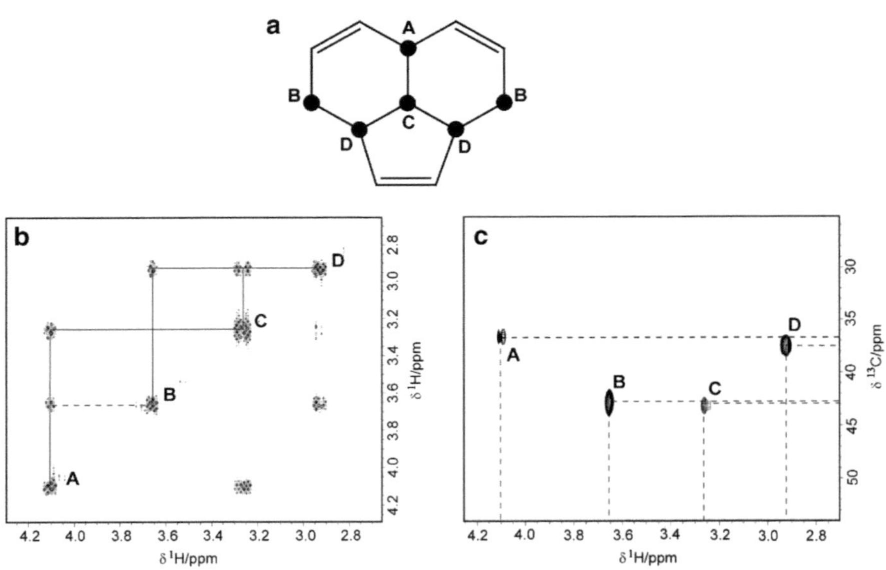

Fig. 9.3 Assignment of protons and sp³-carbons in the C_{3v} isomer of $C_{60}H_{18}$ using a combination of COSY- and ^{1}H-^{13}C HSQC-experiments. (**a**) Representation of one of the three equivalent spin-systems found in the structure. (**b**) COSY spectrum showing the proton connectivity through $^{3}J_{HH}$-couplings (*solid lines*) and the long-range coupling between A and B (*dashed line*). (**c**) The corresponding ^{1}H-^{13}C HSQC spectrum with assignment of the sp³-carbon resonances

These two experiments are very useful to establish the number and sizes of proton spin-systems in higher hydrogenated fullerenes where severe overlap makes the interpretation of an ordinary COSY spectrum difficult (Gakh et al. 2003; Wågberg et al. 2008).

In Fig. 9.3 we show an illustrative example how a combination of COSY and HSQC experiments can be used to determine the connectivity of the protons in the ABCD spin-system of the C_{3v} isomer of $C_{60}H_{18}$ (Briggs et al. 2005; Wågberg et al. 2005) and to assign the ^{13}C chemical shifts of the sp^3-carbons (Wågberg et al. 2005). Strong cross-peaks in the COSY spectrum (Fig. 9.3a) correspond to protons connected via $^3J_{HH}$-couplings (solid lines). The peaks are connected in the order A→C→D→B and the relative peak intensities calculated from the ^1H NMR spectrum are 1:2:1:2. The dashed line corresponds to the $^5J_{HH}$-coupling between protons A and B. The only spin-system compatible with these results is shown in the figure. The ^{13}C-shifts of the carbons directly bonded to these protons could be determined from the HSQC spectrum (b).

9.6 ^{3}He-NMR Spectroscopy

The possibility to trap ^3He inside fullerenes using high temperature and pressure has enabled the use of ^3He NMR to study fullerene structures and also a wide range of fullerene derivatives including hydrogenated fullerenes (Saunders et al. 1993, 1994). Table 9.3 summarizes the ^3He NMR data reported for different hydrogenated fullerenes derived from $^3He@C_{60}$ and $^3He@C_{70}$. ^3He chemical shifts are reported relative to dissolved ^3He gas set at 0 ppm. Although ^3He NMR spectra do not contain as much structural information as ^1H or ^{13}C NMR spectra, the simple appearance with only one peak for each isomer makes it ideal as a probe to determine the number of isomers and their relative abundance (Fig. 9.4). In the synthesis of higher hydrogenated fullerenes this is especially important where a large number of isomers usually are formed and where ^1H and ^{13}C NMR spectra become exceedingly complex.

The ^3He nucleus is well suited for NMR spectroscopy considering the fact that it is a spin ½ nucleus with a high magnetogyric ratio (approximately 75% compared to ^1H). The chemical shift of the ^3He nucleus encapsulated in a fullerene is very sensitive to the substitution pattern on the fullerene surface as the ring current of the residual π-systems exerts a large effect on the ^3He chemical shift (Elser and Haddon 1987; Saunders et al. 1994). As a result, a wide chemical shift range has been observed for different ^3He-containing hydrogenated fullerenes derived from $^3He@C_{60}$ and $^3He@C_{70}$ (see Table 9.3). Even for unsubstituted $^3HeC_{60}$ and $^3He@C_{70}$ there is a large difference in ^3He chemical shifts ($^3He@C_{60}$ $\delta = -6.3$ ppm and $^3He@C_{70}$ $\delta = -28.8$ ppm) (Saunders et al. 1994). Calculated ^3He NMR chemical shifts have shown to agree well with experimental values and could thus be used for tentative assignment of observed ^3He peaks (Boltalina et al. 1999; Bühl et al. 1995; Elser and Haddon 1987; Nossal et al. 2001b; Wang et al. 2003). An interesting trend relating ^3He chemical shifts with the

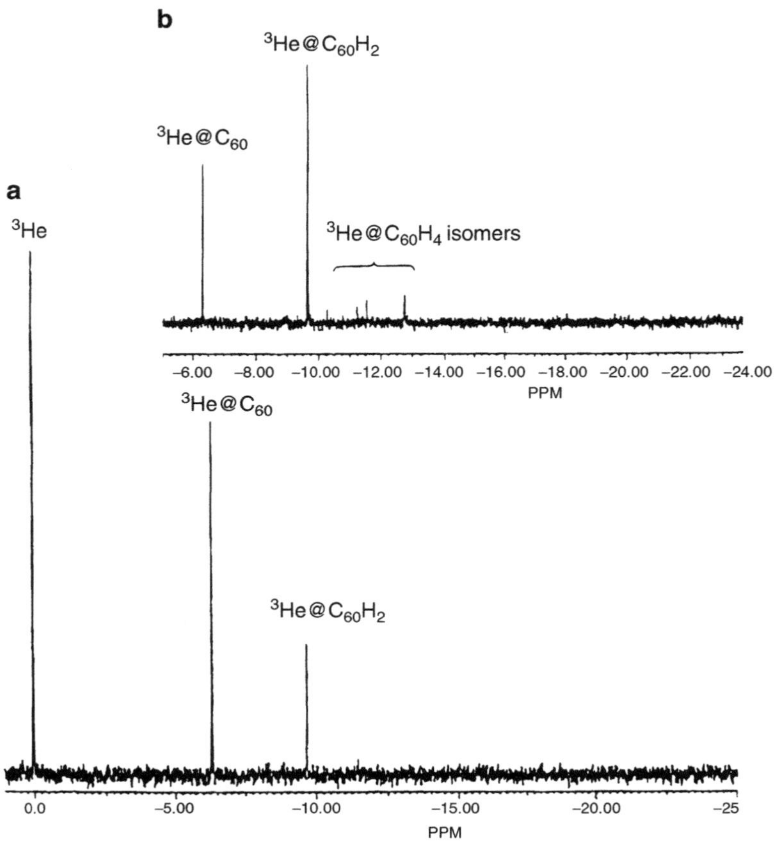

Fig. 9.4 (**a**) ^{3}He NMR spectrum for a reaction mixture of $^3He@C_{60}H_2$ produced from hydroboration of $^3He@C_{60}$. (**b**) ^{3}He NMR spectrum from diimide reduction of $^3He@C_{60}$ showing a signal from $^3He@C_{60}H_2$ as well as from several different isomers of $^3He@C_{60}H_4$ (Cross et al. 1996)

degree of hydrogenation and the subsequent substitution patterns have been observed (Peera et al. 2003). Addition of hydrogen will result in an upfield shift of ^{3}He going from $^3He@C_{60}$ ($\delta = -6.3$ ppm) to $^3He@C_{60}H_{18}$ that has the most upfield chemical shift ($\delta = -16.4$ ppm) (Billups et al. 1997b). This trend can be seen in Table 9.3 (entries 1–6) and is reversed upon further hydrogenation where ^{3}He is shifted downfield all the way to approximately −3 ppm for $^3He@C_{60}H_{44}$ (Peera et al. 2003). Peera and co-workers proposed that these trends results from the fact that hydrogenation of C_{60} distorts the essentially spherical geometry of C_{60} to a more hemispherical appearance for $C_{60}H_{18}$ where ^{3}He is situated directly above the benzenoid ring in the center of the $C_{60}H_{18}$ crown. This will result in significant shielding of ^{3}He and explains the upfield shift observed in $^3He@C_{60}H_{18}$. Further hydrogenation will reduce this shielding effect as the structure re-expands towards a more spherical geometry.

Table 9.3 Observed ^3He chemical shifts for hydrogenated C_{60} and C_{70}

Entry	Structure	Reference	Symmetry	Solvent	^3He chemical shifts
1	1,2-$^3He@C_{60}H_2$	1	C_s	1-methylnaphthalene/CD_2Cl_2	−9.66
2	$^3He@C_{60}H_4$	1	-[a]	1-methylnaphthalene/CD_2Cl_2	−10.3, −10.81, −11.26, −11.56, −12.75, −12.79
3	$^3He@C_{60}H_4$	2	-[a]	C_6D_6/CS_2	−9.3, −10.0, −10.9, −11.21, −12.42, −12.48
4	1,2,18,22,23,36-$^3He@C_{60}H_6$	3	C_3	1-methylnaphthalene/CD_2Cl_2	−16.35
5	1,2,33,41,42,50-$^3He@C_{60}H_6$	3	D_3	1-methylnaphthalene/CD_2Cl_2	−15.36
6	$^3He@C_{60}H_{18}$	2	-[a]	-[b]	−16.45
7	$^3He@C_{60}H_{36}$	2	D_{3d}[c]	-[b]	−7.7
8	$^3He@C_{60}H_{36}$[d]	4	C_1	o-$C_6D_4Cl_2$	−8.01(broad)
9	$^3He@C_{60}H_{36}$[d]	4	C_3	o-$C_6D_4Cl_2$	−8.14(broad)
10	$^3He@C_{60}H_{38}$	5	-[a]	o-$C_6D_4Cl_2$	−6.4(several overlapping peaks)
11	$^3He@C_{60}H_{40}$	5	-[a]	o-$C_6D_4Cl_2$	−3.0(several overlapping peaks)
12	$^3He@C_{60}H_{42}$	5	-[a]	o-$C_6D_4Cl_2$	−3.0(several overlapping peaks)
13	$^3He@C_{60}H_{44}$	5	-[a]	o-$C_6D_4Cl_2$	−3.0(several overlapping peaks)
14	1,2-$^3He@C_{70}H_2$	3	C_s	1-methylnaphthalene/CD_2Cl_2	−27.18
15	1,2,56,57-$^3He@C_{70}H_4$	3	C_2	1-methylnaphthalene/CD_2Cl_2	−25.3
16	7,19,23,27,33,37,44,53-$^3He@C_{70}H_8$	3	C_s	1-methylnaphthalene/CD_2Cl_2	−18.84
17	7,8,19,26,33,37,45,49,53,63-$^3He@C_{70}H_{10}$	3	C_s	1-methylnaphthalene/CD_2Cl_2	−17.17

[a]Not assigned
[b]Not reported
[c]Tentative assignment
[d]Assigned to structures 5a and 5b in accordance with the findings of Gakh et al (Gakh et al. 2003)
1 – (Cross et al. 1996), 2 – (Billups et al. 1997b), 3 – (Wang et al. 2000), 4 – (Nossal et al. 2001b), 5 – (Peera et al. 2003)

Numerous studies have shown the utility of ^{3}He NMR in analysis of hydrogenated fullerenes derived from both $^{3}He@C_{60}$ and $^{3}He@C_{70}$. $^{3}He@C_{60}H_2$ resulting from a reduction of $^{3}He@C_{60}$ by hydroboration was studied by Cross et al. who found a single ^{3}He NMR signal at $\delta = -9.66$ ppm (Table 9.3, entry 1) (Cross et al. 1996). Further reduction of this material resulted in the formation of six different isomers of $^{3}He@C_{60}H_4$ (Table 9.3, entry 2). No assignments of these peaks were attempted. Reduction of $^{3}He@C_{60}$ with hydrazine resulted in a mixture where six of the eight possible isomers of $^{3}He@C_{60}H_4$ could be detected (Table 9.3, entry 3) (Billups et al. 1997a). ^{3}He NMR of $^{3}He@C_{60}H_6$ prepared through Zn(Cu) reduction of $^{3}He@C_{60}$ revealed the presence of three different isomers of $^{3}He@C_{60}H_6$ of which two could be identified as 1,2,33,41,42,50-$^{3}He@C_{60}H_6$ and 1,2,18,22,23,36-$^{3}He@C_{60}H_6$ (Table 9.3, entries 4–5). The third peak at $\delta = -14.24$ ppm was not assigned (Wang et al. 2000). Billups et al. observed formation of $^{3}He@C_{60}H_{18}$ with a peak at $\delta = -16.45$ ppm as a by-product in the synthesis of $^{3}He@C_{60}H_{36}$ (Billups et al. 1997b). $^{3}He@C_{60}H_{36}$ has been prepared through both Birch reduction and reduction using dihydroanthracene (Billups et al. 1997b; Nossal et al. 2001b). Billups et al. only observed two distinct peaks in the ^{3}He NMR spectrum at $\delta = -7.7$ and -7.8 ppm where the isomer at -7.7 ppm was tentatively assigned as having D_{3d} symmetry based on ^{3}He chemical shift calculations (Bühl et al. 1995). Nossal and co-workers found the resulting isomer mixture after Birch reduction to be much more complex with broad ^{3}He peaks even after HPLC separation suggesting a mixture of isomers in the individual fractions (cf. Section 9.7.5 and Table 9.3, entries 8–9). A mixture of $^{3}He@C_{60}H_{38}$, $^{3}He@C_{60}H_{40}$, $^{3}He@C_{60}H_{42}$ and $^{3}He@C_{60}H_{44}$ has been prepared through Benkeser reduction of $^{3}He@C_{60}H_{36}$ (Peera et al. 2003). ^{3}He NMR revealed that a multitude of isomers were formed for each degree of hydrogenation.

A number of hydrogenated $^{3}He@C_{70}$ derivatives have also been studied with ^{3}He NMR (Table 9.3, entries 14–17). Wang and co-workers used Zn(Cu) reduction of $^{3}He@C_{70}$ to produce a mixture of $^{3}He@C_{70}H_2$, $^{3}He@C_{70}H_4$ and $^{3}He@C_{70}H_8$ (Wang et al. 2000). The crude reaction mixture showed ^{3}He resonances at $\delta = -28.81, -27.18, -25.53$ and -17.84 ppm where the signal at $\delta = -28.81$ ppm corresponds to unreacted $^{3}He@C_{70}$ (Saunders et al. 1994). After HPLC separation the peak at $\delta = -27.18$ ppm was assigned to 1,2-$^{3}He@C_{70}H_2$ isomer and the peak at $\delta = -25.3$ ppm to the major $^{3}He@C_{70}H_4$ isomer, i.e. 1,2,56,7-$^{3}He@C_{70}H_4$. Two minor $^{3}He@C_{70}H_4$ isomers at $\delta = -24.77$ and -23.76 ppm were also observed. The peak at $\delta = -18.84$ ppm was assigned to 7,19,23,27,33,37,44,53-$^{3}He@C_{70}H_8$. In the same study, $^{3}He@C_{70}H_{10}$ was also analyzed and the major isomer, 7,8,19,26,33,37,45,49,53,63-$^{3}He@C_{70}H_{10}$ appeared at $\delta = -17.17$ ppm. It can be noted that there is a striking difference between $^{3}He@C_{60}$- and $^{3}He@C_{70}$-derivatives, reflecting the different magnetic environment experienced by ^{3}He in the parent compounds $^{3}He@C_{60}$ and $^{3}He@C_{70}$ (Saunders et al. 1994). Also, based on these studies, the relationship between ^{3}He chemical shift and degree of hydrogenation seems to be opposite for $^{3}He@C_{70}$-derivatives compared to $^{3}He@C_{60}$-derivatives with a downfield ^{3}He shift for each subsequent pair-wise hydrogen addition (Table 9.3, entries 14–17).

9.7 Specific Isomers of Hydrogenated Fullerenes Studied with NMR Spectroscopy

9.7.1 $C_{60}H_2$

$C_{60}H_2$ was the first hydrogenated fullerene that was fully characterized regarding structure and symmetry (Henderson and Cahill 1993). The simplest of all hydrogenated fullerenes, $C_{60}H_2$ can form 23 different thermodynamically stable isomers according to calculations (Henderson 1992; Matsuzawa et al. 1992). Out of these, however, Henderson et al. established that two possible isomers were in agreement with the measured NMR spectra. Taking into account that hydrogenation at the 6,6 ring fusion is far more energetically favorable over the 6,5 ring fusion (Henderson and Cahill 1993) the authors proposed that the formed isomer was 1,2-$C_{60}H_2$ **(1a)** as shown in Fig. 9.5.

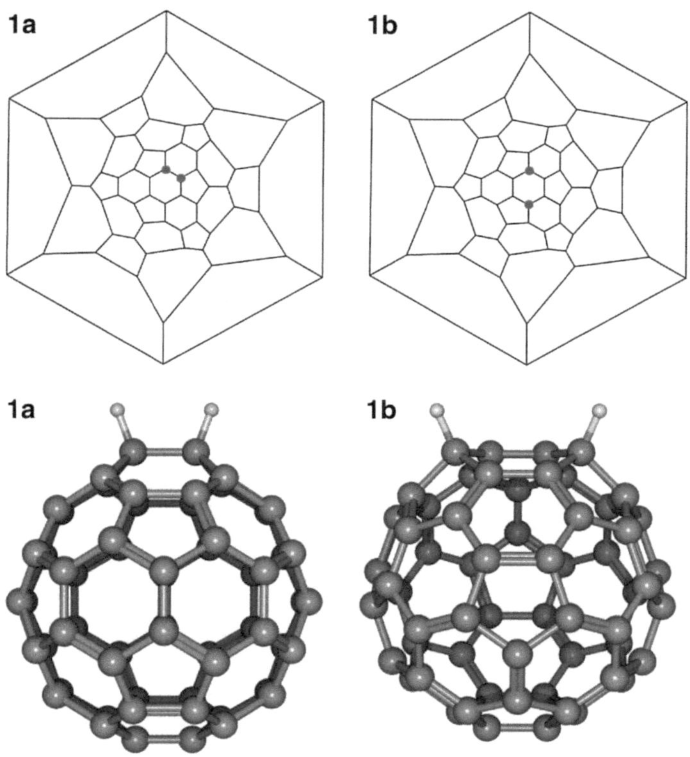

Fig. 9.5 Schlegel diagrams and 3D structures of 1,2-$C_{60}H_2$ **(1a)** and 1,4-$C_{60}H_2$ **(1b)**. All 3D structures in this and the following figures are MM2 minimized in CS Chem3D Ultra v. 7.0.0 by CambridgeSoft

Numerous reports have later confirmed this structure **(1a)** (An et al. 1993; Avent et al. 1994; Becker et al. 1993; Bergosh et al. 1997; Meier et al. 1994, 2002a; Wang et al. 2005). In addition Avent et al. also found the next most stable dihydrofullerene, 1,4-$C_{60}H_2$ **(1b)** (Avent et al. 1994). The fact that both hydrogen are adjacent only to sp^2-carbons in 1,4-$C_{60}H_2$ is manifested by an expected downfield shift of the ^{1}H singlet from 6.97 ppm for 1,2-$C_{60}H_2$ **(1a)** to 7.23 ppm (Table 9.1, entries 3 and 5). While the ^{1}H chemical shifts have been reported for the dihydrofullerenes the complete assignment of the ^{13}C resonances remained undetermined for long time. The main reason for this is the difficulty to establish C–C correlations due to the low natural abundance of ^{13}C in normally produced samples. Meier et al. synthesized ^{13}C enriched (13%) 1,2-$C_{60}H_2$ and they were able to assign the complete ^{13}C NMR spectrum using the INADEQUATE experiment (Meier et al. 2002a). Whereas the ^{1}H NMR spectrum contains only one singlet the ^{13}C NMR spectrum contains one sp^3 resonance at 54.1 ppm (representing two equivalent carbons C1 and C2) and 16 sp^2 resonances in the range 136.5–152.7 ppm (Table 9.2, entry 2). Similar ^{13}C chemical shifts had previously also been observed by Bergosh et al. (Table 9.2, entry 1) although no assignment of all resonances was attempted in that study (Bergosh et al. 1997). The most downfield resonance has been assigned to the sp^2-carbons adjacent to the sp^3 hybridized carbons, a feature that is also observed for $C_{60}H_6$ (Bergosh et al. 1997; Meier et al. 2002a, 2003).

Meier et al. explain the observed ^{13}C chemical shift range by a London methodology (Meier et al. 2002a, 2003) that previously was used for pristine fullerenes (Pasquarello et al. 1993). The London theory explains how paramagnetic ring currents are established on the fullerene surface and these could very well explain the observed ^{13}C chemical shift distribution in 1,2-$C_{60}H_2$. Figure 9.6 shows the net paramagnetic ring current deduced for 1,2-$C_{60}H_2$.

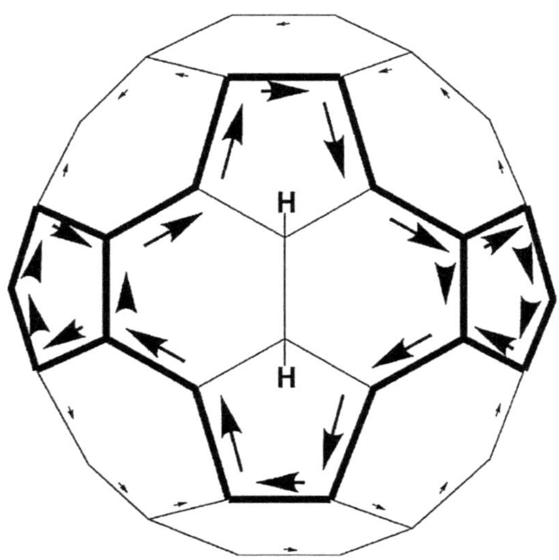

Fig. 9.6 The net paramagnetic ring current in 1,2-$C_{60}H_2$ derived from a London methodology (Meier et al. 2002a)

The almost spherical shape of C_{60} has made it to an interesting test system for studying molecular dynamics. A very powerful method to study molecular dynamics is by T_1 relaxation experiments, which have been used for C_{60} both in solution (Jones and Rodriguez 1992) and in the solid state (Johnson et al. 1992; Tycko et al. 1991a, b; Walton et al. 1993; Yannoni et al. 1991). However, these studies are challenging because the low natural abundance of ^{13}C makes it difficult to establish the contributions of the different relaxation mechanisms. Irwin et al. instead used $C_{60}H_2$ to study the molecular dynamics (Irwin et al. 1994). The advantage of measuring ^1H NMR of the two attached protons is the large natural abundance of ^1H (99.99%) and the fact that the only viable relaxation mechanism is governed by dipolar relaxation which is a well described process (Abragam 1961). The rotational correlation times, τ_c, they derive for $C_{60}H_2$ (at 303 K) was about 10 ps which is similar as for C_{60} after some re-evaluations taking into account that ^{13}C spin lattice relaxation in C_{60} has contributions from both the chemical shift anisotropy and the spin rotation mechanisms. The fast correlation time for $C_{60}H_2$ is somewhat surprising considering the anisotropic magnetic susceptibility of the molecule which could in principle lead to an ordering effect instead of the observed free rotation at room temperature. Indeed Alemany et al. have reported that at high magnetic fields such an alignment of $C_{60}H_2$ can be observed leading to a splitting of the singlet δ at 6.14 ppm by 0.30 Hz when measured at 750 MHz field strength (Alemany et al. 1997a, b). It should be noted, however, that this effect is only observed when the solvent used also is partially aligned in the magnetic field (in this case C_6D_6).

9.7.2 $C_{60}H_4$

Strong evidence that the addition of hydrogen across the 6,6 bonds are thermodynamically most favored was provided by Henderson et al. who showed that also the second hydrogenation step occurs across 6,6 bonds (Henderson et al. 1994b). The most stable isomer following this hydrogenation mechanism is 1,2,3,4-$C_{60}H_4$ (**2a**) which was first obtained through a hydroboration process starting from $C_{60}H_2$ (Henderson and Cahill 1993; Henderson et al. 1994b) (see Fig. 9.7). In this context it could be mentioned that hydrogenation of C_{60} often follows similar mechanisms as fluorination. This is manifested by a high stability of the isostructural compound 1,2,3,4-$C_{60}F_4$ (Boltalina et al. 2002).

The stability of (**2a**) and its characteristic AA′BB′coupling scheme shown in Fig. 9.8 was confirmed also by Avent et al. (Avent et al. 1994) while Bergosh et al. found 1,2,18,36-$C_{60}H_4$ (**2b**) as one of the main products when reducing C_{60} with Zn(Cu) (Bergosh et al. 1997). The latter isomer has been found to co-exists with (**2a**) also in a study by Henderson et al. (1994b), which is in accordance with the small difference in calculated formation energy (Henderson and Cahill 1993; Matsuzawa et al. 1992; Narita et al. 2000).

Bergosh et al. also show the existence of a 1,2,33,50-$C_{60}H_4$ (**2c**) isomer in their product mixture (Bergosh et al. 1997). This isomer is important since it is the only $C_{60}H_4$ isomer that can evolve into one of the most stable $C_{60}H_6$ isomer, 1,2,33,41,42,50-$C_{60}H_6$

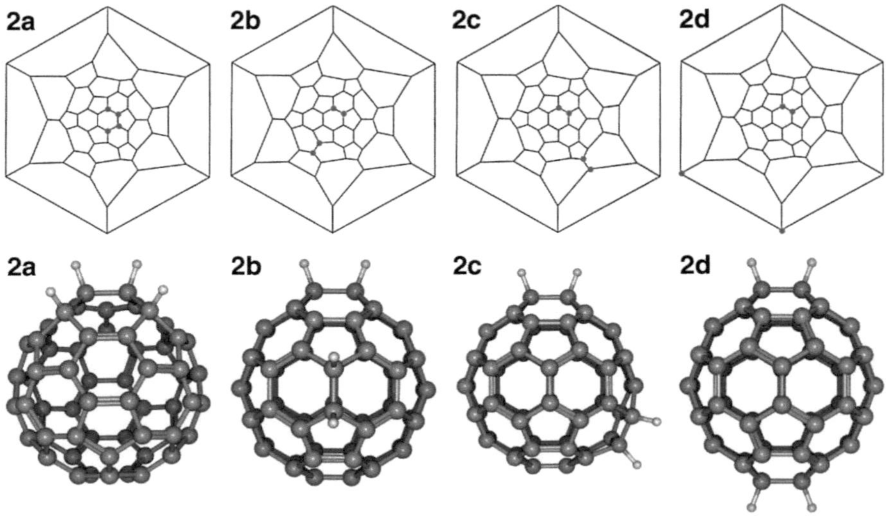

Fig. 9.7 Schlegel diagrams and 3D structures of 1,2,3,4-$C_{60}H_4$ (**2a**), 1,2,18,36-$C_{60}H_4$ (**2b**), 1,2,33,50-$C_{60}H_4$ (**2c**) and 1,2,55,60-$C_{60}H_4$ (**2d**)

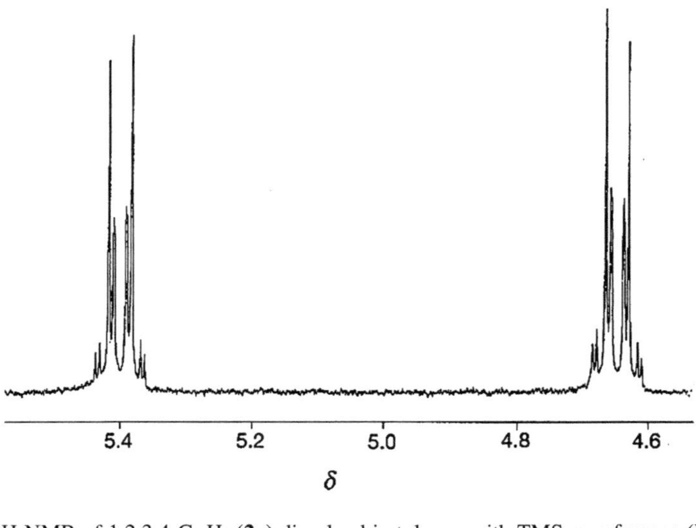

Fig. 9.8 ^{1}H NMR of 1,2,3,4-$C_{60}H_4$ (**2a**) dissolved in toluene with TMS as reference (Henderson et al. 1994b)

(**3a**) that is found at a prolonged reaction times (see structure below). It should be noted that also a few other $C_{60}H_4$ isomers probably can be stable. Henderson et al. find strong indications for the 1,2,55,60-$C_{60}H_4$ (**2d**) isomer as well as the presence of other unassigned structures (Henderson et al. 1994b). Avent et al. have also proposed the formation of a small yield of 1,2,57,58-$C_{60}H_4$ which is a relatively unexpected structure since it involves a hydrogenation along a 5,6 bond (Avent et al. 1994).

9.7.3 $C_{60}H_6$

Using the same reaction conditions as in the synthesis of $C_{60}H_4$ but with extended reaction time, Bergosh et al. were able to synthesize 1,2,33,41,42,50-$C_{60}H_6$ (**3b**) (Bergosh et al. 1997) (see Fig. 9.9). This isomer is presumably formed by addition of two hydrogens to 1,2,33,50-$C_{60}H_4$ (**2c**). On the other hand, despite an equal abundance of 1,2,18,36-$C_{60}H_4$ isomer (**2b**) which could form the 1,2,18,22,23,36-$C_{60}H_6$ (**3a**) isomer very small amounts of this isomer was found. This is rationalized by that the latter probably is removed by either selective destruction or by isomerization during the synthetic process (Bergosh et al. 1997).

In an attempt to find evidence for isomerization during the synthesis, Bergosh et al. also performed Zn(Cu) reduction of C_{60} in the presence of D_2O (Bergosh et al. 1997). The resulting $C_{60}D_4$ mixture with the same isomer ratio as before was thereafter

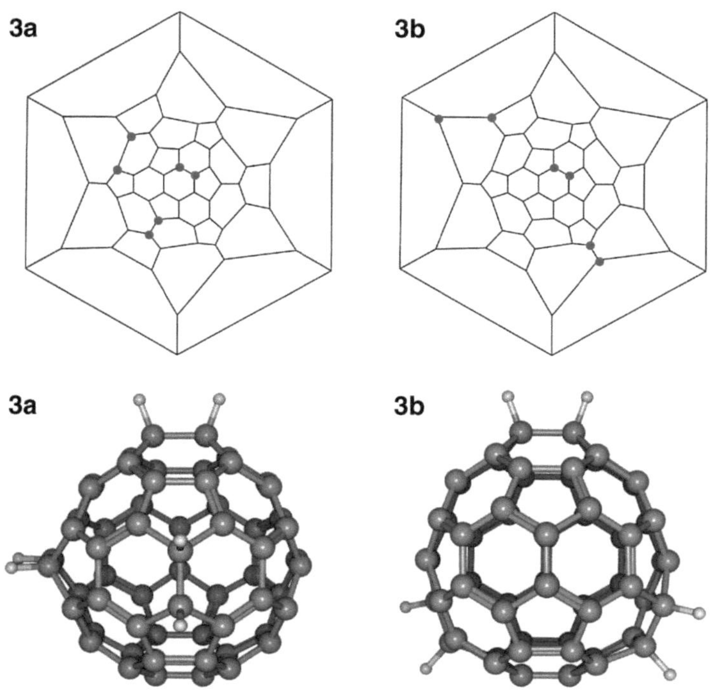

Fig. 9.9 Schlegel diagrams and 3D structures of 1,2,33,41,42,50-$C_{60}H_6$ (**3a**) and 1,2,18,22,23,36-$C_{60}H_6$ (**3b**)

further reduced in the presence of normal water. By following the evolution of the NMR signals they conclude that hydrogenation of **2b** and **2c** produces **3a** and **3b** respectively with no signs of deuterium loss or changes in deuterium positions. Therefore it is believed that under these relatively mild reaction conditions no isomerization takes place. The fact that the **3a** isomer is found in much lower yield can thus rather be rationalized by that this isomer is less stable, in agreement also with other studies (Wang et al. 2000). The mild reducing conditions of Zn(Cu) is confirmed by Meier et al. who observed that irrespective of full access to fresh metal the reaction stops at $C_{60}H_6$ (**3b**) in a 30–40% yield and that further reduction was not achieved (Meier et al. 2000). Isomer (**3b**) is the only $C_{60}H_6$ isomers where all ^{13}C and ^1H resonance have been fully assigned (see Tables 9.1 and 9.2) (Bergosh et al. 1997; Meier et al. 1996, 2002a). As for $C_{60}H_2$ the chemical shifts of the sp^2-carbons are distinctively divided into two classes, one set of signals at 150–160 ppm assigned to sp^2-carbons in pentagons containing a sp^3-carbon and one set in the range 140–150 ppm representing carbons in pentagons containing only sp^2-carbons (Table 9.2) (Meier et al. 2002a).

9.7.4 $C_{60}H_{18}$

Of all the hydrogenated fullerenes the structure of $C_{60}H_{18}$ is probably the most intriguing and fascinating. Despite the huge amount ($1.5 * 10^{13}$) of theoretically predicted possible isomers of $C_{60}H_{18}$ (Balasubramanian 1991) only one have been characterized experimentally (Athans et al. 2007; Briggs et al. 2005; Darwish et al. 1996b; Wågberg et al. 2005). This is mainly attributed to the stepwise addition that has been shown to occur in the hydrogenation process of fullerenes and that increases the possibility to form certain isomers at the cost of others. Darwish et al. claimed that this stepwise addition can occur due to an increased electron localization (and thus reactivity) in the double bonds next to the protonated carbons (Darwish et al. 1996b).

By taking both experimental (Darwish et al. 1996b) and theoretical (Clare and Kepert 1994); (Cahill and Rohlfing 1996) studies into account it is established that the most dominant hydrogenation pathway to $C_{60}H_{18}$ follows the scheme 1–2–3–6 in Fig. 9.10 below.

The resulting isomer which has been shown to be particularly stable has C_{3v} crown symmetry (Fig. 9.11) giving a characteristic ABCD spin system shown in Fig. 9.3. The stability of this isomer is emphasized by its very high yield by both polyamine reduction (Briggs et al. 2005) and direct hydrogenation at high temperature and high pressure (Wågberg et al. 2005). The peculiar C_{3v} symmetry of $C_{60}H_{18}$ raised the question of whether the six-membered ring at the bottom of the $C_{60}H_{18}$ structure has a truly flat aromatic character. Wågberg et al. reported that the NMR chemical shift of the carbons in the six member ring indeed is very close to values reported for relevant aromatic systems such as hexamethyl benzene (Wågberg et al. 2005). This observation strongly supports the presence of a flat aromatic system in $C_{60}H_{18}$ which makes it isostructural to $C_{60}F_{18}$ (Troyanov et al. 2002).

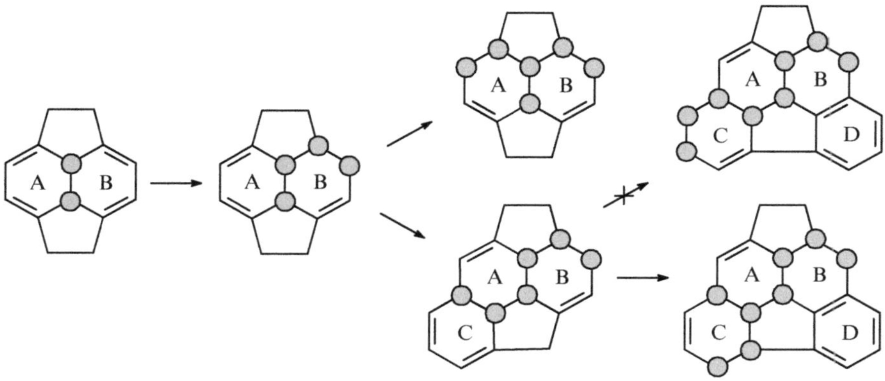

Fig. 9.10 Hydrogenation pathway to $C_{60}H_{18}$ as proposed by (Darwish et al. 1996b)

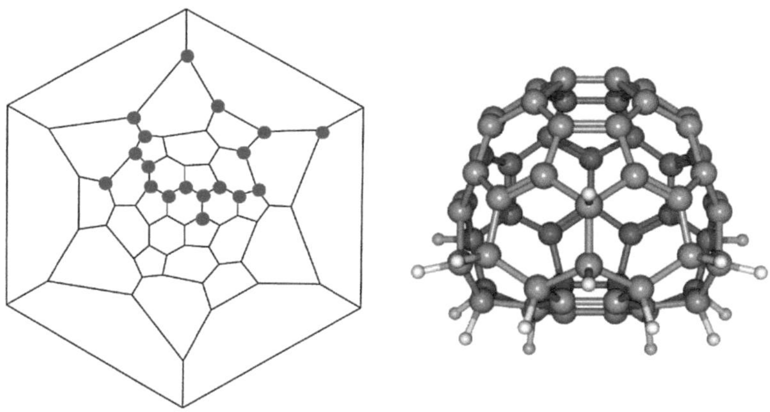

(4) (4)

Fig. 9.11 Schlegel diagrams and 3D structure of $C_{60}H_{18}$, with C_{3v} symmetry **(4)**

9.7.5 $C_{60}H_{36}$

Materials of composition $C_{60}H_{36}$, together with minor amounts of $C_{60}H_{18}$, were the first examples of chemically modified fullerene derivatives (Haufler et al. 1990). This material was formed by Birch reduction of C_{60}. It was early realized that the number of possible isomers of $C_{60}H_{36}$ is enormous; theoretically more than $6 * 10^{14}$ possible isomers have been proposed (Balasubramanian 1991). This has spawned a large number of experimental and theoretical studies of this hydrogenated fullerene, see for example (Gakh et al. 2003) and references therein. The detailed characterization of this material has, however, proven to be extremely difficult both due to the number of isomers formed but also due to the fact that the product tends to decompose if not handled with care. It was early found that the hydrogenation give rise to materials of very complex isomeric composition as reflected by the broad featureless NMR spectra obtained (Attalla et al. 1993; Darwish et al. 1995; Govindaraj 1993; Hall et al. 1993; Haufler et al. 1990; Shigematsu et al. 1993). Even after separation by HPLC, the NMR spectra of materials produced by Birch reduction remain difficult to interpret as a result of broad and overlapping peaks (Nossal et al. 2001b). By the use of ^{13}C solid state NMR spectroscopy, it was observed that the different fractions demonstrated different chemical shifts in the sp^2 region, manifesting that the fractions contain isomers of diverse hydrogenation patterns (Nossal et al. 2001b). In order to get further information about the number and distribution of isomers in the separated fractions, ^{3}He NMR spectroscopy was utilized. Again broad featureless signals were observed but in different spectral regions, once more proving different isomer distributions in the fractions. In order to tentatively establish the structures of the different isomers, the ^{3}He chemical shifts were computed for a selected number of $C_{60}H_{36}$ isomers. It was found that different isomers will show chemical shift in the observed spectral regions but no definitive structural assignments were attained.

In another $C_{60}H_{36}$ sample, prepared by transfer hydrogenation using 9,10-dihydroanthracene at elevated temperatures, fewer isomers were formed (Nossal et al. 2001b). After HPLC separation, the main fraction gave rise to resolved NMR spectra. The ^{3}He spectrum indicates the presence of two major isomers in a 3:1 ratio (see Table 9.3 entries 8 and 9 for tentative assignments). A similar observation was also reported earlier (Billups et al. 1997b). The 1D ^{1}H NMR spectra show relative good resolution albeit with extensive overlap. The presence of at least two major isomers prevented the interpretation of the obtained ^{1}H spectra (Nossal et al. 2001b). The ^{13}C NMR spectrum gave rise to 80 resolved signals of similar intensity. Based on the NMR spectral information it was proposed that the sample consists of one C_1 isomer and one C_3 isomer in a 3:1 ratio. This would account for the number of signals and the comparable signal intensities in the ^{13}C NMR spectrum. It was, however, not possible to deduce the exact structure of these isomers based on the spectral data.

More recently, a report was published where analogous transfer hydrogenation conditions were used (Gakh et al. 2003). The reaction conditions were optimized to prepare a sample that could be analyzed directly by NMR spectroscopy without the need for HPLC purification. The sample was found to contain 60–70% of a C_1 isomer, 25–30% of a C_3 isomer and 2–5% of a T isomer. By a combination of several 2D techniques, (^{1}H-^{13}C-HSQC, ^{1}H-^{13}C-COSY-HSQC and ^{1}H-^{13}C-TOCSY-HSQC) acquired with high spectral resolution, the structures could be established, Fig. 9.12. These isomers are among the structures previously considered in earlier investigations (Clare and Kepert 2002a; Nossal et al. 2001b). As previously speculated, all pentagons have three vicinal hydrogens in these isomers, in accordance with earlier predictions based on calculated stabilities. The most abundant C_1 isomer is however predicted to be the least stable of the observed isomers (Gakh et al. 2003). It was proposed that high temperature production methods give rise to fewer isomers as a result of hydrogen rearrangement on the fullerene surface under these conditions (Gakh et al. 2003).

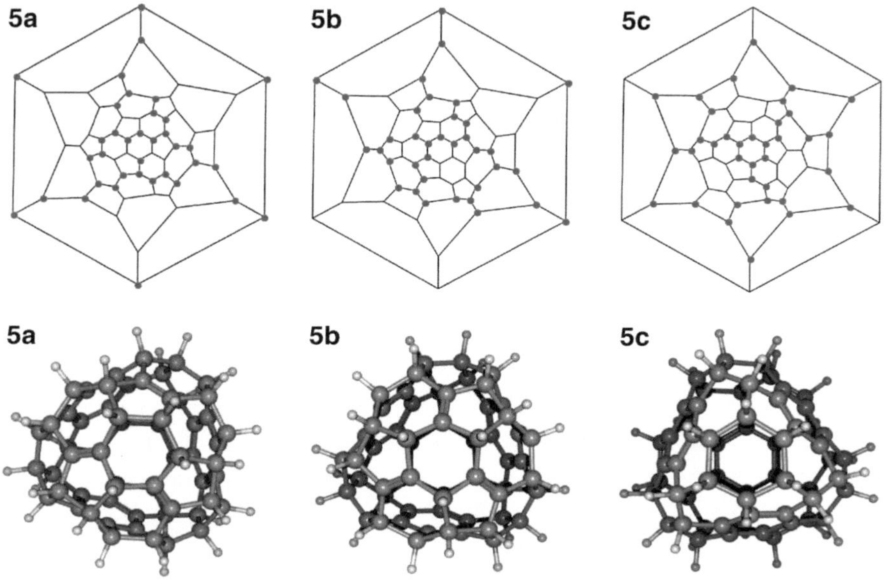

Fig. 9.12 Schlegel diagrams and 3D structures of the C_1 **(5a)**, C_3 **(5b)** and T **(5c)** isomers of $C_{60}H_{36}$ found by (Gakh et al. 2003)

9.7.6 $C_{70}H_2$

One of the most efficient and useful synthesis technique to produce C_{60} and C_{70} of low hydrogenation degree is reduction by Zn(Cu) (Bergosh et al. 1997; Spielmann et al. 1998). In most cases the hydrogenation of C_{70} follows similar patterns and mechanisms as for hydrogenated C_{60} but with some differences. In Table 9.2 it can be seen that in general the ^{1}H chemical shifts for hydrogenated C_{70} are shifted downfield compared to their C_{60} counterparts. Using hydroboration, Avent et al. established that C_{70} is less prone to undergo hydrogenation compared to C_{60} (Avent et al. 1994). In disagreement with this statement is, however, the reduction by Zn(Cu) that has been used to reduce C_{70} to $C_{70}H_{10}$ (Spielmann et al. 2000) whereas the reduction of C_{60} stops already at $C_{60}H_6$ (Meier et al. 2000). In other respects the hydrogenation behavior is similar for the two systems. Dihydrogenated C_{70} is a good example where the most abundant isomer formed under various reduction conditions (Avent et al. 1994; Henderson et al. 1994a; Spielmann et al. 1998) is the 1,2-$C_{70}H_2$ isomer **(6a)** which is analogous to 1,2-$C_{60}H_2$ **(1a)**. In two of these studies minor amounts of 5,6-$C_{70}H_2$ **(6b)** isomers are also found (Fig. 9.13) (Avent et al. 1994; Henderson et al. 1994a). Note that the numbering differs between the reports. The readily formation of 1,2-$C_{70}H_2$ was confirmed also by ^{3}He-NMR (Wang et al. 2000).

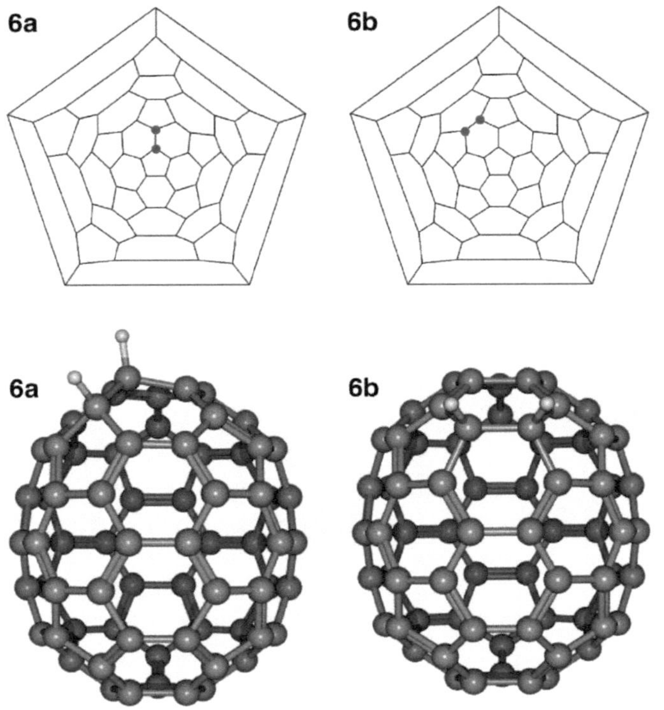

Fig. 9.13 Schlegel diagrams and 3D structures of 1,2-$C_{70}H_2$ **(6a)** and 5,6-$C_{70}H_2$ **(6b)**

9.7.7 $C_{70}H_4$

Prolonged reaction of C_{70} in Zn(Cu) leads to the formation of $C_{70}H_4$. The lower symmetry of C_{70} implies that the numbers of possible isomers of $C_{70}H_n$ are larger than for $C_{60}H_n$. A consequence of this is that complex mixtures can be formed when synthesizing higher hydrogenated $C_{70}H_n$ (Avent et al. 1994).

The second pair of hydrogens are expected to attach at the opposite pole of the C_{70} molecule as observed in the addition of metal complexes (Balch et al. 1992). Spielmann et al. studied several isomers of $C_{70}H_n$ (n = 2, 4, and 8) and took advantage of ^{1}H-coupled ^{13}C NMR spectroscopy to differentiate isomers that are indistinguishable in ^{1}H-decoupled ^{13}C NMR experiments (Spielmann et al. 1998). Six isomers were observed but only few of them could be purified to enable structure determination by NMR spectroscopy. The ^{13}C NMR spectrum of the $C_{70}H_4$ sample gave 35 resonances which ruled out the 1,2,41,58-$C_{70}H_4$ isomer but is consistent with either of the isomers **(7c)** or **(7d)** in Fig. 9.14. After careful analysis Spielmann et al. could assign the major isomer to 1,2,56,57-$C_{70}H_4$ **(7c)**, and the minor isomer to 1,2,67,68-$C_{70}H_4$ **(7d)** which is in agreement with the expectations that the most stable isomers should have hydrogenation pairs on

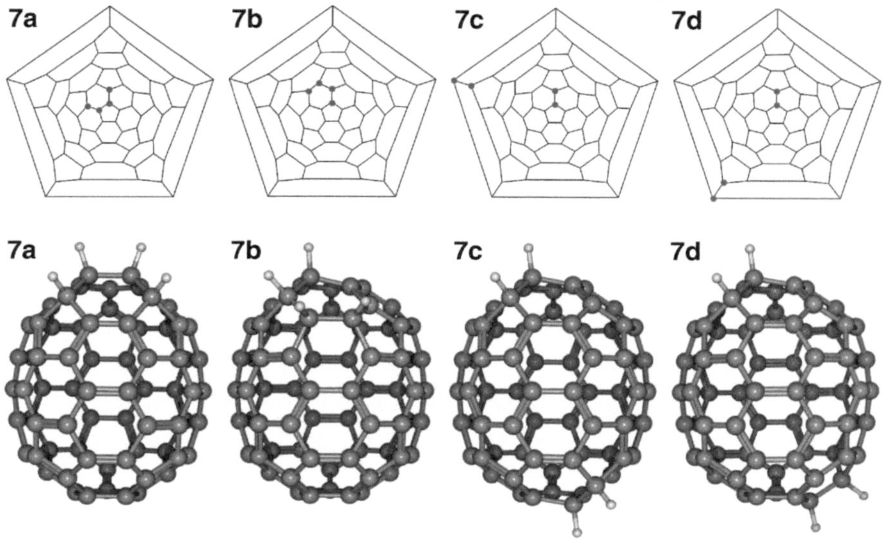

Fig. 9.14 Schlegel diagrams and 3D structures of 1,2,3,4-$C_{70}H_4$ **(7a)**, 1,2,5,6-$C_{70}H_4$ **(7b)**, 1,2,56,57-$C_{70}H_4$ **(7c)** and 1,2,67,68-$C_{70}H_4$ **(7d)**

opposite poles of the molecule (Spielmann et al. 1998). In an earlier study Avent et al. made a detailed analysis of the ^{1}H NMR spectrum for a $C_{70}H_4$ isomer mixture (Avent et al. 1994). By careful analysis of the spin systems they could assign part of the spectra to the 1,2,3,4-$C_{70}H_4$ **(7a)** and the 1,2,5,6-$C_{70}H_4$ isomers **(7b)**. The isomers **(7c)** and **(7d)** observed by Spielmann et al. (1998) was not observed in their study.

9.7.8 $C_{70}H_8$

As mentioned earlier, the reduction of C_{70} by Zn(Cu) could reduce C_{70} to a higher hydrogenation state than C_{60}. By this method Spielmann et al. could synthesize, purify and determine the structure of one of the major $C_{70}H_8$ isomers (Spielmann et al. 1998). The complex ^{13}C spectrum consisted of 37 resonances of which 33 were twice as intense as the remaining 4 signals. A linear ABCD spin system could be identified from a DQF-COSY spectrum. By further analysis of both chemical shifts and the magnitude of J-couplings Spielmann and co-workers could assign the major isomer in their $C_{70}H_8$ synthesis product to be the quite interesting isomer 7,19,23,27,33,37,44,53-$C_{70}H_8$ **(8)** (Fig. 9.15). This structure exhibits several very unique features. In contrast to most other $C_{70}H_n$ structures it is not hydrogenated at the polar regions but rather at the equator. Furthermore, none of the protonated carbons are vicinal to one another which are normally the case.

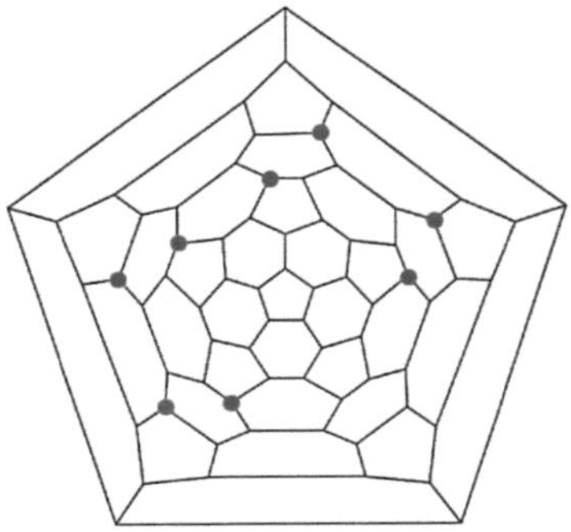

(8)

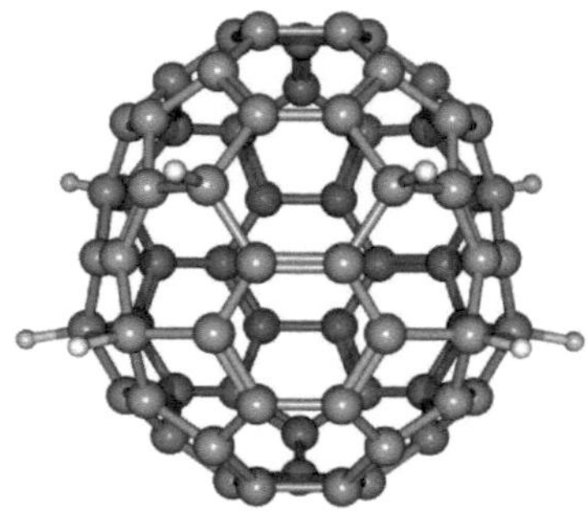

(8)

Fig. 9.15 Schlegel diagram and 3D structure of 7,19,23,27,33,37,44,53-$C_{70}H_8$ **(8)**

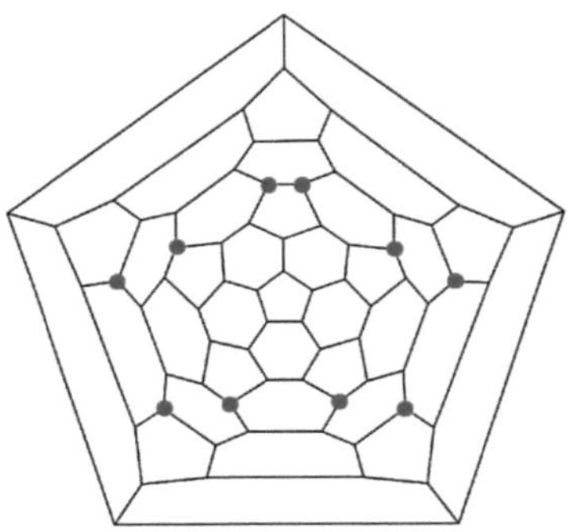

(9)

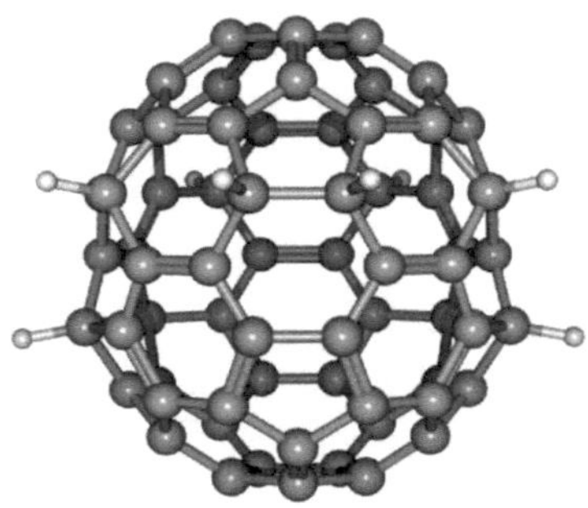

(9)

Fig. 9.16 Schlegel diagram and 3D structure of 7,8,19,26,33,37,45,49,53,63-$C_{70}H_{10}$ **(9)**

9.7.9 $C_{70}H_{10}$

By using isomer **(8)** as a starting material for further Zn(Cu) reduction Spielmann et al. could identify two isomers of $C_{70}H_{10}$, of which one was identified by both ^{1}H and ^{13}C NMR spectroscopy (Spielmann et al. 2000). The identified structure, 7,8,19,26,33,37,45,49,53,63-$C_{70}H_{10}$ **(9)** (Fig. 9.16) is a reduction product of the isomer **(8)** (Fig. 9.15) but it is also the reduction product when starting from C_{70} upon extended reaction times.

9.7.10 $C_{70}H_{38}$

Among the higher hydrogenated $C_{70}H_n$ fullerenes almost no structures have been determined. One exception to this is the structural assignment of $C_{70}H_{38}$ (Wågberg et al. 2008).

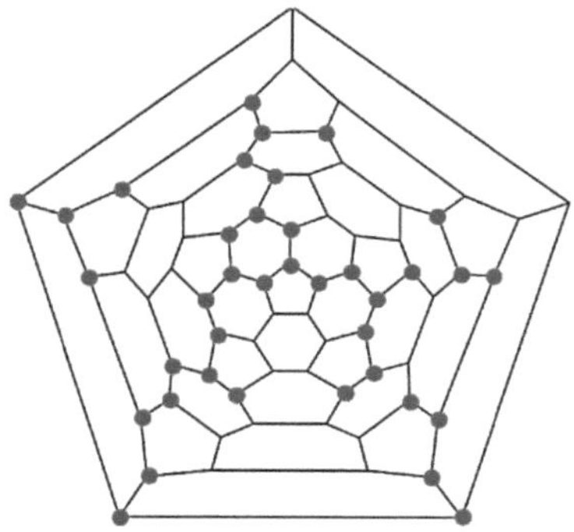

(10)

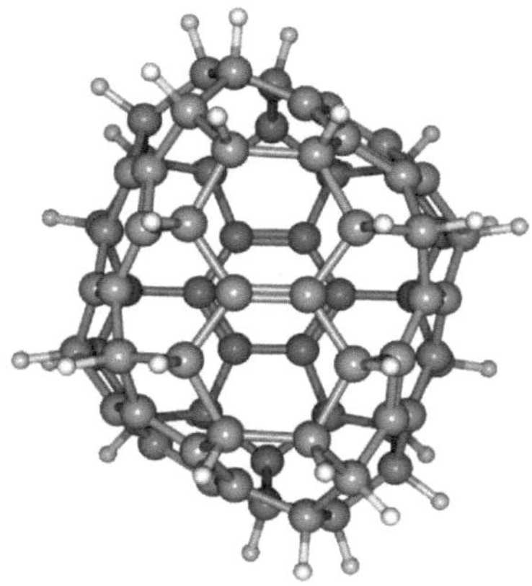

(10)

Fig. 9.17 Schlegel diagram and 3D structure of the $C_{70}H_{38}$ isomer with C_2 symmetry characterized by Wågberg et al. (2008)

By treating C_{70} in hydrogen gas at 100–120 bar and a temperature of 400°C, $C_{70}H_{38}$ consisting of approximately 50% of a single isomer could be obtained (Fig. 9.17). To enable the characterization of this very complex structure, ^{13}C-enriched material was synthesized starting from ^{13}C-enriched C_{70}. The INADEQUATE experiment could subsequently be used to get information about J_{CC}-couplings which proved necessary for the structure determination. The established isomer has some interesting features. For example, it has two protonated carbons at the equator, contrary to earlier predictions (Clare and Kepert 2002b). Furthermore, it contains five benzenoid ring systems. It is interesting to note that **(10)** is isostructural to a fluorinated $C_{70}F_{38}$ structure proposed by Hitchkock et al. to be particularly stable (Hitchcock et al. 2005). However in their study this specific $C_{70}F_{38}$ isomer could not be observed experimentally.

The authors propose that the long treatment time at extreme conditions probably leads to formation of the most stable isomer by a rearrangement of hydrogen on the fullerene surface (Wågberg et al. 2008).

9.8 Conclusions

NMR spectroscopy has been an extraordinary important technique for structure determination of hydrogenated fullerenes ever since the first synthesis of hydrogenated fullerenes almost two decades ago. During this time, technical advances have pushed the limits regarding both sensitivity and spectral resolution. Also, sophisticated 2D NMR experiments have been implemented more and more frequently in recent years. Consequently, NMR spectroscopy has now been used to successfully characterize complex hydrogenated fullerenes whose structure for a long time remained unknown. In contrast to the fluorinated fullerenes as well as other fullerene derivatives, no structures based on X-ray crystallography have as yet been reported and all established structures are based solely upon NMR data. The wealth of structural information gained through NMR spectroscopy ensures that this analytical technique will continue to be crucial for structure determination of hydrogenated fullerenes also in the future. However, even with state of the art spectrometers, structure determination remains a challenging task, especially considering that most synthetic procedures result in complex mixtures of isomers. Successful characterization of hydrogenated fullerenes using NMR spectroscopy is thus closely linked to development of more selective synthetic methods and improved separation techniques. Pure and well characterized substances will enable detailed investigations of the inherent chemical and physical properties of hydrogenated fullerenes. NMR spectroscopy will be one of the cornerstone techniques to achieve these goals.

References

Abragam A (1961) Principles of nuclear magnetism. Oxford University Press, Oxford
Alemany LB, Gonzalez A, Billups WE, Willcott MR, Ezell E, Gozansky E (1997a) J Org Chem 62:5771
Alemany LB, Gonzalez A, Luo WM, Billups WE (1997b) J Am Chem Soc 119:5047
An YZ, Anderson JL, Rubin Y (1993) J Org Chem 58:4799
Athans AJ, Briggs JB, Jia WL, Miller GP (2007) J Mater Chem 17:2636
Attalla MI, Vassallo AM, Tattam BN, Hanna JV (1993) J Phys Chem 97:6329
Avent AG, Darwish AD, Heimbach DK, Kroto HW, Meidine MF, Parsons JP, Remars C, Roers R, Ohashi O, Taylor R, Walton DRM (1994) J Chem Soc, Perkin Trans 2:15
Balasubramanian K (1991) Chem Phys Lett 182:257
Balch AL, Lee JW, Olmstead MM (1992) Angew Chem – Int Edit 31:3
Ballenweg S, Gleiter R, Kratschmer W (1993) Tetrahedron Lett 34:3737
Banks MR, Dale MJ, Gosney I, Hodgson PKG, Jennings RCK, Jones AC, Lecoultre J, Langridgesmith PRR, Maier JP, Scrivens JH, Smith MJC, Smyth CJ, Taylor AT, Thorburn P, Webster AS (1993) J Chem Soc – Chem Commun 1149
Bax A, Summers MF (1986) J Am Chem Soc 108:2093
Bax A, Freeman R, Kempsell SP (1980) J Am Chem Soc 102:4849
Becker L, Evans TP, Bada JL (1993) J Org Chem 58:7630
Berger S (1984) Org Magn Res 22:47
Bergosh RG, Meier MS, Cooke JAL, Spielmann HP, Weedon BR (1997) J Org Chem 62:7667
Billups WE, Luo WM, Gonzalez A, Arguello D, Alemany LB, Marriott T, Saunders M, Jimenez-Vazquez HA, Khong A (1997a) Tetrahedron Lett 38:171
Billups WE, Gonzalez A, Gesenberg C, Luo WM, Marriott T, Alemany LB, Saunders M, Jimenez-Vazquez HA, Khong A (1997b) Tetrahedron Lett 38:175
Boltalina OV, Bühl M, Khong A, Saunders M, Street JM, Taylor R (1999) J Chem Soc – Perkin Trans 2:1475
Boltalina OV, Darwish AD, Street JM, Taylor R, Wei XW (2002) J Chem Soc – Perkin Trans 2:251
Briggs JB, Montgomery M, Silva LL, Miller GP (2005) Org Lett 7:5553
Bucsi I, Szabo P, Aniszfeld R, Prakash GKS, Olah GA (1997) Chromatographia 48:59
Bühl M, Thiel W, Schneider U (1995) J Am Chem Soc 117:4623

Cahill PA, Rohlfing CM (1996) Tetrahedron 52:5247
Clare BW, Kepert DL (1994) Theochem – J Mol Struct 109:1
Clare BW, Kepert DL (2002a) Theochem – J Mol Struct 589:195
Clare BW, Kepert DL (2002b) Theochem – J Mol Struct 583:45
Cross RJ, Jimenez-Vazquez HA, Lu Q, Saunders M, Schuster DI, Wilson SR, Zhao H (1996) J Am Chem Soc 118:11454
Darwish AD, Abdulsada AK, Langley GJ, Kroto HW, Taylor R, Walton DRM (1995) J Chem Soc –Perkin Trans 2:2359
Darwish AD, AbdulSada AK, Langley GJ, Kroto HW, Taylor R, Walton DRM (1996a) Synth Met 77:303
Darwish AD, Avent AG, Taylor R, Walton DRM (1996b) J Chem Soc –Perkin Trans 2:2051
Drelinkiewicz A, Byszewski P, Bielanski A (1996) React Kinet Catal Lett 59:19
Dunlap BI, Brenner DW, Schriver GW (1994) J Phys Chem 98:1756
Elser V, Haddon RC (1987) Nature 325:792
Gakh AA, Romanovich AY, Bax A (2003) J Am Chem Soc 125:7902
Govindaraj A (1993) Curr Sci 65:868
Hall LE, McKenzie DR, Attalla MI, Vassallo AM, Davis RL, Dunlop JB, Cockayne DJH (1993) J Phys Chem 97:5741
Haufler RE, Conceicao J, Chibante LPF, Chai Y, Byrne NE, Flanagan S, Haley MM, Obrien SC, Pan C, Xiao Z, Billups WE, Ciufolini MA, Hauge RH, Margrave JL, Wilson LJ, Curl RF, Smalley RE (1990) J Phys Chem 94:8634
Henderson CC, Cahil PA (1992) Chem Phys Lett 198:570
Henderson CC, Cahill PA (1993) Science 259:1885
Henderson CC, Rohlfing CM, Gillen KT, Cahill PA (1994a) Science 264:397
Henderson CC, Rohlfing CM, Assink RA, Cahill PA (1994b) Angew Chem – Int Ed Engl 33:786
Hirsch A (1994) The chemistry of fullerenes. Thieme Verlag, New York/Stuttgart
Hitchcock PB, Avent AG, Martsinovich N, Troshin PA, Taylor R (2005) Org Lett 7:1975
Irwin AD, Assink RA, Henderson CC, Cahill PA (1994) J Phys Chem 98:11832
Jameson CJ (1996) Annu Rev Phys Chem 47:135
Johnson RD, Yannoni CS, Dorn HC, Salem JR, Bethune DS (1992) Science 255:1235
Jones VK, Rodriguez AA (1992) Chem Phys Lett 198:373
Kay LE, Keifer P, Saarinen T (1992) J Am Chem Soc 114:10663
Kintigh J, Briggs JB, Letourneau K, Miller GP (2007) J Mater Chem 17:4647
Lerner L, Bax A (1987) Carbohydr Res 166:35
Lobach AS, Shul'ga YM, Roshchupkina OS, Rebrov AI, Perov AA, Morozov YG, Spector VN, Ovchinnikov AA (1998) Full Sci Technol 6:375
Marion D, Wüthrich K (1983) Biochem Biophys Res Commun 113:967
Matsuzawa N, Fukunaga T, Dixon DA (1992) J Phys Chem 96:10747
Meier MS, Corbin FS, Vance VK, Clayton M, Mollman M (1994) Tetrahedron Lett 35:5789
Meier MS, Weedon BR, Spielmann HP (1996) J Am Chem Soc 118:11682
Meier MS, Spielmann HP, Haddon RC, Bergosh RG, Gallagher ME, Hamon MA, Weedon BR (2000) Carbon 38:1535
Meier MS, Spielmann HP, Bergosh RG, Haddon RC (2002a) J Am Chem Soc 124:8090
Meier MS, Bergosh RG, Gallagher ME, Spielmann HP, Wang ZW (2002b) J Org Chem 67:5946
Meier MS, Spielmann HP, Bergosh RG, Tetreau MC (2003) J Org Chem 68:7867
Narita S, Morikawa T, Shibuya TI (2000) Theochem – J Mol Struct 528:263
Nossal J, Saini RK, Alemany LB, Meier M, Billups WE (2001a) Eur J Org Chem 4167
Nossal J, Saini RK, Sadana AK, Bettinger HF, Alemany LB, Scuseria GE, Billups WE, Saunders M, Khong A, Weisemann R (2001b) J Am Chem Soc 123:8482
Pasquarello A, Schluter M, Haddon RC (1993) Phys Rev A 47:1783
Peera A, Saini RK, Alemany LB, Billups WESaunders M, Khong A, Syamala MS, Cross RJ (2003). Eur J Org Chem 4140
Rance M, Sørensen OW, Bodenhausen G, Wagner G, Ernst RR, Wüthrich K (1983) Biochem Biophys Res Commun 117:479

Rüchardt C, Gerst M, Ebenhoch J, Beckhaus HD, Campbell EEB, Tellgmann R, Schwarz H, Weiske T, Pitter S (1993) Angew Chem – Int Ed Engl 32:584
Saunders M, Jimenezvazquez HA, Cross RJ, Poreda RJ (1993) Science 259:1428
Saunders M, Jimenezvazquez HA, Cross RJ, Mroczkowski S, Freedberg DI, Anet FAL (1994) Nature 367:256
Shigematsu K, Abe K, Mitani M, Tanaka K (1993) Fullerenes Sci Technol 1:309
Spielmann HP, Wang GW, Meier MS, Weedon BR (1998) J Org Chem 63:9865
Spielmann HP, Weedon BR, Meier MS (2000) J Org Chem 65:2755
Troyanov SI, Boltalina OV, Kouvytchko IV, Troshin PA, Kemnitz E, Hitchcock PB, Taylor R (2002) Fullerenes Nanot Carbon Nanostruct 10:243
Tycko R, Dabbagh G, Fleming RM, Haddon RC, Makhija AV, Zahurak SM (1991a) Phys Rev Lett 67:1886
Tycko R, Haddon RC, Dabbagh G, Glarum SH, Douglass DC, Mujsce AM (1991b) J Phys Chem 95:518
Unkefer CJ, London RE, Whaley TW, Daub GH (1983) J Am Chem Soc 105:733
Wågberg T, Johnels D, Peera A, Hedenström M, Schulga YM, Tsybin YO, Purcell JM, Marshall AG, Noreus D, Sato T, Talyzin AV (2005) Org Lett 7:5557
Wågberg T, Hedenström M, Talyzin AV, Sethson I, Tsybin YO, Purcell JM, Marshall AG, Noreus D, Johnels D (2008) Angew Chem – Int Ed 47:2796
Walton JH, Kamasaquashie AK, Joers JM, Gullion T (1993) Chem Phys Lett 203:237
Wang GW, Weedon BR, Meier MS, Saunders M, Cross RJ (2000) Org Lett 2:2241
Wang GW, Zhang XH, Zhan H, Guo QX, Wu YD (2003) J Org Chem 68:6732
Wang GW, Li YJ, Li FB, Liu YC (2005) Lett Org Chem 2:595
Yannoni CS, Johnson RD, Meijer G, Bethune DS, Salem JR (1991) J Phys Chem 95:9

Chapter 10
Low Temperature Infrared Spectroscopy of C_{60} and C_{70} Fullerenes and Fullerane $C_{60}H_{18}$

Franco Cataldo[1], Susana Iglesias-Groth[2], and Arturo Manchado[2]

Abstract The FT-IR spectra of the fullerenes C_{60} and C_{70}, the fullerane $C_{60}H_{18}$ and the fullerane mixture $C_{60}H_x$ (77%) and $C_{70}H_y$ (22%) with $x \approx y > 36$–38 have been taken in the range of temperatures comprised between +250°C (523 K) and −180°C (93 K). For all the samples studied, as general rule, it has been observed a shift of the infrared absorption bands toward higher frequencies at lower temperatures although some exceptions have been identified. The C–H stretching bands of fulleranes appear almost independent from the temperature and this is of extreme importance from the astrochemical point of view for the search of these molecular species in space. As expected, at 93 K the infrared spectra appear better resolved with absorption bands more sharp and of higher intensity than the same bands measured at higher temperature.

All the infrared spectra of the present study have been made on samples embedded in KBr matrix and all data were extrapolated to 0 K. These spectral data at extremely low temperatures are of paramount importance for astrochemical search of these molecules in space. By comparing the gas phase spectra of both C_{60} and C_{70} fullerenes extrapolated to 0 K with the data taken in KBr matrix, it has been found that at 0 K the entity of the band shift due to matrix effect is 5–10 cm^{-1} toward lower frequencies. In other words, the gas phase spectral bands are systematically shifted 5–10 cm^{-1} toward higher frequencies than the same bands recorded in KBr. Instead, the matrix effect becomes quite negligible when the spectral data taken in KBr are extrapolated to >1,000 K. In such case there is a fair agreement between the band position of C_{60} and C_{70} fullerenes in the gas phase and the extrapolation from data taken in KBr matrix.

[1]Istituto Nazionale di Astrofisica, Osservatorio Astrofisico di Catania, Via S. Sofia 78, 95123 Catania, Italy
and
Actinium Chemical Research, Via Casilina 1626/A, 00133 Rome, Italy
e-mail: franco.cataldo@fastwebnet.it

[2]Instituto de Astrofisica de Canarias, Via Lactea s/n, E-38200 La Laguna, Tenerife, 2 CSIC, Spain
e-mail: sigroth@iac.es; amt@iac.es

F. Cataldo and S. Iglesias-Groth (eds.), *Fulleranes: The Hydrogenated Fullerenes*, Carbon Materials: Chemistry and Physics 2, DOI 10.1007/978-1-4020-9887-1_10,

10.1 Introduction

The vibrational spectroscopy of C_{60} and C_{70} fullerenes has been investigated in numerous papers and reviewed by Kuzmany et al. 1995; Kuzmany and Winter 2000. Much less attention has been dedicated to the low temperature spectra of these two molecules and there are no reports at all concerning the low temperature spectra of fulleranes in general and in particular to the most stable fullerane $C_{60}H_{18}$.

From the astrochemistry point of view, it is extremely important to know the low temperature spectra of molecules which can be the object of research in the space. In particular fullerenes are thought to be present in the circumstellar environment of late-type carbon-rich stars probably mixed or embedded with other forms of elemental carbon. Particularly remarkable sources of fullerenes are thought to be the *R Coronae Borealis* class of stars. These stars are helium-rich carbon giants and present a circumstellar environment particularly favorable for fullerenes formation. These stars are pulsating like the Cepheides but very irregularly. Thus, it may happens that their magnitude drops suddenly from fifth to the faint magnitude of 14, recovering sometimes quickly, other times taking months. The dimming comes from irregular clouds of carbon dust ejected by the stars, perhaps as a result of pulsation (Kaler 2006). Once ejected in the space fullerenes are extremely reactive for instance with the ubiquitous atomic hydrogen, leading to the formation of hydrogenated derivatives knows as fulleranes.

The present chapter is therefore dedicated to review the low temperature infrared spectroscopy of C_{60} and C_{70} fullerenes, $C_{60}H_{18}$ fullerane (one of the most stable hydrogenated derivative of fullerenes) and a mixture of hydrogenated fullerenes referred collectively as $C_{60}H_x/C_{70}H_y$. Additionally, also the gas phase spectra of C_{60} and C_{70} fullerenes will be shortly reviewed.

10.2 Experimental

10.2.1 Materials and Equipment

C_{60} and C_{70} fullerenes were 99% pure chemicals from MTR Ltd (USA), $C_{60}H_{18}$ was synthesized by hydrogenating C_{60} with hydrogen iodide (Cataldo et al. 2010) and fullerene hydrides $C_{60}H_x/C_{70}H_y$ were obtained from ABCR GmbH (Germany). Infrared spectrophotometric grade KBr was obtained from Aldrich (USA).

The FT-IR spectra were recorded on samples embedded in KBr pellets at a resolution of 4 or 1 cm^{-1} on a spectrometer Nicolet IR-300 from Thermo-Fischer Corp.

The low temperature apparatus consisted of a variable temperature cell from Specac model P/N 21525 equipped with KBr windows and sample holder which is able to work in the range comprised between +250°C and −196°C.

10.2.2 *Experimental Procedure*

The sample to be analyzed, say C_{60} fullerene, is mixed with an appropriate amount of KBr in an agate mortar and then transferred into a press and compressed at 4,000 Kg into a pellet with a diameter of 1.2 cm and a thickness of 0.2 cm. The pellet was mounted into the sample holder of the Specac variable temperature cell and inserted into the cell. The cell was then evacuated with the aid of a pump to a vacuum of 0.1 torr and then heated gradually at 120°C in order to permit the humidity absorbed on the internal surfaces of the cell and in the KBr pellet to evaporate. The sample was then cooled to the desired temperature to record the infrared spectrum. In order to go below room temperature, use was made of liquid nitrogen, added cautiously and in small amount in the cavity present inside the cell. Such cavity is connected with the sample holder and permits to cool the sample to the desired temperature. The temperature of the sample was monitored with adequate thermocouples. The lowest temperature reached with this apparatus was −180°C (93K) while the highest temperature was +250°C. Heating is provided by the Joule effect and an external thermal control unit.

All the spectral processing operations, change of scale from cm^{-1} to μm were made through the Omnic software of our spectrometer.

The Figs. 10.1 and 10.2 show the variable temperature cell from Specac outside and mounted inside the FT-IR spectrometer.

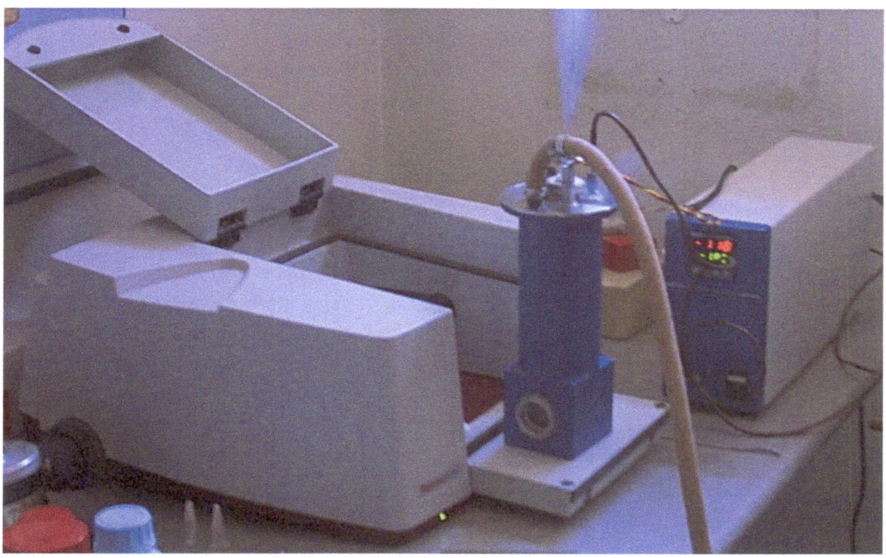

Fig. 10.1 The variable temperature cell ready to be installed inside the FT-IR spectrometer. Behind the cell, on the right, there is the temperature control unit

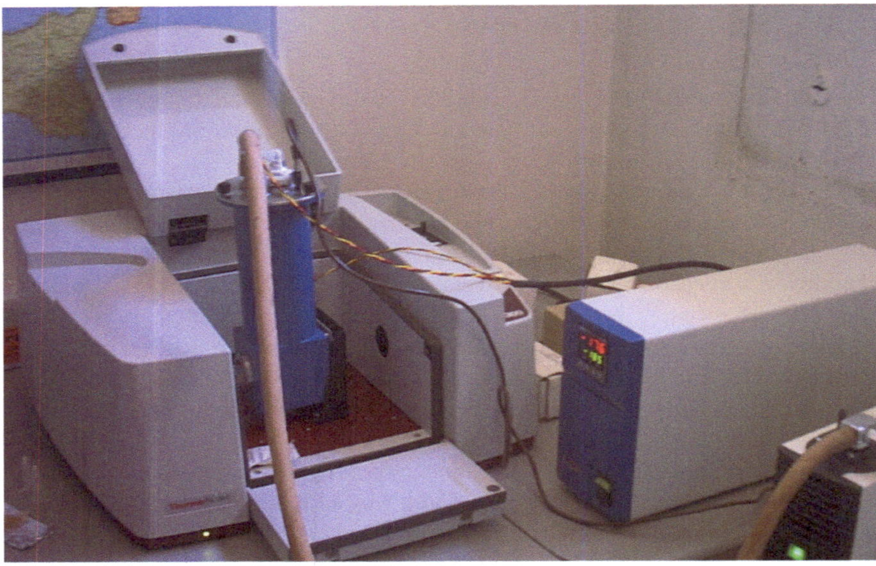

Fig. 10.2 The variable temperature cell installed inside the FT-IR spectrometer. The temperature control reports –176°C (97 K)

10.3 Results and Discussion

10.3.1 The Low Temperature and High Temperature Gas Phase FT-IR Spectra of C_{60} Fullerene

Fullerene C_{60} has the highest symmetry of any known molecule. Although there are 174 vibrational degrees of freedom (3N-6) for each C_{60} molecule, the icosahedral symmetry of the fullerene C_{60} gives rise to a number of degenerate modes, so that only 46 mode frequency are expected for this molecule. Of these, four are infrared-active and 10 are Raman-active, whereas the remaining modes are optically inactive.

The infrared spectrum of C_{60} is very simple consisting of four modes with F_{1u} symmetry observed at frequencies of 527 ($F_{1u}(1)$), 576 ($F_{1u}(2)$), 1182 ($F_{1u}(3)$) and 1429 ($F_{1u}(4)$) cm^{-1}. The 527 and 576 cm^{-1} modes are associated with a primarily radial motion of the carbon atoms, while the 1182 and 1427 cm^{-1} modes are essentially associated with a tangential motion of the carbon atoms (Kuzmany et al. 1995; Kuzmany and Winter 2000). The most characteristic vibrational mode is the pentagonal "pinch" mode at 1427 cm^{-1}. When C_{60} is cooled to –13°C (260 K) a phase transition occurs to a state with high degree of orientational order referred as "ratchet phase" (Graja and Swietlik, 1995). On further cooling the C_{60} crystals between –123°C (150 K) and –183°C (90 K), a glass transition can be observed due to the fact that C_{60} molecules shuffle into two nearly degenerate orientations. On further decreasing temperature, the population of C_{60} molecules with energetically less favorable orientation decrease but not vanish completely. Indeed, even below −183°C (90 K) about 17% of the C_{60} is found to be frozen in the energetically less favorable orientation leading to a glass-like behaviour (Graja and Swietlik 1995; Kuzmany et al. 1995; Kuzmany and Winter 2000). Such transitions can be followed by plotting the relative intensity or the peak height of the four infrared absorption bands as function of the temperature (Kamaras et al. 1993; Graja and Swietlik 1995).

In Fig. 10.3 are reported the FT-IR spectra of C_{60} measured with our apparatus at +250°C (523 K), +75°C (351 K) and at −180°C (93 K). The well known phenomena of band sharpening and increase in their intensity can be easily observed. In other words at low temperature the infrared spectrum results better resolved. Such phenomenon can be better appreciated in Fig. 10.4 where the synthetic spectrum derived from the subtraction of the low temperature (−180°C) spectrum of C_{60} from that recorded at +250°C is reported. The peaks pointing downward are due to the low temperature spectrum: 6.988, 8.443, 17.316, 19.044 μm. The peaks pointing upward are due to the high temperature spectrum: 6.946 and 7.052 μm, 8.422 and 8.506 μm, 17.210 and 17.484 μm, 18.833 and 19.191 μm. It is evident from the spectrum in Fig. 10.4 that the bandwidth was larger at higher temperature and became narrower at low temperature. This behavior of the C_{60} infrared absorption

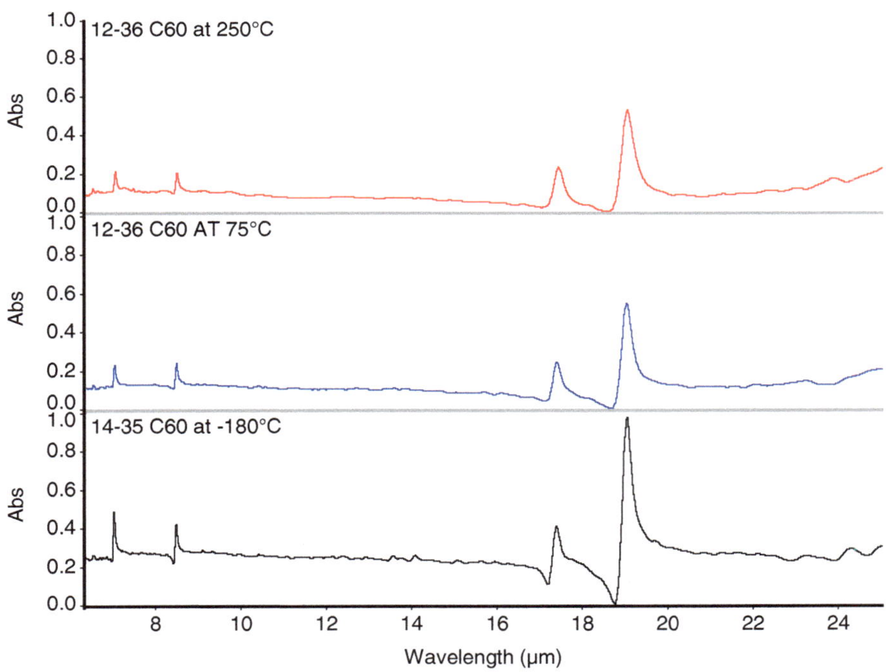

Fig. 10.3 FT-IR spectra of C_{60} in KBr taken respectively (From top to bottom) at +250°C, +75°C and at −180°C

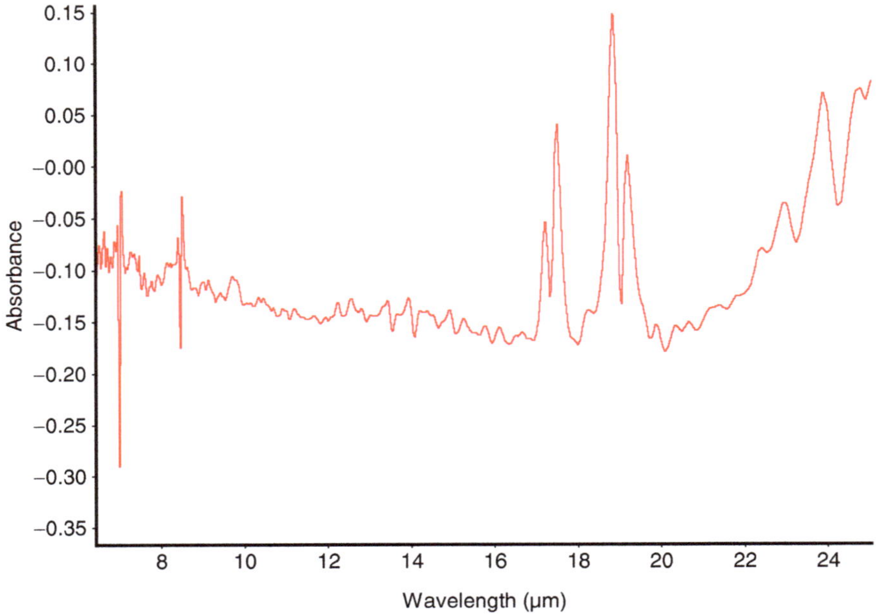

Fig. 10.4 Spectrum derived from the subtraction of the low temperature (−180°C) spectrum of C_{60} from that recorded at +250°C. The peaks pointing downward are due to the low temperature spectrum: 6.988, 8.443, 17.316, 19.044 μm. The peaks pointing upward are due to the high temperature spectrum: 6.946 and 7.052 μm, 8.422 and 8.506 μm, 17.210 and 17.484 μm, 18.833 and 19.191 μm. It is evident from such spectrum the bandwidth was larger at higher temperature and became narrower at low temperature

spectrum toward temperature was already reported and analyzed by Frum et al. (1991) and Nemes et al. (1994).

An additional feature of the low temperature spectra of C_{60} is the shift of the infrared band position. Such phenomenon is illustrated in Fig. 10.5 on the $F_{1u}(3)$ band of C_{60}. On lowering the temperature from +250°C (523 K) to −180°C (93 K) the absorption peak originally located at 1178 cm^{-1} shifts to higher frequencies i.e. 1182 cm^{-1}. Additionally, the bandwidth is reduced from 8.95 to 7.97 cm^{-1} respectively from +250°C to −180°C and the peak area passes from 2.275 to 2.400 since at lower temperature there is a sharpening of the absorption band but also an increase in intensity. It is known that even in the solid state the C_{60} molecules rotate at high temperature and there is an interaction between adjacent molecules; the coupling of the vibration with the diffuse rotation is the cause the band broadening effect. On cooling the rotation is inhibited and also there is a decay of the vibration into a lower lying optical mode and a lattice mode (Kuzmany and Winter 2000).

By measuring the infrared peak position of C_{60} at different temperature in the range of +250°C (523 K) to −180°C (93 K) it is possible to know by extrapolation the peak position a 0 K, an important information for the search of the fullerene

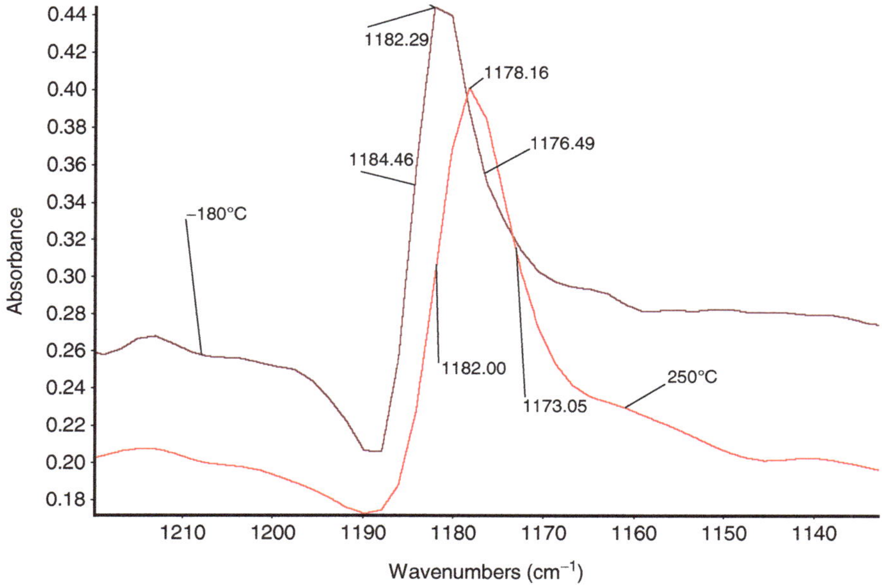

Fig. 10.5 FT-IR spectra of C_{60} (in KBr). On lowering the temperature from +250°C to −180°C the absorption peak originally located at 1178 cm^{-1} shifts at higher frequencies i.e. 1182 cm^{-1}. Additionally, the bandwidth is reduced from 8.95 to 7.97 cm^{-1} respectively from +250°C to −180°C and the peak area passes from 2.275 to 2.400 since at lower temperature there is a sharpening of the absorption band but also an increase in intensity

infrared "signature" in the interstellar or circumstellar medium. Figure 10.6 illustrates the procedure for the pentagonal "pinch" vibrational mode of C_{60}: the peak position at three (or more) different temperatures are plotted against the temperature and the resulting linear dependence of peak position with temperature is employed to determine the infrared absorption peak position of C_{60} at 0 K. As shown in Fig. 10.6 the expected peak position of the pentagonal "pinch" vibrational mode of C_{60} at 0 K is 1430.9 cm^{-1}, about 9.7 cm^{-1} higher frequency than the band position at +250°C. In Table 10.1 are reported the peak position of all the 4 absorption bands of C_{60} extrapolated to 0 K. As already reported previously, our infrared measurements on C_{60} were made in a KBr matrix.

Thus, in Table 10.1 are reported also the infrared peak positions of C_{60} in other environments. It is interesting to compare the infrared data taken at 83 K in an Ar matrix with our data recorded at 93 K in KBr matrix. The four infrared band frequencies of C_{60} are shifted toward higher wavenumbers from 2.5 to 6.1 cm^{-1} for the C_{60} sample embedded in Ar matrix suggesting lower interaction with matrix in comparison to the sample imbedded in KBr matrix but also lower degree of freedom of the molecules and perhaps a much more C_{60}–C_{60} lattice interaction. Furthermore, also the values of full-width at half maximum (fwhm) of the absorption bands reported in parenthesis in Table 10.1 show significant differences, being

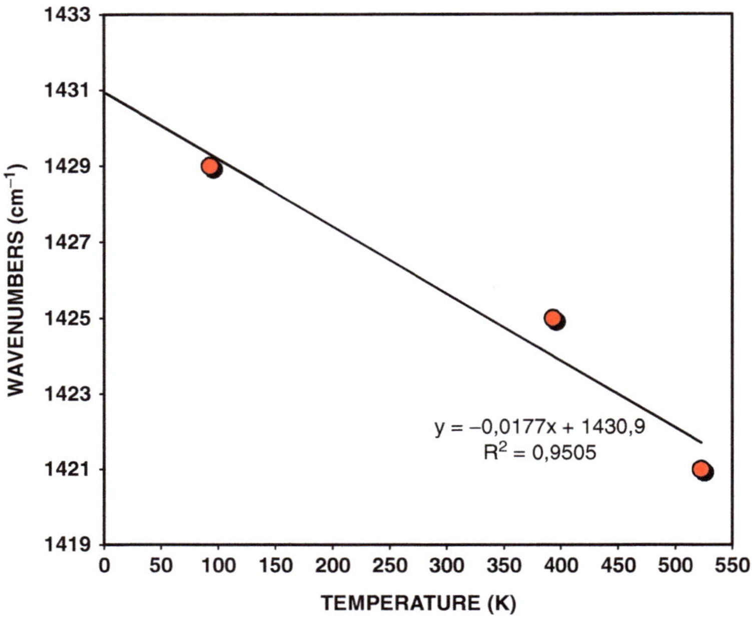

Fig. 10.6 Extrapolation to 0 K of the C_{60} infrared absorption band position. This is the case of the $F_{1u}(4)$ band, the pentagonal "pinch" mode. The intercept at 0 K occurs at 1430.9 cm^{-1}

significantly smaller for the C_{60} sample in Ar matrix, in line with the explanation already given for the frequency shift.

Frum et al. 1991 as well as Nemes et al. 1994 have studied the gas phase spectra of C_{60} above 1,000 K. Their data are reported in Table 10.1. At the mentioned temperatures is not surprising to find fwhm values above 10 cm^{-1} attributable to the high rotational freedom of the molecules at these temperatures. It is interesting to note that fwhm of the same order of magnitude were found in our infrared study on C_{60} embedded in a KBr matrix and heated at 523 K. By extrapolating the vibrational frequencies measured at 523 K in KBr to 1,083 K using the following equations:

$$\nu \text{ F1u(4)} = -0.0177 \text{ T} + 1430.9$$

$$\nu \text{ F1u(3)} = -0.00884 \text{ T} + 1183.0$$

$$\nu \text{ F1u(2)} = -0.0063 \text{ T} + 576.8$$

The expected infrared frequencies at 1,085 K in KBr are: 570.0, 1173.4 and 1411.7 cm^{-1}. By comparing these values with the gas phase values reported in Table 10.1, it can be observed that at high temperature the matrix effect of KBr affects only the bands at 1173.5 and 1412.0 cm^{-1} causing a shift to higher frequencies respectively of 4.5 and 7.6 cm^{-1} in comparison to the gas phase bands. The band at 570 cm^{-1} appears independent from temperature.

Table 10.1 Summary of C_{60} infrared absorption bands and width (fwhm in parenthesis) in cm^{-1}

0 K[a]	0 K[b]	0 K	83 K[b]	93 K[a]	523 K[a]	1,065 K[b]	1,083 K[c]	1,083 K	1,083 K
KBr matrix	Gas phase	Gas/KBr	Ar matrix	KBr matrix	KBr matrix	Gas phase	Gas phase	KBr matrix	Gas/KBr
Extrap. to 0 K	Extrap. to 0 K	Difference						Extrap. to 1,083 K	Difference
1,430.9	1,435.5	4.6	1,431.9 (4)	1,428.9 (10.8)	1,421.2 (10.7)	1,406.9 (12)	1,407.2 (13.5)	1,411.7	−4.5
1,183.0	1,190.7	7.7	1,184.8 (2)	1,182.3 (7.9)	1,178.2 (8.9)	1,169.1 (13)	1,165.8 (n.d.)	1,173.4	−7.6
576.8	574.4	−2.4	579.3 (2)	574.7 (6.9)	572.7 (10.0)	570.3 (13)	570.0 (13.7)	570.0	0.0
525.0	531.2	6.2	530.1 (1)	524.0 (6.9)	524.0 (8.8)	527.1 (11)	527.5 (11.6)		

[a]This work
[b]Frum et al. 1991
[c]Nemes et al. 1994

The vibrational frequencies of the gas phase spectra of C_{60} taken in the range of temperature comprised between 879 and 1,212 K were extrapolated to 0 K to give the theoretical position of these bands: an approach of astrochemical interest (Nemes et al. 1994). The results of that extrapolation are reported in Table 10.1. It can be observed that the vibrational frequencies at 0 K derived from the gas phase data are shifted toward higher wavenumbers in comparison to the data derived from C_{60} in KBr matrix (and extrapolated to 0 K) by 6.2 cm^{-1} for the $F_{1u}(1)$ mode, by 7.7 cm^{-1} for the $F_{1u}(3)$ and by 4.6 cm^{-1} for the $F_{1u}(4)$ mode. Instead only for the $F_{1u}(2)$ mode the band extrapolated to 0 K is shifted to shorter wavenumbers by 2.4 cm^{-1} in comparison to the data in KBr matrix extrapolated at 0 K.

10.3.2 The Low Temperature and High Temperature Gas Phase FT-IR Spectra of C_{70} Fullerene

In C_{70}, because of its lower D_{5h} symmetry, there are five kinds of non-equivalent atomic sites and eight kinds of non-equivalent bonds. This means that the number of normal vibrations increases for C_{70} in comparison to C_{60}. Although there are now 204 vibrational degrees of freedom for the 70-atom molecule, the symmetry of C_{70} gives rise to a number of degenerate modes so that only 122 modes are expected. Of these 31 are infrared-active and 53 are Raman-active.

Owing to the elongated form and reduced symmetry of C_{70} molecule, the orientational transition and crystal phases are more complicated than in C_{60}. A variety of measurements have shown that C_{70} crystal can be prepared in either f.c.c. or hexagonal, close-packed h.c.p. form. Since the two forms are almost isoenergetic the can co-exist under determinate conditions. The transition from a fully disordered phase to a partially ordered phase occurs at +64°C (337 K) but the low temperature ordered phase can be reached at +3°C (276 K) (Varma et al. 1993). Another transition in C_{70} occurs at −73°C (200 K).

In Fig. 10.7 are reported two FT-IR spectra of C_{70} recorded respectively at −180°C and at +250°C as in the previous case of C_{60}. In the wavelength (μm) abscissa scale it is difficult to appreciate the band shift due to temperature change occurred in C_{70} embedded in KBr. Therefore, Figs. 10.8 and 10.9 show the details of the spectra in wavenumbers putting in evidence the small band shift measured at the two temperatures employed.

As already observed for C_{60}, by passing from high (523 K) to low temperatures (93 K) there is a systematic shift of the vibrational bands to higher frequencies although there are also bands which appear completely insensitive to the temperature change. For C_{60} an example of infrared band not sensitive to temperature change, at least in KBr matrix, is that at 524 cm^{-1} (see Table 10.1). As shown in Table 10.2, a similar phenomenon can be observed also in the infrared spectra of C_{70} where a series of vibrational absorption bands are quite insensitive to the temperature change of the sample in KBr matrix. However, there are also other

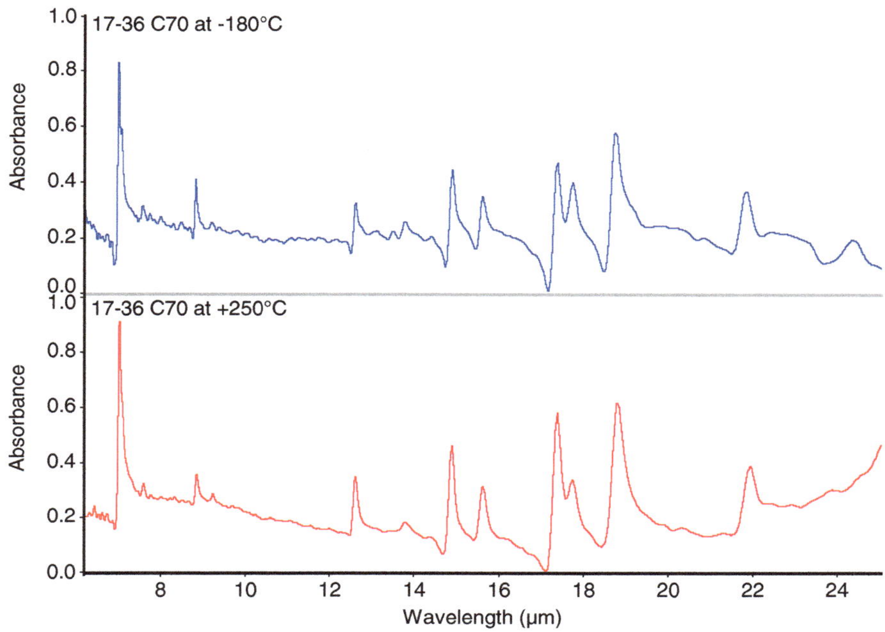

Fig. 10.7 The FT-IR spectrum of C_{70} (in KBr)

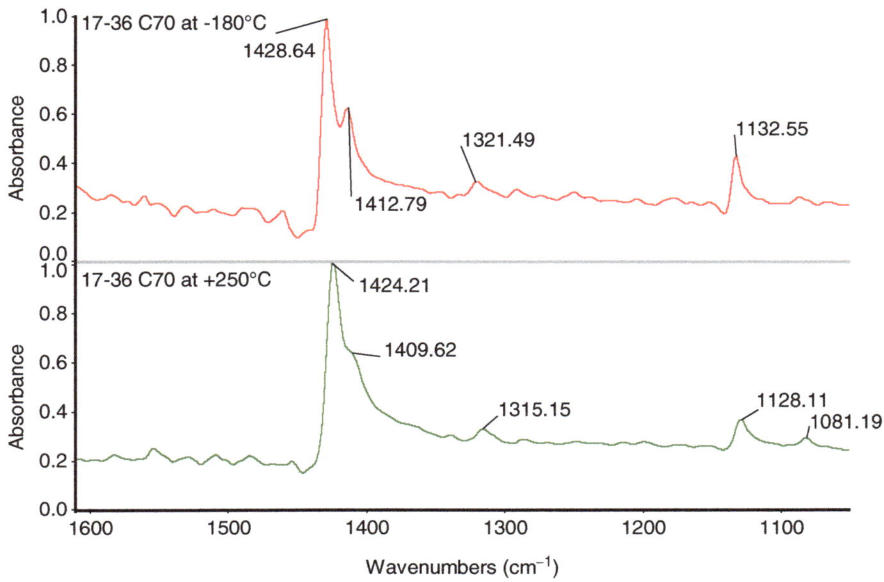

Fig. 10.8 FT-IR spectrum of C_{70} in KBr taken at −180°C and at +250°C detail in the range between 1600 and 1000 cm^{-1}

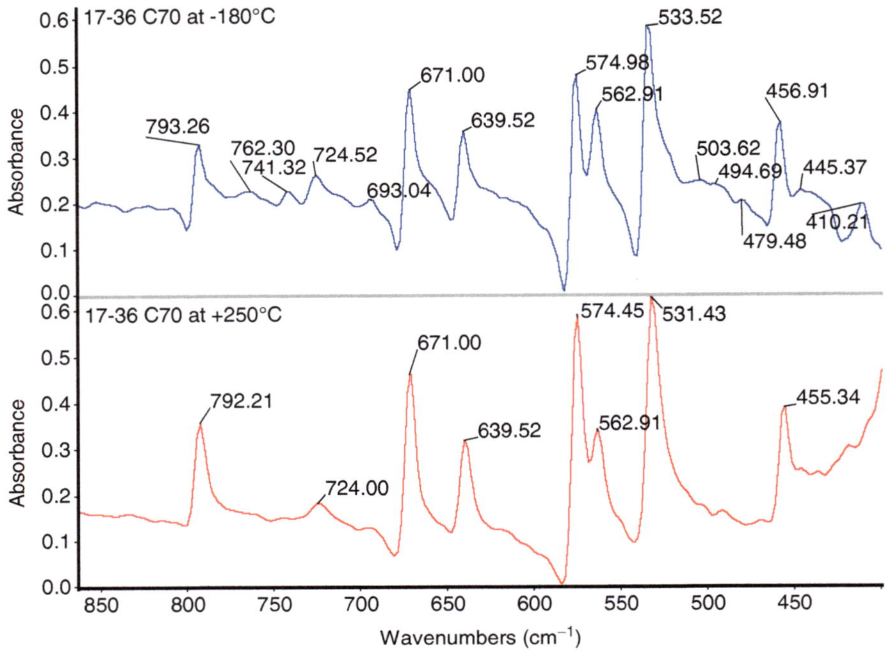

Fig. 10.9 FT-IR spectrum of C_{70} in KBr taken at −180°C and at +250°C detail in the range between 900 and 400 cm^{-1}

vibrational bands of C_{70} whose frequency is dependent from the temperature and this has allowed us to extrapolate their position at 0 K. The results of such extrapolations are reported in Table 10.2 in comparison with the data concerning the C_{70} absorption bands extrapolated to 0 K from gas phase infrared spectral data taken from Nemes et al. (1994). As expected in both cases the temperature-sensitive bands tend to shift to the highest possible frequencies and the shift appears more pronounced with the extrapolation from the gas phase data rather than for the data taken in KBr and this is certainly due to a matrix effect.

The infrared vibrational frequencies of C_{70} taken in KBr matrix in the range between 93 to 523 K were also extrapolated to very high temperature, i.e. 1,083 K and compared with the gas phase spectrum of C_{70} (Nemes et al. 1994). High temperature causes a shift toward lower frequencies and band broadening. Surprisingly, in this case the agreement between the infrared peak position taken from C_{70} in KBr matrix and extrapolated to 1,083 K show a general good agreement with the experimental spectrum measured on C_{70} in the gas phase. Another interesting aspect reported in Table 10.2 regards the fact that the emission infrared band position of the C_{70} in the gas phase are almost coincident with the absorption bands (Nemes et al. 1994).

Table 10.2 Summary of C_{70} infrared absorption bands and width (fwhm in parenthesis) in cm^{-1}

0 K[a]	0 K[b]	0 K	93 K[a]	523 K[a]	1,083 K	1,083 K[b]	1,083 K[b]	1,083 K
KBr matrix	Gas phase	Gas/KBr	KBr matrix	KBr matrix	KBr matrix	Gas phase	Gas phase emission	Gas/KBr
Extrap. to 0 K	Extrap. to 0 K	Difference			Extrap. to 1,083 K			Difference
1,429.6	1,432.0	2.4	1,428.6	1,424.2	1,418.5	1,411.8 (18.2)	1,411.8	6.7
1,413.5			1,412.8	1,409.6	1,405.5			
1,322.9			1,321.5	1,315.1	1,306.8			
1,133.5	1,144.1	10.6	1,132.5	1,128.1	1,122.4	1,121.7 (14,7)	1.122.2	0.2
1,085.8	1,099.6	13.8	1,085.0	1,081.2	1,076.3	1.077	1,077.2	−0.9
793.5	793.0	−0.5	793.3	792.2	790.7		7,93.3	−2.6
724.6			724.5	724.0	723.3			
			671.0	671.0				0.0
	638.0		639.5	639.5			638,5	
	583.1		574.5	574.4		574,9	575,3	
	567.7		562.9	562.9		556,9	553,3	
534.0	545.0	11.0	533.5	531.4	528.6	527,6 (10,0)	528,2	0.4
457.2			456.9	455.3	453.2			

[a]This work
[b]Nemes et al. 1994

10.3.3 The Low Temperature FT-IR Spectra of $C_{60}H_{18}$ Fullerane

Among the numerous hydrogenated fullerenes synthesized till now, the most stable and accessible is $C_{60}H_{18}$. Fulleranes with higher degree of hydrogenation tend to decompose thermally to $C_{60}H_{18}$ or to undergo oxidation and hydrolysis (Darwish et al. 1995; Cataldo 2003). Instead $C_{60}H_{18}$ is easy accessible through the high pressure hydrogenation reaction or through the reaction between hydrogen iodide and C_{60} followed by a moderate thermal annealing of the product under inert atmosphere (Cataldo et al. 2010). Once produced, $C_{60}H_{18}$ shows a remarkable hydrolytic stability and oxidation resistance in air. This has permitted us to study its infrared absorption spectrum from +250°C (523 K) to −175°C (98 K). The infrared spectra of $C_{60}H_{18}$ recorded at the two extreme temperatures are shown with μm scale in abscissa in Fig. 10.10. To appreciate the band shift caused by the temperature change, in Figs. 10.11 and 10.12 are reported the details of the two spectra with the wavenumber scale in abscissa. Figure 10.11 shows the C–H stretching region of $C_{60}H_{18}$; the position of the infrared absorption bands underlines the aliphatic/naphthenic character of the C–H bond of $C_{60}H_{18}$. The most common trend of the

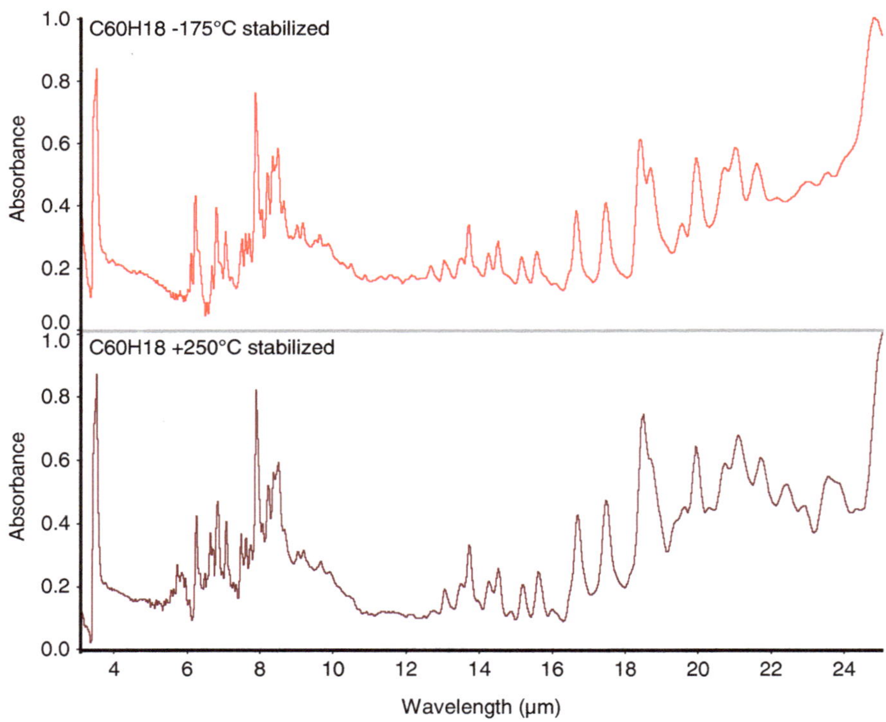

Fig. 10.10 FT-IR spectra of $C_{60}H_{18}$ in KBr taken at −175°C and at +250°C

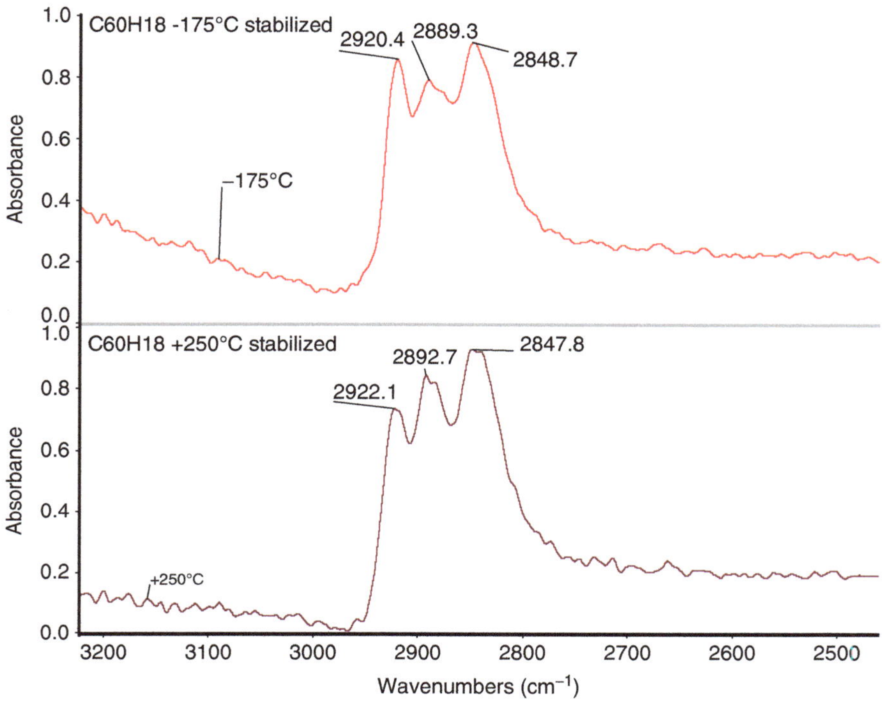

Fig. 10.11 FT-IR spectra of $C_{60}H_{18}$ in KBr taken at –175°C and at +250°C. The C–H stretching band region is shown in detail. The stretching bands underline the aliphatic/naphthenic nature of the C–H bond in fullerane

infrared absorption bands by passing from high temperature to low temperature is the shift toward higher frequencies of the absorption bands although in some cases there are bands that appears almost insensitive to the temperature change. In Fig. 10.11 it can be observed that the C–H stretching modes of $C_{60}H_{18}$ are insensitive to temperature or, in countertendency, are slightly shifted to lower frequencies at lower temperatures. However, this cannot be interpreted as an anomaly since at lower temperatures the phenomenon of band splitting combined with the narrowing of the absorption bands may lead to wrong conclusions. Indeed in the fingerprint region of the infrared spectra of $C_{60}H_{18}$ (see Fig. 10.12), the band shift occurs always toward higher frequency by passing from high to low temperature. In Table 10.1 a summary of the most important absorption bands of $C_{60}H_{18}$ and the entity of shift to higher frequency are reported.

A simplified model of the stretching frequency of 2 atoms and their connecting bond is offered by the harmonic oscillator of two masses connected by a spring (Herzberg 1950).

$$\nu_n = (2\pi c)^{-1} (k/\mu)^{1/2} = (2\pi c)^{-1} (F/x\mu)^{1/2}$$

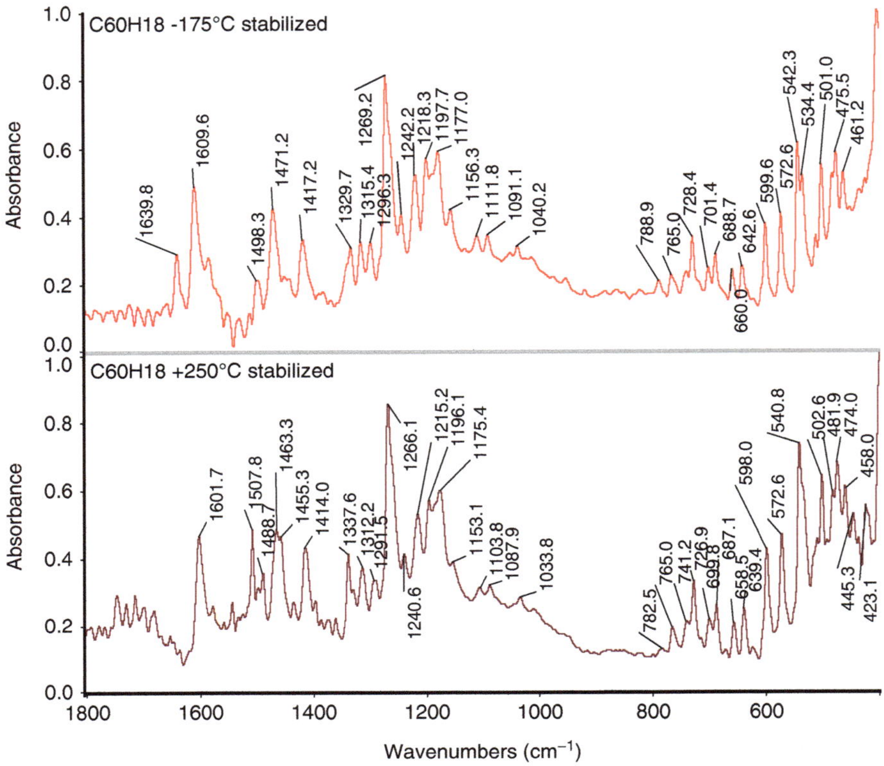

Fig. 10.12 FT-IR spectra of $C_{60}H_{18}$ in KBr taken at −175°C and at +250°C. The "fingerprint" region is shown in detail

Where ν_n is the vibrational frequency in wavenumber, c the velocity of light, k the force constant and μ the reduced mass of the system. Applying the Hooke's law, the force constant k = F/x with F the force acting on the two masses and x their displacement. By lowering the temperature of the system the displacement of the masses is minimized, the force constant increased and thus the vibrational frequency is increased. It can be calculated for the C–H band that an increase of the force constant of 0.2% leads to a band shift of ≈3.5 cm^{-1}, similar to the values observed experimentally.

In the previous sections we have shown also our approach to extrapolate to 0 K the infrared absorption bands measured at +250°C and at −175°C. The results of such extrapolation are shown in Table 10.3. The limitations of such approach have already been evidenced in the previous sections: the matrix effect plays an important role. In fact in Tables 10.1 and 10.2 it has been shown that the extrapolation to 0 K from gas phase spectral data leads systematically to band position shift at higher frequencies than those extrapolated from spectra of fullerenes embedded in

Table 10.3 Summary of $C_{60}H_{18}$ infrared absorption bands in cm^{-1}

0 K[a]	98 K[a]	523 K[a]	
KBr matrix	KBr matrix	KBr matrix	Shift from
Extrap. to 0 K			523 to 98 K
2,920.0	2,920.4	2,922.1	
2,888.6	2,889.3	2,892.7	
2,848.9	2,848.7	2,847.8	0.9
	1,639.8		
1,611.3	1,609.6	1,601.7	7.9
		1,507.8	
1,500.4	1,498.3	1,488.7	9.6
1,472.9	1,471.2	1,463.3	7.9
		1,455.3	
1,417.9	1,417.2	1,414.0	3.2
		1,337.6	
1,330.4	1,329.7	1,326.6	3.1
1,316.1	1,315.4	1,312.2	3.2
1,297.3	1,296.3	1,291.5	4.8
1,269.9	1,269.2	1,266.1	3.1
1,242.5	1,242.2	1,240.6	1.6
1,219.0	1,218.3	1,215.2	3.1
1,198.0	1,197.7	1,196.1	1.6
1,177.3	1,177.0	1,175.4	1.6
1,157.0	1,156.3	1,153.1	3.2
1,113.5	1,111.8	1,103.8	8.0
1,091.8	1,091.1	1,087.9	3.2
1,041.6	1,040.2	1,033.8	6.4
790.3	788.9	782.5	6.4
765.0	765.0	765.0	0
728.7	728.4	726.9	1.5
701.7	701.4	699.8	1.6
689.0	688.7	687.1	1.6
660.3	660.0	658.5	1.5
643.3	642.6	639.4	3.2
599.9	599.6	598.0	1.6
572.6	572.6	572.6	0
542.6	542.3	540.8	1.5
	534.4		
500.7	501.0	502.6	−1.6
475.8	475.5	474.0	1.5
461.9	461.2	458.0	3.2
		445.3	
		423.1	

[a]This work

KBr matrix. The shift toward higher frequency is from 5 to 10 cm^{-1} by passing from KBr to gas phase data extrapolated to 0 K. Thus, the data of $C_{60}H_{18}$ reported in Table 10.3 and extrapolated to 0 K should be corrected for an additional shift of 5–10 cm^{-1} toward higher frequencies.

10.3.4 The Low Temperature FT-IR Spectra of Mixture of Fulleranes $C_{60}H_x$ (77%) and $C_{70}H_y$ (22%)

When a mixture of C_{60} and C_{70} fullerenes are subjected to dry hydrogenation under high pressure and temperature a mixture of products generally referred as fullerene hydrides or fulleranes $C_{60}H_x$ (77%) and $C_{70}H_y$ (22%) are obtained. The chromatographic analysis of these products reveals that they consist of a complex mixture of products (Cataldo and Iglesias-Groth 2010). In addition to fulleranes with higher degree of hydrogenation than that achievable by "wet" synthesis ($C_{60}H_x$ and $C_{70}H_y$ with $x \approx y > 36$–38) which are the main components, other products have been detected and thought to be derived from the fullerane cage breakdown (Cataldo and Iglesias-Groth 2010).

The mixture of fulleranes $C_{60}H_x$ (77%) and $C_{70}H_y$ (22%) show a remarkable stability in humid air and this has permitted us to record their spectra in the range of temperatures between +250°C (523 K) and −180°C (93 K). The infrared spectra taken at the two extreme temperatures are shown in Fig. 10.13 with μm scale in abscissa. The spectra at both temperature are dominated by the strong C–H stretching band.

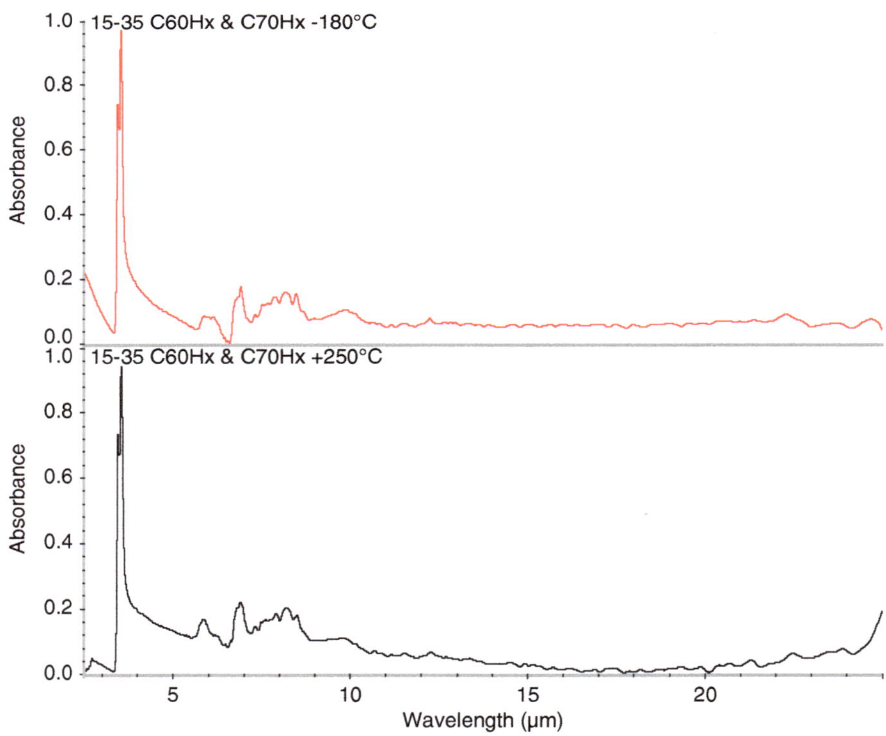

Fig. 10.13 FT-IR spectra in KBr of a mixture of fullerene hydrides $C_{60}H_x$ and $C_{70}H_y$ with $x > 36$–38

As already reported for $C_{60}H_{18}$, also in the case of the mixture of fulleranes $C_{60}H_x$ (77%) and $C_{70}H_y$ (22%) the position of the C–H stretching bands appears almost independent from the temperature and can be followed in Table 10.4. Instead, the bands in the fingerprint region of the spectrum of the $C_{60}H_x$ and $C_{70}H_y$ mixture show dependence from temperature as shown both in Fig. 10.14 and in Table 10.4. The shift toward higher frequencies at lower temperature appears quite large for certain bands. There are just three absorption bands that in countertendency show an inverse dependence of the frequency with temperature but this could also be due to phase transitions and formation of new phases.

In any case, Table 10.4 shows also the extrapolation to 0 K of the infrared absorption bands of the fulleranes $C_{60}H_x$ (77%) and $C_{70}H_y$ (22%). An extrapolation which should be taken with the usual limitation derived from the fact that we are measured the spectra in a KBr matrix. Thus, the real values at 0 K in the gas phase are expected with a further correction of 5–10 cm^{-1} toward higher frequencies.

The extrapolation of the infrared absorption bands to high temperature has not been made since the fulleranes $C_{60}H_x$ (77%) and $C_{70}H_y$ (22%) start to decompose under inert atmosphere at 478°C (751 K) (Cataldo and Iglesias-Groth 2009a).

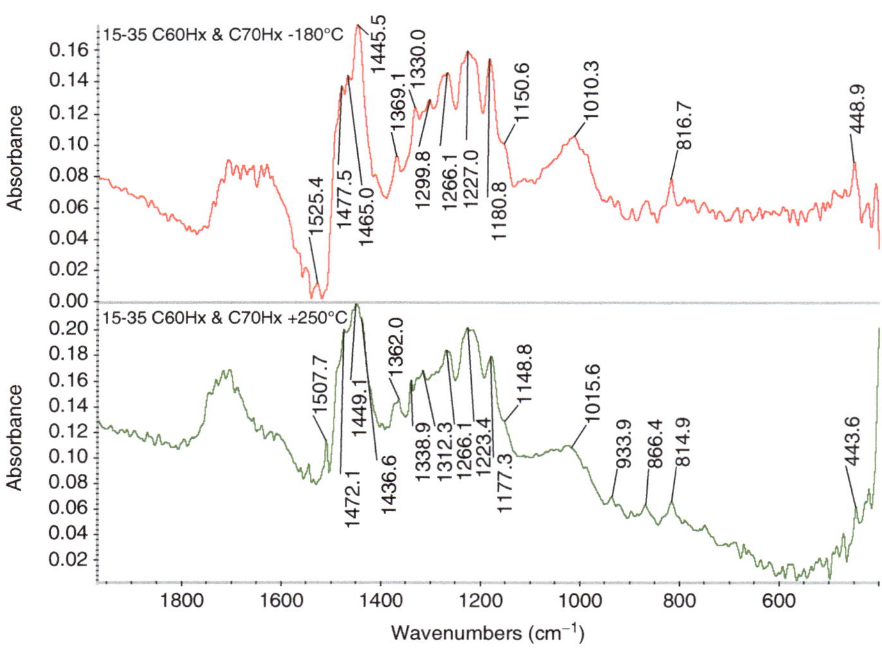

Fig. 10.14 FT-IR spectra of in KBr of a mixture of fullerene hydrides $C_{60}H_x$ and $C_{70}H_y$ with x > 36–38. Detail of the fingerprint region

Table 10.4 Summary of $C60H_x + C70H_y$ infrared absorption bands in cm^{-1}

0 K[a]	98 K[a]	523 K[a]	
KBr matrix	KBr matrix	KBr matrix	Shift from
Extrap. to 0 K			523 to 98 K
2,908.1	2,908.0	2,907.5	0.5
2,848.6	2,849.0	2,850.8	−1.8
	2,839.9		
2,837.2	2,837.2	2,837.2	
2,826.8	2,826.8	2,826.8	0.0
1,529.2	1,525.4	1,507.7	17.7
1,478.7	1,477.5	1,472.1	5.4
1,468.4	1,465.0	1,449.1	15.9
1,447.4	1,445.5	1,436.6	8.9
1,370.6	1,369.1	1,362.0	7.1
1,328.1	1,330.0	1,338.9	−8.9
1,297.1	1,299.8	1,312.3	−12.5
1,266.1	1,266.1	1,266.1	0.0
1,227.8	1,227.0	1,223.4	3.6
1,181.6	1,180.8	1,177.3	3.5
1,151.0	1,150.6	1,148.8	1.8
1,009.2	1,010.3	1,015.6	−5.3
937.3	936.7	933.9	2.8
	867.4	866.4	1.0
817.1	816.7	814.9	1.8
450.0	448.9	443.6	5.3

[a]This work

10.4 Conclusions

For the first time have been reported the low temperature spectra of fullerane $C_{60}H_{18}$ and fulleranes mixture $C_{60}H_x$ (77%) and $C_{70}H_y$ (22%). The position of the infrared absorption bands have been extrapolated to 0 K, an important key tool for searching such molecules in space.

Additionally, the infrared absorption spectra of C_{60} and C_{70} have been recorded in KBr matrix at 93 K and were extrapolated to 0 K and above 1,000 K. The results of such extrapolation have been compared with gas phase spectral data taken on C_{60} and C_{70} above 1,000 K and extrapolated to 0 K. The entity of the matrix effect exerted by KBr has been quantified at 0 K and it has been found that at 1,000 K the matrix effect of KBr becomes negligible.

Acknowledgements The present research work has been supported by grant AYA2007-64748 ofthe Spanish Ministerio de Ciencia e Innovacion.

References

Cataldo F (2003) Fullerenes Nanot Carbon Nanostruct 11:295

Cataldo F, Iglesias-Groth S (2010) Fullerenes Nanot Carbon Nanostruct 18: in press

Cataldo F, Iglesias-Groth S, Manchado A (2010) Fullerenes Nanot Carbon Nanostruct 18 (in press)

Darwish AD, Abdul-Sada AK, Langley GJ, Kroto HW, Taylor R, Walton DRM (1995) J Chem Soc Perkin Trans 2:2359

Frum CI, Engleman R, Hedderich HG, Bernath PF, Lamb LD, Huffman DR (1991) Chem Phys Lett 176:504

Graja A, Swietlik R (1995) Synth Metals 70:1417

Herzberg G (1950) Molecular spectra and molecular structure. Van Nostrand Reinhold, New York

Kaler JB (2006) The Cambridge encyclopedia of stars. Cambridge University Press, Cambridge, p 203

Kamaras K, Akselrod L, Roth S, Mittelbach A, Honle W, Von Schnering HG (1993) Chem Phys Lett 214:338

Kuzmany H, Winter J (2000) Vibrational properties of fullerenes and fullerides. In: Andreoni W (ed) The physics of fullerenes and related materials. World Scientific, Singapore, p 208

Kuzmany H, Winkler R, Pichler (1995) J Phys Condens Matter 7:6601

Nemes L, Ram RS, Bernath PF, Tinker FA, Zumwalt MC, Lamb LD, Huffman DR (1994) Chem Phys Lett 218:295

Varma V, Sashandri R, Govindaraj A, Sood AK, Rao CNR (1993) Chem Phys Lett 203:54

Chapter 11
High-Pressure Hydrogenated Carbon Nanostructures

A.V. Bazhenov[1], I.O. Bashkin[1], and K.P. Meletov[1]

Abstract High-pressure hydrogenation of the single-walled carbon nanotubes, graphite nanofibers and fullerenes C_{60} was developed. Produced samples have been studied by their combustion, gas thermodesorption, mass-spectroscopy, X-ray, IR and Raman scattering spectroscopes.

Synthesized carbon nanotubes, graphite nanofibers with the hydrogen content corresponding to the chemical formula $CH_{0,8\div0,9}$ and fullerenes $C_{60}H_x$ with x from x = 36 to the unbelievable x = 60 were produced and elucidated. Physisorption takes place only for small percent of hydrogen in this case. Dominant amount of hydrogen forms the strong covalent C–H bonds thermally stable up to 600°C. As a result of hydrogenation high-frequency conductivity of free carriers in nanotubes decreases by one order of magnitude and electron transitions between van Hove singularities in the density of electron states of nanotubes disappear.

We have shown that $C_{60}H_{36}$ is a set of isomers in our case. Combination of the vibrational modes of $C_{60}H_{60}$ and, for example, $C_{60}H_{48}$ can explain the emission and absorption spectra of interstellar and circumstellar clouds: spectral positions of not only narrow lines, but broad backgrounds as well.

Hydrogen interaction with the carbon nanostructural materials (nanotubes, nanofibers, fullerenes C_{60} and C_{70} has been intensively studied over the last years. A developed surface of nanotubes and nanofibers induced a considerable applied interest aimed at hydrogen storage and reduced consumption of organic fuel in modern industry. For the academic studies, of interest is the nature of the hydrogen interaction with the carbon nanomaterials.

In view of practical application, the carbon nanotubes or nanofibers were saturated with molecular hydrogen under relatively mild conditions: the hydrogen pressure did not exceed 10 ÷ 12 MPa at room or liquid nitrogen temperatures. The data of the application research were reviewed, e.g., by Dillon and Heben (2001).

[1]Institute of Solid State Physics, Chernogolovka, Moscow District, Russia
e-mail: bazhenov@issp.ac.ru; bashkin@issp.ac.ru; mele@issp.ac.ru

F. Cataldo and S. Iglesias-Groth (eds.), *Fulleranes: The Hydrogenated Fullerenes*, Carbon Materials: Chemistry and Physics 2, DOI 10.1007/978-1-4020-9887-1_11,

The processes of hydrogen absorption and desorption in the compression/decompression cycles were found to be reversible and reproducible at any temperature. It was generally assumed on the basis of the absorption/desorption data, that the mechanism of the hydrogen absorption is physisorption of the H_2 molecules on the graphene layers. The data on the electrical properties and Raman spectra of the single-walled nanotubes measured in the hydrogen absorption/desorption cycles at pressures up to 0.8 MPa and temperatures from 4 to 500 K (Pradhan et al. 2002) supported the assumption of hydrogen physisorption. There was, however, a remarkable discrepancy as concerns the hydrogen capacity: the maximal amount of absorbed hydrogen reported by different authors showed a scatter by two orders of magnitude, from several tenths to several tens of wt% H. Numerous studies have shown that maximum amount of adsorbed hydrogen is ≈6 wt% H.

An alternate approach was the study of hydrogen chemical reaction with fullerenes producing strong chemical bonds. There were some works on interstitial solution of molecular hydrogen in fullerenes (e.g., Assink et al. 1992), but the main efforts were made to their chemical modification. There have been proposed several hydrogenation reactions where fullerenes C_{60} or C_{70} acted as traps of hydrogen (discussed in many reviews, e.g., Taylor and Walton 1993; Hirsch 1994; Goldshleger and Moravskii 1997).

Fullerenes could be hydrogenated in organic solvents due to the metal-acid reaction, or due to the hydrogen transfer from a hydrogen-rich compound like 9,10-dihydroanthracene, or in the presence of hydrogenation catalysts and so on. Hydrogen binding in the reaction products was stronger compared to physisorption because of formation of the strong covalent C–H bonds. There were prepared many compounds $C_{60}H_x$ and $C_{70}H_x$ with even values of x. The most complete study of the properties was carried out on $C_{60}H_{36}$ and $C_{60}H_{18}$. Maximal content of hydrogen was x = 36. We know only one publication where more hydrogen-rich compounds with x = 38–50 have also been observed in the mass spectra of the products of catalytic hydrogenation (Shigematsu et al. 1993).

Direct fullerene hydrogenation was accomplished at hydrogen pressures elevated to 50–85 MPa and T = 573–623 K (Jin et al. 1994). Later this method was widely used to prepare hydrofullerenes of the some compositions as after the chemical reactions. The chemical potential of hydrogen increases under pressure, and there are many examples of the increase in hydrogen solubility or the formation of hydride phases in the metal–hydrogen systems at high pressures. The technique used in our work provides hydrogen pressures up to 9 GPa at temperatures from 77 to 900 K. With these treatment parameters, it was possible to prepare new hydrogen-rich compounds on the basis of the carbon nanomaterials with covalently bound hydrogen.

High hydrogen pressures were generated using the toroid-type quasi-hydrostatic devices. A general view of the high-pressure cell is shown in Fig. 11.1. A sample of about 100 mg mass and a pellet of AlH_3 (or AlD_3) were placed into a copper ampoule and separated with a thin Pd foil. The ampoule was tightly plugged with a copper lid using gallium as solder. Both Cu and Ga are less permeable to hydrogen and retain it in the ampoule. Thermally unstable AlH_3 rapidly decomposed under

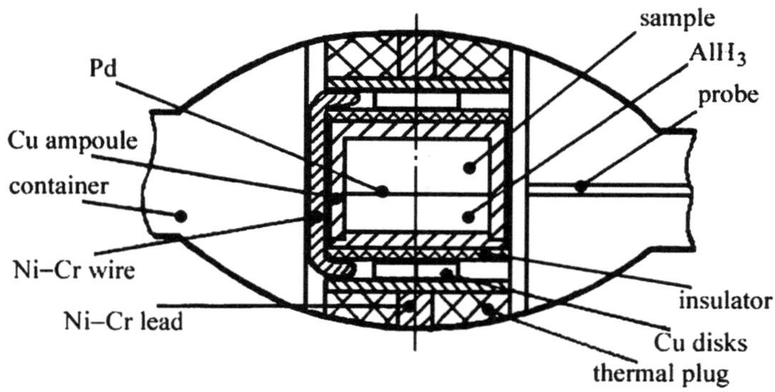

Fig. 11.1 Layout of the high-pressure cell

high pressure on heating above 250°C, permeated through the Pd foil and reacted with the sample. The ampoule was mounted in the channel of a pipestone container together with the Ni–Cr wire heater, a thermocouple entered through a special probe hole. The container also served as a pressure-transmitting medium.

The procedure of the thermobaric treatment consisted of several steps. Initial compression to 1.5 GPa and heating to 270°C resulted in the hydrogen atmosphere in the ampoule. Then pressure was increased to the final value, and the sample was maintained at 350°C overnight (about 20 h), then at 450–490°C for 6–10 h. After cooling to room or liquid-nitrogen temperature and decompression, the recovered samples were stored without access of the ambient atmosphere until measurements.

11.1 Experimental Methods for Testing of Samples

The chemical analysis of the products was carried out using combustion of a 3 mg sample in the oxygen flow at 1,400°C and weighting of the combustion products, CO_2 and H_2O (or D_2O).

The X-ray diffraction patterns were measured at room temperature on a Siemens D500 diffractometer with the monochrome $CuK_{\alpha 1}$ irradiation.

The thermal stability of the hydrogenated materials was estimated by measurements of gas thermodesorption. A piece of the sample cooled to liquid nitrogen under pressure was placed, in a liquid nitrogen bath, into a non-hermetic copper container, and the latter was moved into a quartz ampoule externally cooled with liquid nitrogen. The ampoule was connected to a vacuum system with the calibrated volume, the system was evacuated to 10^{-6} MPa, and then the ampoule was heated at a rate of 20 K/min. The pressure P of a gas evolved upon heating to 650°C was registered manually; the sample was weighed after measurements. We call these dependences as a "manometric curves".

The mass spectra of the gases evolved from the deuterated SWNT sample heated in vacuum were measured with the MI 1201V mass spectrometer. Gas ionization in the ion source of the spectrometer was produced with a 70-eV electron beam. To obtain the gas phase, the sample was placed in a quartz ampoule of a pyrolyzer that was connected to the injection system of the mass spectrometer through a fine control valve. Then the ampoule was evacuated to a pressure of about $2 \cdot \times 10^{-5}$ Pa in order to remove the surface and weakly bound impurities from the sample. After the evacuation, the ampoule was isolated from the vacuum system and the sample was heated to 550°C in five steps. At each step, the sample was kept at a fixed temperature for 3 h; then the fine control valve was open and the mass-spectrometric analysis of the gas collected in the ampoule was performed. After the analysis, the quartz ampoule was again evacuated, the valve was closed, and the sample was heated to the next temperature. The measurements were carried out over the range $1 \leq m/z \leq 90$, where m is the atomic mass and z is the ion charge. The spectrometer resolution of about 0.08% ensured a reliable determination of the gas-phase components.

IR spectra were measured by Bruker IFS-113v IR Fourier spectrometer: transition spectra and diffuse reflection at T = 295 K. IR microscope of the Fourier spectrometer was used for the measurements of the transmission spectra in the spectral range 600–9,000 cm^{-1} Transmission spectra of the thin films were also measured in the spectral range from 4,000 to 18,000 cm^{-1} using a standard double-beam spectrometer.

Raman spectroscopy: Raman spectra from small SWNT pieces with typical dimensions of 100 μm were recorded in the back-scattering geometry using two different micro-Raman setups comprised of a triple monochromator DILOR XY and a CCD detector system, cooled either to liquid nitrogen temperature or –100°C. The 488 or 514.5 nm line of an Ar^+ laser, as well as the 647.1 nm line of a Kr^+ laser, were used for excitation, while the beam intensity on the sample was ≅0.5 mW. The laser line was focused on the sample by means of a 100× objective with a spatial resolution of ~1 μm.

11.2 High-Pressure Hydrogenated Single-Walled Carbon Nanotubes and Nanofibers

Our first results about the high-pressure hydrogenation of single-walled carbon nanotubes (SWNTs) and graphite nanofibers (GNFs) were published in (Bashkin et al. 2004). Starting GNFs were synthesized in a direct-flow quartz reactor in a CO/H_2 = 4/1 gas mixture at 600°C for 6 h using a mixed Fe/Cu = 7/3 catalyst. Scanning electron microscopy showed that the GNF length was, on the average, of 30 μm and the diameter ranged from 100 to 300 nm. The content of graphite nanofiber in the prepared material was about 90%.

Carbon black containing 15–20% SWNTs was synthesized by the electric arc method in helium atmosphere at a pressure of 0.86 atm using a metallic Co/Ni = 3/1 catalyst (Loutfy et al. 1999). To remove impurities from SWNTs, carbon black was subjected to ultrasonic treatment in a concentrated hydrochloric acid, and then to the multistage treatment with hydrochloric acid alternating with oxidation in air at temperatures up to 540°C. The content of SWNTs in the product was estimated using scanning and transmission electron microscopy and was found to be equal to 50–60%.

In the experiments, a GNF or SWNT sample with a mass of about 60 mg was placed in a high-pressure chamber and saturated with hydrogen obtained by thermal decomposition of AlH_3. The sample was held under a hydrogen pressure of 9 GPa first for 18 h at t = 350°C and then for another 6 h at 450°C. At the end of holding, the chamber was cooled to −140°C and unloaded to atmospheric pressure at this temperature. Then, the hydrogen-saturated material was taken out from the chamber and further held in liquid nitrogen.

Hydrogenated GNF and SWNT powders with a mass of several milligrams were chosen to determine their thermal stability, hydrogen content and to study them by X-ray diffraction, IR spectroscopy and so on.

The typical manometric curves for hydrogenated GNFs and SWNTs are shown in Fig. 11.2. The right axis of the graph indicates the amount of liberated hydrogen x as calculated under the assumption that the gas consisted only of H_2 molecules. The x(t) dependences for hydrogenated GNFs and SWNTs are closely similar to each other. In the interval from T = 77 K to t = 0°C, the amount of liberated gas increases with temperature rather slowly, a small jump is observed near t = 0°C, and the gas release is terminated near room temperature. The total amount of hydrogen released upon heating to room temperature is 0.15–0.5 wt%. Gas is virtually not evolved from the samples upon heating from room temperature to 450°C, but the second stage of intense release begins near 600°C, and about 5 wt% H is collected at 600–650°C, i.e., an order of magnitude greater than upon heating to room temperature. The rate of gas release is low, so that the process is not terminated up to 600–650°C, and, as is shown in Fig. 11.2 by the example of GNF–H_x, an additional amount of gas is liberated upon repeated sample heating to 600°C at the same rate.

To determine the total hydrogen content and estimate the composition of liberated gas, the hydrogenated GNF and SWNT samples heated to room temperature were burned out in an oxygen flow at 1,400°C and the combustion products H_2O and CO_2 were weighed. These measurements gave x = 6.3 wt% H for GNF and x = 6.8 wt% H for SWNTs (this corresponds to the chemical formulas $CH_{0.81}$ and $CH_{0.88}$, respectively), with a spread in data less than 0.05 wt%. The data obtained agree satisfactorily with the estimate x ≈ 5 wt%H derived from the gas release between room temperature and 650°C, taking into account that the gas release was incomplete during the first run. Such an agreement is the evidence that hydrogen was liberated predominantly in the form of H_2 molecules rather than of hydrocarbons (e.g., if methane CH_4 were released, the amount of its molecules and, correspondingly, pressure would be twice as low in the gas-release experiments). A comparison of the burning results with the gas-release data allows the conclusion to be drawn that, after measurements with heating to 600–650°C presented in Fig. 11.2, about 1.2 and 1.7 wt% H remained in the GNF and SWNT samples, respectively.

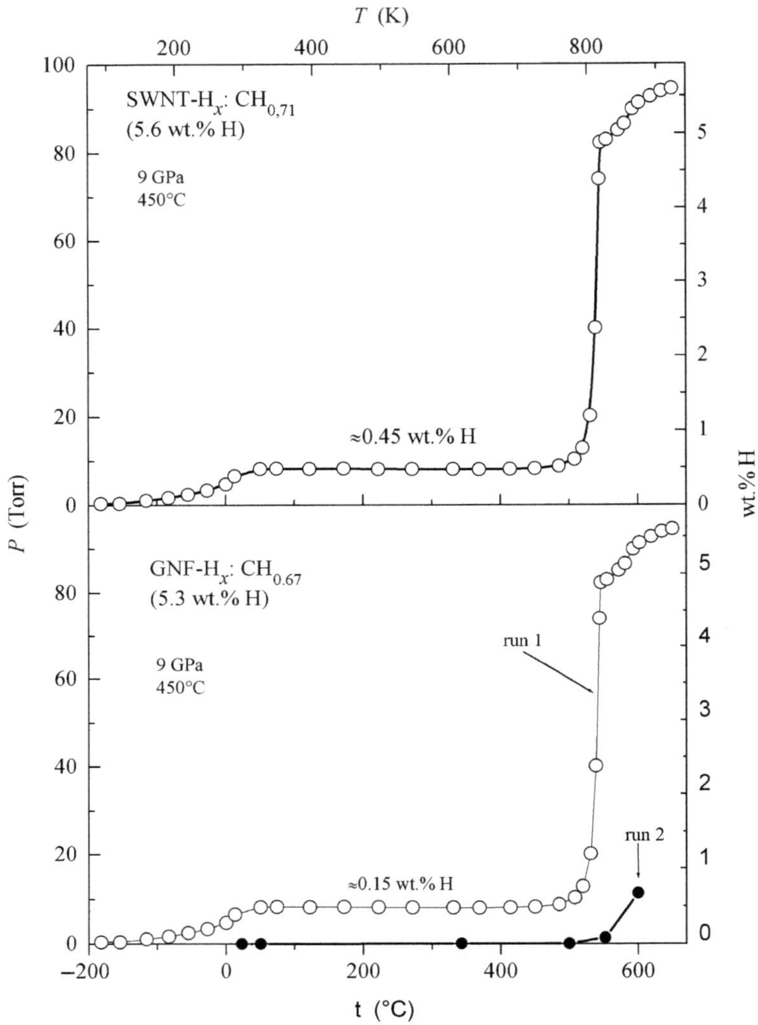

Fig. 11.2 Temperature dependence of the gas pressure in a preliminarily evacuated volume (*left vertical scale*) and its recalculation into the amount of hydrogen evolved from the sample (*right scale*) upon heating at a rate of 20 K/min for single-walled carbon nanotubes (SWNTs) and graphite nanofibers (GNFs, two heating cycles) saturated with hydrogen at a pressure of 9 GPa and temperatures up to 450°C

Mass-spectrometer date of the hydrogenated carbon nanotubes we have published in (Shulga et al. 2004). These nanotubes for the gas mass-spectrometry were synthesized by the electric-arc method using a nickel–yttrium catalyst. To remove amorphous carbon and the metal catalyst, the primary products containing 10–15 wt% of carbon SWNTs were purified by repeated oxidation in air at temperatures up to 550°C alternated with washing in hydrochloric acid. The microprobe analysis showed that final content of the metal impurities in SWNT was less than 1 wt%, with the ratio of the detected metals Ni:Y:Cu:Zn = 1:1.5:0.6:0.7. The incombustible residue was consistent with the total oxide amount within ±30%. The purified products were studied using the high-resolution electron microscopy (Krestinin et al. 2003a, b). The sensitivity of the absorption spectra in the near-IR region to the purity of products (Krestinin et al. 2003b; Chiang et al. 2001a) was used for quantitative determination of the SWNT content in this batch, it amounted to 80–85 wt%. The nanotubes had a narrow diameter distribution with an average value of 1.5 nm and were strongly aggregated in bundles, microcrystalline films, and polycrystalline covers. The main carbon impurities, according to the electron-microscopic data, were graphitized black particles and graphite blocks up to 10–15 μm in size.

To reduce or to take into account the effect of the ambient atmosphere and other experimental factors, we used the heavier hydrogen isotope, deuterium, in this study. SWNTs deuterated under the final pressure of 5 GPa during a two-step exposure at T = 350°C for 21 h and at T = 460–490°C for 9 h. According to the data of two combustion tests, deuterated SWNT contained 10.8 ± 0.1 wt% D.

The mass spectra of gases evolved from deuterated SWNT at various heating steps are shown in Fig. 11.3. The main constituent of the gas phase at temperatures to 400°C was a mixture of the hydrocarbon molecules and radicals. The mixture consisted of both deuterated hydrocarbons, as evidenced by the high intensities of the peaks with m/z = 17–20 and 31–36, and compounds including the light isotope, as follows from the occurrence of the peaks with the odd mass numbers. The latter fact is indicative of the impurity of the light hydrogen isotope, protium. The occurrence of a minor impurity of protium both in parent SWNT and in AlD_3 was found in other experiments. The origin of protium in AlD_3 is quite clear: our AlD_3 contains approximately 8% admixture of the AlH_3. Parent SWNT had a minor impurity of solvents and media used for their preparation, diethyl ether and acetone, whose fragmentation products were ascertained by the analysis of the fine structure of the peaks at m/z = 28, 31, 44, 45, 59, and 74 (Shulga et al. 2004).

The most prominent feature of the spectra in Fig. 11.3 is a dramatic difference between the compositions of the gas phase below and above 400°C. The concentration of the D_2 molecules in the gas phase increased as the temperature was increased from 100°C to 400°C. In the spectra measured after heating to 500°C and 550°C, the peaks with m/z = 4 and 3 became dominant. Consequently, D_2 and HD molecules were the main components of the gas phase at these temperatures. This remarkable fact shows that the main step in the manometric curves in Fig. 11.2 is actually due to evolution of molecular hydrogen whereas light hydrocarbon molecules contribute mainly at lower temperatures when the gas desorption is small.

Figure 11.4 shows the X-ray diffraction patterns of GNFs and SWNTs in the initial and hydrogenated states and after various annealing procedures. Partial annealing with the removal of ≈40% of hydrogen absorbed under pressure

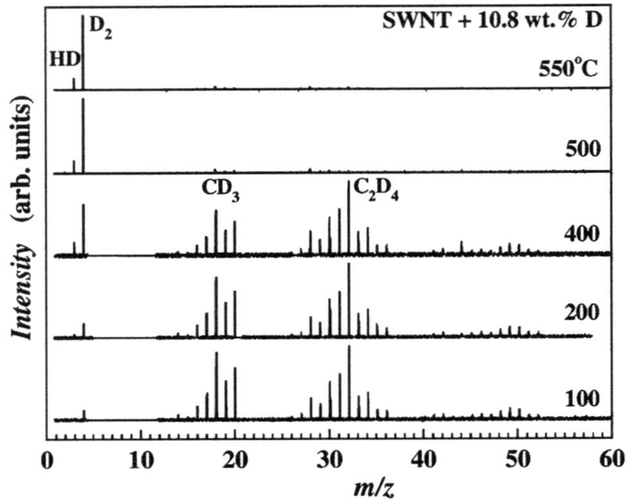

Fig. 11.3 Mass spectra of the gas phase over a sample of deuterated SWNT measured in heating steps at indicated temperatures. The spectra are restricted to the m/z = 60 value, because peaks of the heavier ions have negligible intensities. The most intense peaks are identified according to their principal components

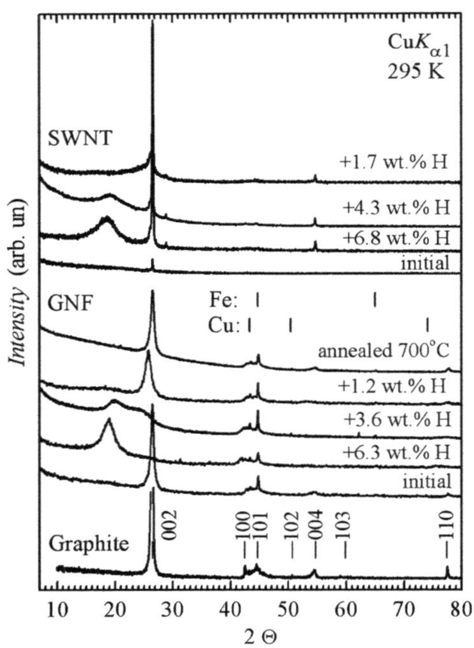

Fig. 11.4 X-ray diffraction patterns of single-walled carbon nanotubes and graphite nanofibers: in the initial state, after saturation with hydrogen at 9 GPa (6.8 and 6.3 wt% H, respectively), after removal of about 40% of absorbed hydrogen (4.3 and 3.6 wt% H), after degassing annealing at 600–650°C (1.7 and 1.2 wt% H), and after prolonged annealing at 700°C (GNFs). For comparison, the diffraction pattern of a GDG-6 graphite powder is also shown. Bar diagrams of Fe and Cu indicate the catalyst admixture in GNFs, T = 295 K

(the residual contents were 3.6 wt% H in GNFs and 4.3 wt% H in SWNTs) was carried out by holding the sample in an evacuated volume at a temperature of about 500°C. Annealing of GNFs at 700°C was performed in a dynamic vacuum of $P < 10^{-5}$ mmHg for 6 h. The X-ray diffraction pattern of the ground high-density graphite GDG-6 is presented in Fig. 11.4 for comparison.

The diffraction patterns of initial GNFs include a strong reflection near the graphite (002) line at $2\Theta = 26.5°$, a number of weak reflections near the other graphite lines, and the lines of catalyst components. After the saturation of the GNFs with hydrogen, a broad intense peak appears near 19° instead of a very strong graphite reflection at $2\Theta = 26.5°$, the other graphite reflections undergo a marked shift to the small angles, and only the catalyst lines do not change their positions. After two heating cycles to 600°C shown in Fig. 11.2, the peak at 19° disappears and the diffraction pattern mainly regains its initial shape. Nevertheless, the most intense reflection remains markedly shifted to small angles, in agreement with the fact that hydrogen is removed incompletely. A prolonged vacuum annealing at 700°C results in a complete recovery of the initial diffraction pattern. A comparison of the diffraction pattern of the sample having 3.6 wt% H after partial annealing with the diffraction patterns of the extreme states shows that it represents the diffraction pattern of a two-phase state, in which the strongest reflections are shifted toward each other ($2\Theta \approx 19.9°$ and 24.0°) and are strongly broadened. A change in the X-ray diffraction patterns indicates that the crystal lattice of graphite nanofibers swells upon hydrogenation and that the structure is recovered after the removal of hydrogen. A change in the diffraction pattern of the multilayer GNF structure upon hydrogenation, 6.3 wt% H, can be considered as being caused mainly by an increase in the spacing between the graphene layers by approximately 40% from 3.36 to 4.67 Å.

In the diffraction pattern of the initial SWNTs, no reflections are seen from the single-walled nanotubes (a weak reflection at small angles $2\Theta \sim 6°$ can be, in principle, observed because of the triangular SWNTs packing in bundles (see e.g. Sharma et al. 2001). A sharp weak reflection at the position of the graphite (002) line should be caused by the presence of an admixture of graphitized particles in the material. The catalyst reflections are not seen. After the hydrogenation of the SWNTs, a broad peak appears near 18.5°. The narrow graphite (002) and (004) reflections are markedly strengthened. We think that it is just the result of the crystallization of the amorphous graphite micro inclusions. The removal of 2.5 wt% H results in weakening of the broad peak 18.5° and its shifting by ≈0.8° to larger angles, and this peak disappears after heating to 650°C. The sharp graphite (002) and (004) reflections are retained and their intensity relative to the background change only slightly. It was reported in the literature that SWNT is not destroyed at hydrostatic pressures below 13 GPa (Sharma et al. 2001). For this reason, a change in the diffraction pattern of SWNTs after the thermal treatment in hydrogen should be assigned to the graphitization of the amorphous carbon particles that are the impurity addictions in the initial material and to the hydrogenation and dehydrogenation of a certain fraction of graphitized particles.

Weakly bonded hydrogen, which is released in an amount of less than 0.5 wt% upon heating to room temperature, can reasonably be assigned to physisorption, which is considered to be the dominant mechanism of hydrogen absorption by carbon materials at pressures below 12 MPa and not too high temperatures.

To elucidate the nature of the bonded states of the main hydrogen mass that is retained up to high temperatures, the IR diffuse reflection spectra were measured for GNF and SWNT powders in the initial state, after the treatment under hydrogen pressure, and after the degassing annealing. Measurements were performed in the range 400–5,000 cm^{-1} at room temperature. The results are presented in Fig. 11.5.

It is not the conventional reflection but the diffuse reflection of powder. Light in this sample is scattered over 2π solid angle. The spectra of initial GNFs and SWNTs (curves 1 of Fig. 11.5) are characterized by a monotonic decrease in intensity with the increase of the photon energy; such a spectral behavior is typical of the spectra of purified nanotubes (Bazhenov et al. 1998). As in the case of nanotubes exhibiting properties of a strongly imperfect metal or semimetal, the light absorption by free carriers caused by their high-frequency conductivity is the main reason for a decrease in the GNFs and SWNTs diffuse reflection with increasing photon energy.

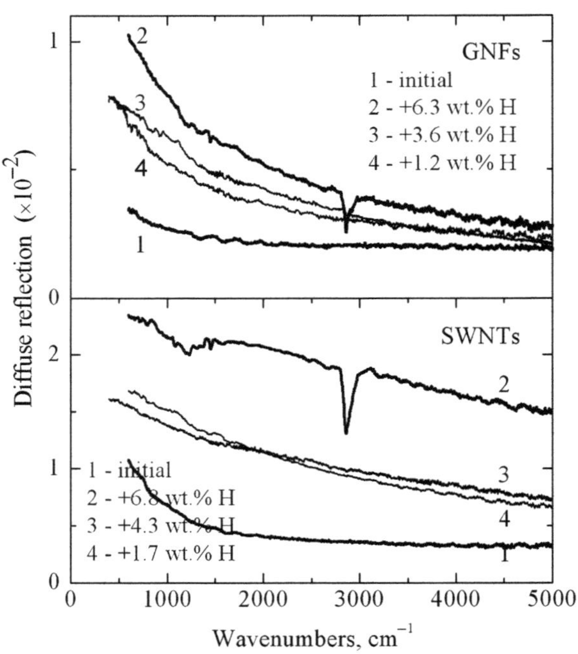

Fig. 11.5 IR diffuse reflection spectra of graphite nanofibers and single-walled carbon nanotubes: in the initial state, after saturation with hydrogen at 9 GPa, after removal of about 40% of absorbed hydrogen, and after degassing annealing. T = 300 K

After hydrogenation, the GNF diffuse reflection increases substantially over the entire range of measurements (curves 2 of Fig. 11.5) and a narrow absorption band at 2,860 cm^{-1} with a halfwidth of 38 cm^{-1} and a weaker band at 2,920 cm^{-1} appear in the spectrum. These energies are typical of the stretching C–H vibrations, which are strong chemical bonds. In the spectra of hydrogenated SWNTs, the diffuse reflection also increases; an asymmetric band with a maximum at 2,860 cm^{-1} and a halfwidth of 95 cm^{-1} appears in the spectrum due to the stretching C–H vibrations, as well. The asymmetry of the SWNT–H_x line can be explained by the overlap between the bands at 2,860 and 2,920 cm^{-1}.

After annealed for 6 h in vacuum at 700°C and removal of the main hydrogen mass, the C–H absorption lines disappear and the spectra become somewhat closer to the spectra of initial samples (curves 4 in Fig. 11.5). Nevertheless, the background diffuse reflection is regained in part and occupies the intermediate position between the spectra of the initial and hydrogenated samples.

Diffuse reflection from powder sample is a complex combination of transmission, internal and external reflections, and scattering. It is dependent on the particle size, absorption and refractive indices of the studied material. The case of proper prepared powder diffuse reflection R carries the information primarily about the transmission spectrum of the sample (Willey 1976; Fuller and Griffiths 1978). The traditional method of the absorption spectra (K) calculation on the base of the diffuse reflection R is the Kubelka–Munk equation $K = (1 - R)^2S/2Rc$, where S is the scattering coefficient, concentration of the studied material is $c = 1$ in our case.

Is it a real absorption? It is well known that analysis on the base of the Kubelka–Munk equation is applicable at diffuse reflection R not much less than $R \approx 30\%$. The case of low diffuse reflection the deviations from linearity should be taken into account. We have R is near 1%. So, we should be careful! The case of strongly absorbing samples it is possible to dilute them in nonabsorbent powder, for example in KBr powder. We have not used this traditional method because were afraid of possible chemical reactions at nigh temperature treatment of the mixture of the hydrogenated SWNTs with KBr.

We have elucidated the problem of the low diffuse reflection in (Bazhenov et al. 2004). Our hydrogenated carbon nanostructures are bulk polycrystalline materials. They are not responsible for IR measurements. We cleaved polycrystalline samples into thin plates and compare their optical properties with grinded samples. Absorption lines associated with vibration modes of the C–H bonds were well detected in the transmittance spectrum of $C_{60}H_{36}$ plates with the help of IR microscope of the Fourier spectrometer. This method was not applicable, however, for the study of SWNTs and NFs due to the strong free carrier absorption and variation of the plate thickness. The grinding of $C_{60}H_{36}$ does not result in a loss of the strongly bonded hydrogen (≈2,900 cm^{-1}). We are shure that grinding is applicable for strongly bonded hydrogen both for NFs and SWNTs as well.

To understand, are the spectra of Fig. 11.6 real absorption or not, we have studied the transmission spectra of the SWNT powder pressed in KBr pellets (Bazhenov

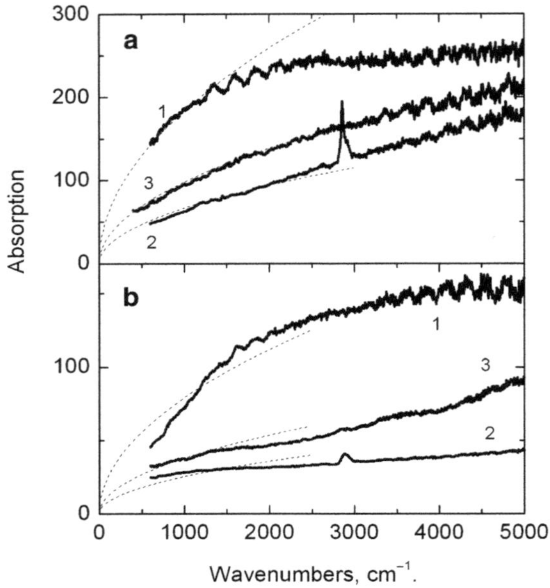

Fig. 11.6 Absorption spectra K restored from diffuse reflection R by using the Kubelka–Munk equation. Spectra 1, 2, and 3 are initial nanostructures, hydrogenated and annealed at 700°C during 6 h, respectively. (**a**) Spectra of NFs, (**b**) SWNTs. Dashed curves are Drude approximation of the absorption spectra, $K = A_i * \nu^{0,5}$

et al. 2004). For both cases we have used analysis on the base of Druder approximation of the free carrier conductivity in the spectral range $\nu < 1{,}500$ cm^{-1}. Using Drude formulas for high-frequency conductivity of the free carries we have got an absorption coefficient K of the free carries $K^2 = 16\pi^2\sigma\nu/c$ at low wave numbers ν, $\nu \ll \gamma$, where γ is the damping factor of free carriers, σ – high-frequency conductivity, c – velocity of light. These dependencies are shown in Fig. 11.6 by dashed curves. They have shown the ten times decrease of the free carrier absorption and, respectively, high-frequency conductivity of the free carries as a result of hydrogenation of SWNTs in both cases. So, we have qualitative concurrence of the results of these two methods.

At $\nu > 2{,}000$ cm^{-1} spectral dependence of monotonies increase of the absorption calculated from our diffuse reflection spectra is strongly different of the absorption spectra of SWNT powder pressed in KBr pellets. This is a result of strong absorption of light by SWNT nanoclusters. To study spectra in this spectral range we have used more complicate method: we have prepared SWNT thin films (Bazhenov et al. 2006). It was not an easy procedure, because it is well known that common organic solvents are not applicable in the case of carbon nanotubes. By date, there are no methods of synthesis of individual nanotubes of certain type. Nanotubes are generally combined into either ropes with hexagonal packing (bundles). This circumstance impedes their practical use. Grinding or treating samples with ultrasound in an organic solvent (acetone, toluene) or an aqueous solution of a surfactant, can give a liquid suspension of nanotubes. However, such a suspension is unstable in time due to the van der Waals interaction between nanotubes. It leads to aggregation of nanotubes and precipitation of aggregates.

To prepare thin films of hydrogenated SWNTs we have dispersed them in acetone by a 10-min ultrasonic treatment. The prepared suspension was then dropped on a CaF_2 or KBr substrates and dried (Bazhenov et al. 2006). Suspension and drying at room temperature resulted in inhomogeneous distribution of nanotubes, because SWNTs got significantly aggregated while drying. The rate of aggregation increased by an order of magnitude for the hydrogenated SWNTs. To get homogeneous films, the suspension was pulverized on the substrates heated to ≈100°C. The preheating efficiently suppressed aggregation of both initial and hydrogenated SWNTs. Films of the initial SWNTs thus prepared were free of acetone, whereas some amount of this solvent (presumably, physisorbed) still remained in the films of the hydrogenated SWNTs. To remove acetone, films of the hydrogenated SWNTs were heated to 150°C in vacuum.

The optical density spectra of thin films (Fig. 11.7) of the initial and hydrogenated SWNTs (5.4 wt% H) show a monotonic increase in the absorption with increasing photon energy in the spectral range 400–4,000 cm^{-1}. Taking into account that our samples are a mixture of metallic and semiconducting SWNTs (Mintwire et al.

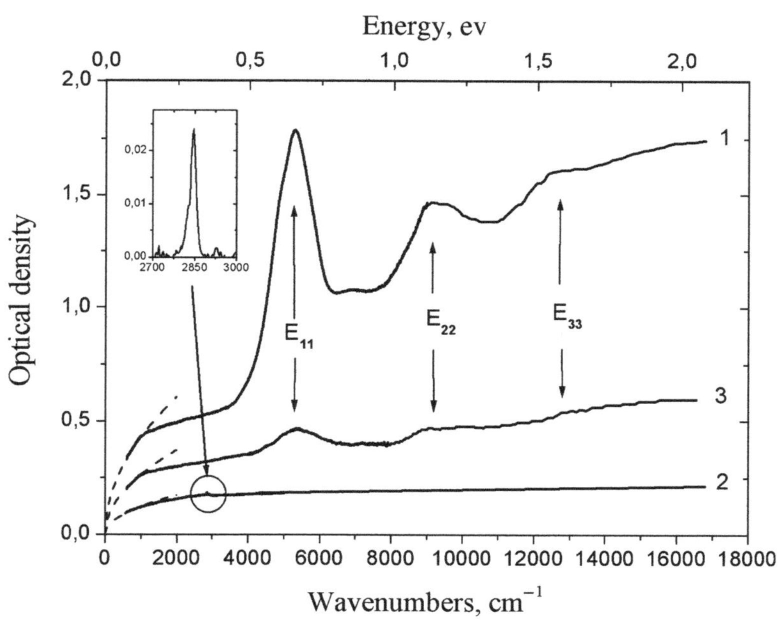

Fig. 11.7 Optical density spectra of thin films of the initial, hydrogenated and outgassed SWNTs (spectra 1, 2 and 3, respectively). The dashed curves in the low-energy part of the spectra represent the Drude approximation of the absorption spectra. The inset in Fig. 11.7 shows the absorption lines of the C–H bonds in a thin film of the hydrogenated SWNTs

1992; Hamada et al. 1992; Saito et al. 1992), this increase of absorption is caused by high-frequency conductivity of the free carriers in metallic nanotubes. Relative intensities of the spectra of Fig. 11.7 we have found as a result of the diffuse reflection measurements of powders at low wavenumbers. The discussed above Drude approximations of the low-energy part of the absorption spectra are shown by dashed curves in Fig. 11.7. Comparison of the spectra 1 and 2 shows that hydrogenation decreases high-frequency conductivity of the SWNTs by one order of magnitude.

Optical density spectra of thin films of the initial SWNTs also demonstrate distinct absorption peaks at 0.62, 1.13 and 1.65 eV (spectrum 1 of Fig. 11.7). These absorption lines result from the van Hove singularities in the density of electron states due to the one-dimensional nature of nanotubes (Lin and Shung 1994).

The first two lines are the transitions in the semiconducting SWNTs, the third line is the transition in the metallic SWNTs (Chiang et al. 2001b). The spectral position E_{11} = 0.62 eV of the first peak permits evaluating of the average diameter d of the semiconducting SWNTs as $d = 2a\delta/E_{11}$ = 1.2 nm in good agreement with the value 1.5 nm derived from the electron-microscopy data.

Peaks E_{11}, E_{22} and E_{33} disappear after hydrogenation (Fig. 11.7, spectrum 2) and a new narrow line arises at 2,845 cm^{-1} (0.353 eV), inset of Fig. 11.7. This line indicates the formation of the strong covalent C–H bonds in the samples as in the diffuse reflection spectra (Figs. 11.5 and 11.6). Study of the thin films has shown a liner dependence of the C–H peak intensity as a function of the hydrogen content. Such proportionality have shown that the most amounts of H atoms were covalently bonded to carbon atoms. The intensity of the line 2,845 cm^{-1} calculated from the diffuse reflection steeply decreases on outgassing and significantly deviates from the linear dependence. This discrepancy can be explained assuming that powder particles in the partly outgassed SWNTs were inhomogeneous: their outer layers were depleted in hydrogen. In fact, diffuse reflection can only feel a very thin surface layer of the particles due to the large value of the free carrier absorption coefficient $K \cong 10^4$ cm^{-1} for SWNTs, while the measured hydrogen content is related to the mean hydrogen content in the material.

The term “covalent” we should use carefully. It is perfect for the Si- or Ge-single crystals. In our case of carbon nanotubes with H we should take into account the transfer of electron between hydrogen and carbon cage, inevitably. So, some iconicity we should take into account. The transfer of electron between hydrogen and carbon cage may explain the disappearance of the E_{11}, E_{22} and E_{33} peaks as a result of hydrogenation of the SWNTs. Transfer of electrons between carbon cage and hydrogen will change the filling of the van Hove singularities (Fig. 11.8). It can explain the observed disappearance of the electronic transitions in the spectra of hydrogenated SWNTs.

The decrease of the free-carrier absorption upon hydrogenation can be associated with localization of free carries near C–H bonds in SWNTs, or increase in the rate of their scattering by these defects. The complete removal of hydrogen by vacuum annealing at 700°C only partly restored the intensity of the

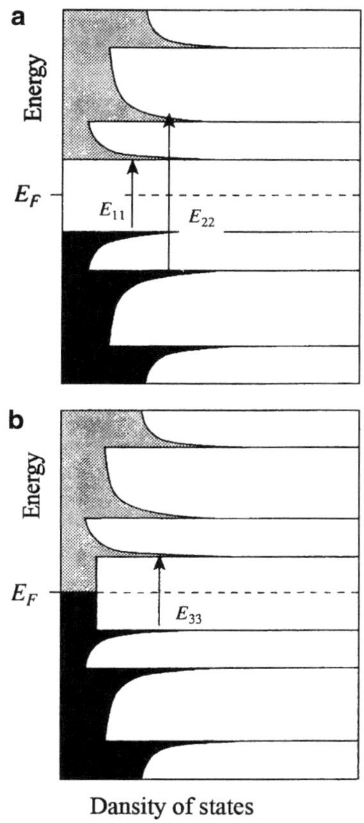

Fig. 11.8 Sketch of the energy structura and IR-active optical transitions E_{ii} in semiconducting (**a**) and metallic (**b**) SWNTs. Painted on the base of the theoretical calculations (Lin and Shung 1994; Chiang et al. 2001b)

electronic transitions and free carrier absorption characteristic of the initial SWNTs. We think, some defects appear in the structure of SWNTs as a result of their hydrogenation at high pressure and removal of hydrogen. We have made a control experiment: SWNTs were exposed to the same power-temperature conditions but without hydrogen. This experiment has shown that electronic transitions vanish in the IR transmission spectrum and free carrier absorption strongly decreases as well. No doubt, mechanical interaction between SWNTs is different of the case of hydrogenation. It follows from the restoration (not complete) of the electron transitions and free-carrier absorption in SWNTs after degassing. Nevertheless, this control experiment permits to suppose the formation of some defects in the structure of SWNTs as a result of high-pressure hydrogenation.

Micro-Raman spectroscopy of the SWNTs hydrogenated at P = 5.0 GPa and T ≅ 500°C have shown a giant structureless hot luminescence background that screens all vibration modes (Meletov et al. 2007). Among them, the most interesting for our study were those with frequencies close to the Raman frequencies of the C–H stretching vibrations of the covalently bonded hydrogen (2,800–3,000 cm^{-1}) and of the H–H stretching vibration of the molecules H_2 (4,130–4,160 cm^{-1}). But, we could not detect them at excitation of the Raman spectra with the help of Ar^+, or Kr^+ lasers. The intensity of the luminescence background is about a 100 times stronger than that for pristine SWNTs. High intensity and frequency dependence of this luminescence are typical for hot luminescence: intensity of the spectrum and frequency dependence have no changed when we have used 488 and 514.5 nm line of an Ar^+ laser, as well as the 647.1 nm line of a Kr^+ laser for excitation. Hot luminescence was attributed to the random binding of the hydrogen atoms on the surface of SWNTs. Hot luminescence vanishes after annealing of the hydrogenated SWNTs in vacuum or in air.

Besides, we have measured Raman spectra of the SWNTs treated under the same high pressure and temperature (HPHT) conditions as those used for the hydrogenation of SWNTs (P = 5.0 GPa, T = 500°C) but without hydrogen atmosphere (spectrum (b) of Fig. 11.9) and spectrum of the hydrogenated SWNTs after their complete degassing at 500°C in vacuum (spectrum (c) of Fig. 11.9). The spectra (b) and (c) were compared with the spectrum (a) of the pristine SWNTs. Note that, the luminescence background in spectrum (b) of the HPHT treated SWNTs is not as high as in the spectra of hydrogenated SWNTs. The main difference between the spectra (a), and (b), (c) is in the broadening of all the Raman peaks in comparison with the spectrum of pristine SWNTs (spectrum (a)), the increase of the D-band intensity, and the appearance of the second-order $D+G^+$ peak at 2,945 cm^{-1}. Raman spectra of pristine SWNTs had shown narrow peaks and a very small intensity of the

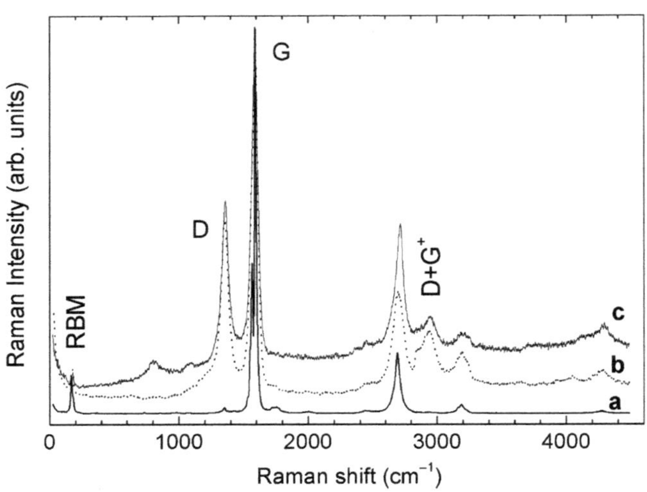

Fig. 11.9 Raman spectra of (**a**) pristine SWNTs, (**b**) SWNTs treated at high pressure and high temperature in the absence of a hydrogen atmosphere, and (**c**) hydrogenated SWNTs after annealing in vacuum at ≅550°C for 1 h

D-band at 1,350 cm^{-1} typical for SWNTs of the high structural order. It is known that the D-band in graphite-related materials is associated with structural disorder while a double-resonance process was employed for their interpretation (Thomsen and Reich 2000). The intense Raman G-band at a higher energy corresponds to the C–C stretching vibrations in tangential and axial directions of the SWNT that splits to G^- (tangential) and G^+ (axial) bands located at 1,567 and 1,592 cm^{-1}, respectively. The shape of the G^--band is sensitive to the electronic properties, strongly related to the diameter and chirality of SWNTs: the Lorentzian lineshape is characteristic of the semiconducting SWNTs, whereas the Breit-Wigner Fano lineshape is typical for metallic SWNTs (Venkateswaran 2004). In our case, the Lorentzian lineshape of the G^--band indicates that the probed SWNTs are semiconducting. The low-frequency Raman band is related to the radial breathing modes (RBM) of the tubes. Its spectral position is using for calculation of the diameter of carbon nanotubes.

Taking into account the results of the IR and Raman scattering measurements, we may conclude that not complete restoration of the IR spectra after outgassing of the hydrogenated SWNTs is a result of appearance of structural disorder in SWNTs due to high pressure/high temperature hydrogenation.

11.3 High-Pressure Hydrogenated C_{60}

As we told above, there were prepared compounds $C_{60}H_x$ and $C_{70}H_x$ with even values of x by chemical reaction of hydrogen with C_{60}. The most complete study of their properties was carried out on $C_{60}H_{36}$ and $C_{60}H_{18}$ and discussed in many reviews (e.g. Taylor and Walton 1993; Hirsch 1994; Goldshleger and Moravskii 1997).

Maximal content of hydrogen was x = 36.

Hear we want to discuss properties of the $C_{60}H_{36}$ prepared by high-pressure technique. The $C_{60}H_{36}$ molecule may exist in a great number of isomeric forms, but only a small number of them are stable (Bühl et al. 1995). The isomeric form with the highest symmetry, T_h, has 12 double bonds, arranged as far apart as possible on the surface of the C_{60} cage, while the form having the double bonds in four isolated aromatic six-membered rings lacking hydrogen atoms and located at the corners of a tetrahedron has a T symmetry structure. Between these two extremes are the isomers with symmetry D_{3d} and S_6, which have two six membered rings at the three-fold axis poles of the molecule, with the other six double bonds isolated in six pentagons. The presence of various isomers in the $C_{60}H_{36}$ specimens depends most likely on the preparation method and on the kinetic parameters controlling hydrogen addition. For example, $C_{60}H_{36}$ prepared by hydrogen transfer from a hydrogen-rich compound like 9,10-dihydroanthracene contains a mixture of the principal isomers D_{3d} and S_6, while $C_{60}H_{36}$, prepared by zinc reduction of C_{60} in aromatic solvents, contains the S_6 isomer as the most abundant (Bini et al. 1998; Bensasson et al. 1997; Darwish et al.1995).

Concerning the solid-state phase of $C_{60}H_{36}$, Hall et al (1993) have suggested the body-centered cubic structure (bcc) with cell parameter 11.785 ± 0.015 Å for the packing of the molecules in the crystalline state. Furthermore, they suppose, at least for the D_{3d} isomer, that at low temperatures the **bcc** crystal structure would be transformed to a body-centered tetragonal one.

The connection of the 36 hydrogen atoms to the C_{60} cage lowers the molecular symmetry and activates Raman scattering from a variety of initially forbidden phonon modes (Bini et al. 1998). In addition, the appearance of the C–H stretching and bending modes and those related to various isomers of $C_{60}H_{36}$, results in a very rich Raman spectrum. The comparison of the phonon frequencies for five principal isomers of $C_{60}H_{36}$, obtained by molecular dynamics calculations, with experimentally observed phonon frequencies has led to the conclusion that the material prepared by the transfer hydrogenation method contains mainly two isomers, those with symmetries D_{3d} and S_6 (Bini et al. 1998).

In review (Meletov and Kourouklis 2005) we have analyzed optical Raman and photoluminescence spectra of the hydro- and deuterofullerene, $C_{60}H_{36}$ and $C_{60}D_{36}$, respectively, prepared by high-pressure hydrogenation. The X-ray analysis, of the obtained material, shows that it has the **bcc** structure, typical for $C_{60}H_{36}$ (Hall et al. 1993), with lattice parameter 11.83 Å. The aim of research was to identify the phonon and electron energy spectra of the high-pressure hydrogenated fullerene, to clarify the isomer composition and homogeneity of samples, as well as to study the isotopic effects in the vibrational spectra. The Raman spectra of the high-pressure hydrogenated samples were compared with those obtained by transfer hydrogenation and with the molecular dynamics calculations data, five low energy isomers were studied (Bini et al. 1998). The Raman data show the presence of all principal isomers in the high-pressure hydrogenated fullerenes, large isotopic shift for the C–H stretching mode, whereas the shift of the modes related to the fullerene molecular cage is negligible. We have also studied the pressure behavior of the Raman and photoluminescence spectra of the $C_{60}H_{36}$ at pressure up to 12 GPa in order to get information about the structural and chemical stability of the material at high pressure. All observed features are reversible with pressure and $C_{60}H_{36}$ is stable in the investigated pressure region.

The comparison of our experimental data with those of Ref. (Bini et al. 1998; Bensasson et al. 1997) shows that the Raman spectrum of the high-pressure hydrogenated $C_{60}H_{36}$ is richer more than five times than that of the transfer hydrogenated $C_{60}H_{36}$. The majority of the experimentally observed Raman peaks (86 peaks from a total number of 126) are very close, with an accuracy of ~5 cm^{-1}, to the calculated frequencies and cross-sections of the Raman active modes (their total number is 400) (Bini et al. 1998). The peaks, which are close to the calculated frequencies, are assigned to all principal isomers, but the majority of them belong to the isomers with the symmetry S_6, T and D_{3d}.

Now we want to discuss IR optical spectra of the $C_{60}H_{36}$ synthesized at high-pressure. Results of this study were published in Bazhenov et al. (2008). There are a lot of publications devoted to theoretical and experimental study of $C_{60}H_{36}$. We should pay attention on the existing discrepancies in the results of theoretical calculations of the dipole-active spectra $C_{60}H_{36}$, compare, for example, papers Bini et al. (1998) and Bulusheva et al. (2001). There were used different theoretical models. Semiempirical method of the MNDO type (Dewar and Thiel 1977) was used in (Bini et al. 1998). Ab initio Hartree–Fock self-consistent field approximation was used in (Bulusheva et al. 2001).

To analyze structure of $C_{60}H_{36}$ produced by our method we elucidated absorption spectra near 2,900 and 1,650 cm^{-1}. It was found that Lorenz curves are the best approximations of the spectra determined by the stretching vibrations of the C–H bonds in $C_{60}H_{36}$. Figure 11.10 shows that the structure of stretching vibrations strongly depends on hydrogen content x. Experimental spectrum of stretching vibrations in $C_{60}H_{48}$ consists of two broad lines 2,914 and 2,841 cm^{-1} and of four lines 2,912, 2,860, 2,830, 2,802 cm^{-1} in $C_{60}H_{36}$. We denote these lines in $C_{60}H_{36}$ as 1, 2, 3 and 4, respectively. The order of their relative intensities is $1 > 2 > 3 > 4$. Infrared and Raman spectra of five low energy isomers of $C_{60}H_{36}$ of the T_h, D_{3d}, D_{3d}(C-K), T and S_6 symmetries were simulated in (Bini et al. 1998) by semiempirical quantum chemical calculations. In accordance with (Bini et al. 1998) the main part of our $C_{60}H_{36}$ is close to the T symmetry. Two discrepancies with experiment we have found: (1) experimental and theoretical intensity orders are $1 > 2 > 3$ and $1 > 2 \gg 3$, respectively; (2) experimental and theoretical intervals of three high energy lines are 82 and 39 cm^{-1} (for T symmetry), respectively. Essentially more close to our experimental frequencies and intensity order were calculated for $C_{60}H_{36}$ of T symmetry using ab initio Hartree–Fock self-consistent field (Bulusheva et al. 2001). Tree high energy lines in our high-pressure hydrogenated $C_{60}H_{36}$: 2,912, 2,860, 2,830 cm^{-1} and their relative intensities are close to the $C_{60}H_{36}$ of the T symmetry 2,912, 2,850, 2,831 cm^{-1} in the chemically synthesized $C_{60}H_{36}$ (Bulusheva et al. 2001).

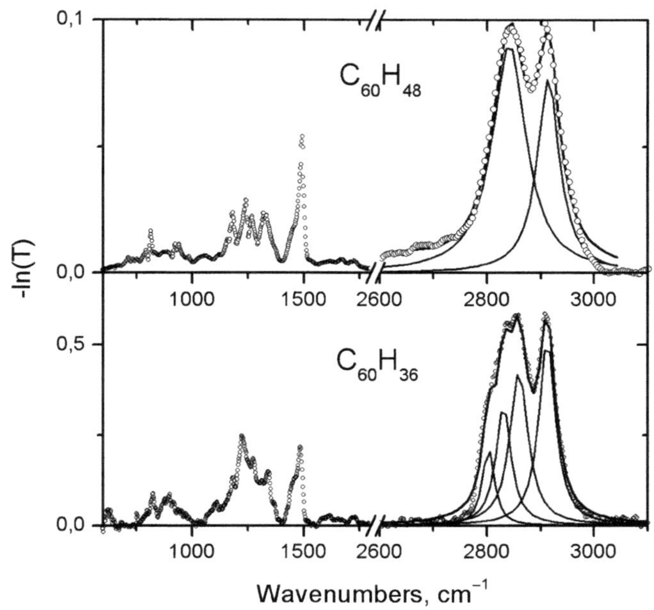

Fig. 11.10 Absorption spectra of $C_{60}H_{36}$, $C_{60}H_{48}$ and their deconvolution: points are experimental spectra; thin solid lines designate the Lorenzian curves and the sum of these curves

In accordance with (Bini et al. 1998) absorption line 4 in $C_{60}H_{36}$ may be explained by admixture of the isomer of the D_{3d} symmetry. Absorption spectrum of $C_{60}H_{36}$ near 1,650 cm^{-1} confirms an admixture of isomers in our case. Theoretical calculations (Bini et al. 1998) have shown absorption lines at 1,634, 1,700 and 1,731 cm^{-1} in isomers of D_{3d}, T_h, D_{3d}(C-K) symmetries, respectively, and absence of absorption in this spectral range in isomers of S_6 and T symmetries. But T-isomer should have absorption line near 1,700 cm^{-1} in accordance with calculations (Bulusheva et al. 2001). We have observed absorption lines at 1,584, 1,611 and 1,723 cm^{-1}. It means that we have not only admixture of the isomer of the D_{3d} symmetry, but isomer of the D_{3d}(C-K) or T_h symmetry as well.

So, we may conclude that we have combination of isomers in our $C_{60}H_{36}$ synthesized at high-pressure and high temperature in accordance with IR and Raman measurements.

Authors of the course of fullerene chemistry (Fullerenes, Ed. "Examen" 2005) are sure that carbon cage of the hydrogenated fullerene with x > 36 should be destroyed as a result of steric strains. Experiment (Talyzin et al. 2006a) has shown that increase of the hydrogenation time at hydrogen pressure 120 bar and temperature T = 673 K results in amorphization of the structure at hydrogen content more than 4.6 wt% (formally $x \geq 36$). The sample with the strongest hydrogenation (5.3 wt%) was completely amorphous. Nevertheless, high-resolution mass spectroscopy of these samples has shown some amount of highly reduced fullerenes up to $C_{60}H_{50}$ and fragments of $C_{57-59}H_x$ with variation of the hydrogen content x from 38 to 48 (Talyzin et al.(2006). The change of the conditions of hydrogenation (20 kbar, T = 773 K, 90 min) has shown that it is possible to produce hydrofullerides $C_{60}H_x$ with $44 \leq x \leq 52$ without noticeable fragmentation or collapse of the carbon cage (Talyzin et al. 2006b). It is important that in the case of high pressure of hydrogenation mass spectroscopy has shown essentially more narrow $C_{60}H_x$ distribution, $44 \leq x \leq 52$ (Talyzin et al. 2006b), in comparison with the hydrogenation at low pressure ($16 \leq x \leq 50$) (Talyzin et al. 2006a). The resultant product is strongly dependent on the experimental method of hydrogenation of C_{60}, and further study of the strongly reduced fullerenes is necessary.

Our method of high-pressure/high temperature is essentially different of the chemical methods of hydrogenation of C_{60}. We have synthesized $C_{60}H_x$ with $36 \leq x \leq 60$ at hydrogen pressure 50 ± 5 kbar, temperatures up to 773 K and time of hydrogenation ≅10 h. Hydrogenated samples were of white color. The value of hydrogen content x was dependent on the experimental conditions: time, temperature, pressure. Hydrogen content in products was estimated by combustion of fulleranes in the oxygen flow and weighting of the resultant products, CO_2 and H_2O. Average hydrogen content was measured in our case with accuracy $\Delta x = \pm(0.5 \div 1)$. We do not know the exact formula for every C_{60} molecule in the product of hydrogenation: the case of one experiment we have got x = 57.5, second experiment has shown x = 60.1. We think that we have $C_{60}H_{60}$ molecules with some admixture of impurity.

IR transmission spectra (T) of the thin polycrystalline specimens or powders were measured using IR microscope of the Fourier-spectrometer at room temperature in the spectral range of 600 ÷ 5,000 cm^{-1}. Optical absorption spectra were calculated as –ln(T). Figure 11.11 demonstrates normalized absorption spectra of fulleranes $C_{60}H_x$ with x = 36, 42, 48 and 60 together with the well known spectrum of fullerit

C_{60} (bottom picture). Stretching and banding vibrations of the C–H bonds appear in the spectra of the hydrogenated C_{60} near 2,900 and 1,450 cm^{-1}, respectively. Also, the well-known increase of a number of dipole-active modes determined by vibrations of the carbon cage was observed when we got over from highly-symmetrical C_{60} to fulleranes $C_{60}H_x$ with x = 36, 42, 48. This is a result of the lowered molecular symmetry and existence of isomers. For example, fullerit C_{60} has 4 dipole-active modes; $C_{60}H_{36}$ of the T symmetry should have 42 dipole-active modes in accordance with theoretical calculation (Bini et al. 1998). A number of the dipole-active vibration modes decreased again in $C_{60}H_{60}$. This indicates their higher symmetry in comparison with $C_{60}H_x$, x = 36 ÷ 48. Our IR spectrum of $C_{60}H_{60}$ is similar to that one measured in highly reduced fullerit with x = 44 ÷ 52 (Talyzin et al. 2006b). Taking into account results of paper (Talyzin et al. 2006b). it is possible to conclude that carbon cage of C_{60} is not destroyed in our $C_{60}H_{60}$.

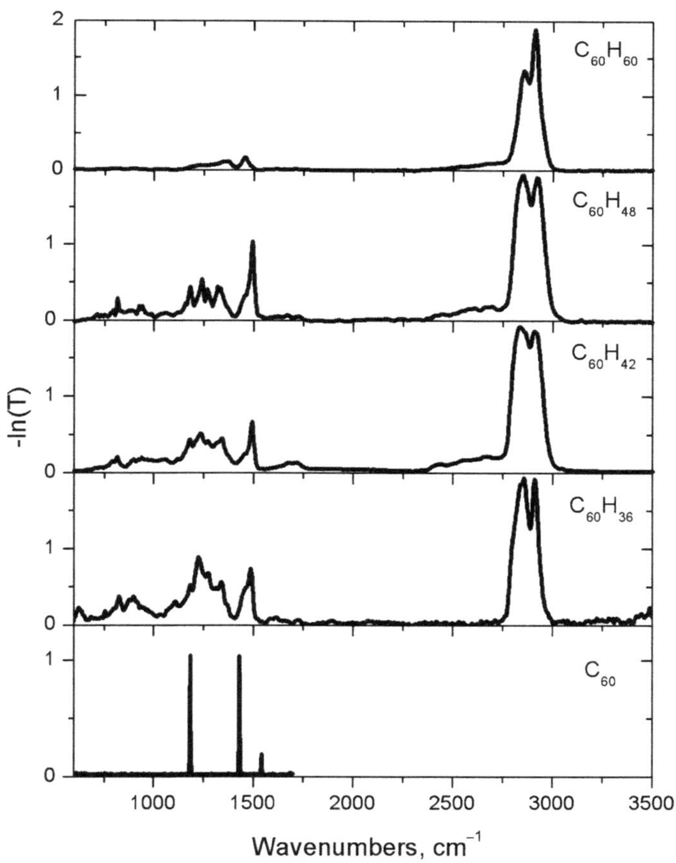

Fig. 11.11 Normalized absorption spectra of fullerit C_{60} and fulleranes $C_{60}H_x$ with x = 36, 42, 48 and 60, T = 300 K

It is interesting that a combination of the vibrational modes of $C_{60}H_{60}$ and, for example, $C_{60}H_{48}$ may explain the emission and absorption spectra of interstellar and circumstellar clouds. Millimeter-wave measurements have led to the detection of over 120 inorganic, organic molecules, radicals, cyclic molecules, and cyanopoynes in the interstellar medium (Kwok 2004). In addition to molecules, solid-state particles were found in the interstellar and circumstellar medium. Solid particles have a high opacity due to strong absorption and scattering of the visible light. These clouds of dust may completely obscure the stars. Clouds may be studied by infrared absorption spectroscopy as well. Infrared emission spectrum of the carbon-rich source NGC 7027 (Russel et al. 1977) is shown in the upper part of Fig. 11.12.

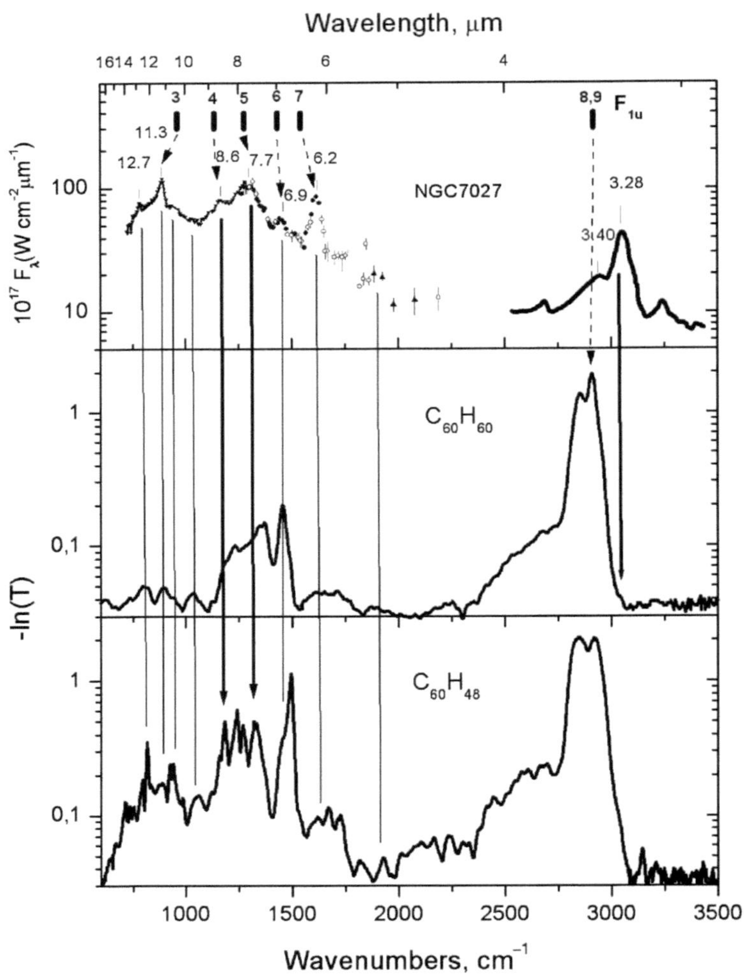

Fig. 11.12 Emission spectrum of NGC 7027 with the wavelengths of the unidentified features marked and normalized absorption spectra of $C_{60}H_{60}$ and $C_{60}H_{48}$

This emission is a result of illumination of the interstellar and circumstellar clouds excited by ultraviolet radiation of stars. A lot of experiments were performed in different laboratories to fit the C–H starching features observed towards the Galactic center: complex molecular mixture formed by irradiation of ices, hydrogenated amorphous carbon, quenched carbonaceous composite produced by quenching the plasma of methane gas, organic materials, such as microorganisms (see references in (Pendelton et al. 1994)). None of widely accepted materials were satisfactory for explanation of the emission spectrum of NGC 7027 (Webster 1991). Adrian Webster (Webster 1991, 1993) proposed that this spectrum is the emission of $C_{60}H_{60}$ and $C_{60}H_x$ with small x. He calculated vibrational spectrum of $C_{60}H_{60}$, using a force-field model. He supposed that carbon atoms in $C_{60}H_{60}$ are located at the vertices of a truncated icosahedron as in usual C_{60}, and the bound hydrogen atoms positioned randomly outward from them (I_h symmetry, the icosahedral grope with inversion). Of the possible 354 vibrational modes in $C_{60}H_{60}$, only 27 belonging to the irreducible representation F_{1u} are IR-active. This representation is triply degenerate, so there are nine IR active frequencies. The result of theoretical calculation is shown by short vertical lines in the upper part of Fig. 11.12. Tentative assignment of calculated frequencies of $C_{60}H_{60}$ to unidentified infrared features is shown by arrows in this figure. Six of the infrared active frequencies match unidentified lines to within 4%, and a seventh differ from an observed line by 8%. Line 787 cm^{-1} (12.7 μm) was not identified. Experimental spectrum of $C_{60}H_{60}$ much better describes unidentified lines. Moreover, not identified in theory (Webster 1991, 1993) line 787 cm^{-1} (12.7 μm) is clearly seen in experimental spectrum.

Two features of the unidentified emission cannot be explained with the help of theoretical and experimental spectra of $C_{60}H_{60}$: broad backgrounds in the spectral range 700 ÷ 1,700 cm^{-1}, and emission line at 3049 cm^{-1} (3.28 μm). It was supposed in (Webster 1991, 1993) that background is an emission of lightly hydrogenated fulleranes $C_{60}H_x$ as a result of decreased molecular symmetry and isomerism. As an example we show spectrum of $C_{60}H_{48}$ in Fig. 11.12. Strong background in the spectral range 700 ÷ 1,700 cm^{-1} takes place in this spectrum. As we have shown above, our fulleranes with hydrogen content x = 36 are a mixture of isomers. We are sure that the same will take place at x = 48. Figure 11.10 shows an increase of background in the spectrum of $C_{60}H_{48}$ in comparison with the spectrum of $C_{60}H_{36}$.

Cataldo F. (2003) the first have shown that the absorption spectrum of $C_{60}H_{36}$ synthesized by chemical method is able to match several IR emission lines detected from interstellar carbon dust. Background in the spectral range 700 ÷ 1,700 cm^{-1} was not detected in this paper.

Stretching vibrations of C–H bonds in experimental absorption spectrum near 2,920 cm^{-1} correspond to the sp^3 bonding (Stoldt et al. 2001) (Fig. 11.3). They are observed in emission spectrum of NGC 7027 as a broad band overlapping with line 3,049 cm^{-1} (3.28 μm). This line was never observed in absorption spectra of interstellar and circumstellar clouds (Russel et al. 1977). Frequency of this line corresponds to the sp^2 bonding (Stoldt et al. 2001). We should note that very weak line 3,050 cm^{-1} was observed in the absorption spectra of synthesized $C_{60}H_{42}$ and $C_{60}H_{60}$ (Fig. 11.13). Taking into account strong steric strains in highly reduced fulleranes, we suppose that a high concentration of short time leaving sp[2] bonds appear in $C_{60}H_x$ with $x \geq 42$ at ultraviolet excitation by stars. The absence of this line in the absorption spectra of interstellar and circumstellar clouds may be explained by emission of only thin boundary layer of cloud as a result of absorption of the ultraviolet light of stars by dust of cloud and shot lifetime of the sp[2] states.

The absorption spectrum of stretching vibrations of the C–H bonds consists of three lines 2,857, 2,910 and 2,942 cm^{-1} (Fig. 11.13). Analogous structure is seen in the absorption spectrum of the cloud IRS6E (Fig. 11.13).

Fig. 11.12 shows that the emission intensity of the low energy states is stronger than that of the high energy lines. The absorption spectra of Fig. 11.12 have reverse relation. Absorption is proportional to the density of the vibrational states. Emission intensity is proportional to the Boltzmann population of the vibrational states at ultraviolet excitation. This may be a reason for high emission intensity of the low energy states in clouds.

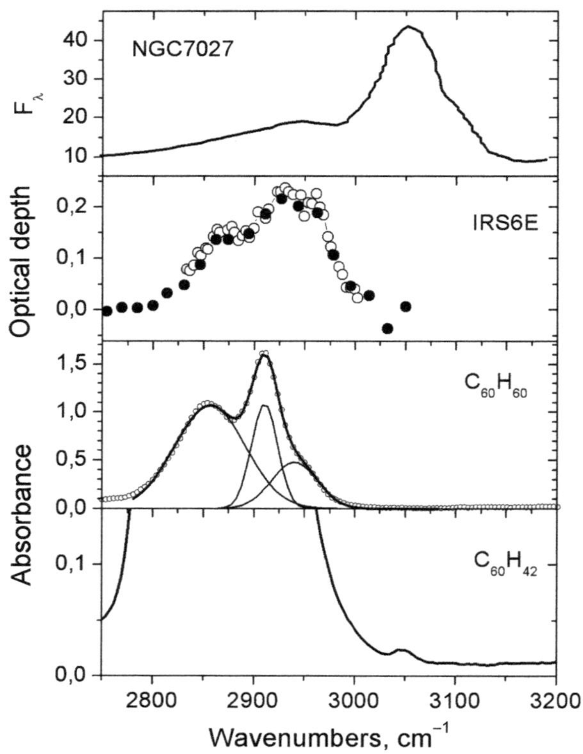

Fig. 11.13 Emission spectrum of NGC 7027, absorption spectrum of the interstellar cloud IRS6E and absorption spectra of synthesized $C_{60}H_{60}$ and $C_{60}H_{42}$ in the range of the stretching vibrations of the C–H bonds

Acknowledgments This work was supported by the program of the Russian Academy of Sciences "Quantum Properties of the Condense Matter", by the Russian Foundation for Basic Research of the Russian Ministry of Science.

References

Assink RA, Schirber JE, Loy DA, Morosin B, Carlson GA (1992) Mater J Res 7:2136
Bashkin IO, Antonov VE, Bazhenov AV, Bdikin IK, Borisenko DN, Krinichnaya EP, Moravsky AP, Harkunov AI, Shul'ga Yu M, Ossipyan Yu A, Ponyatovsky EG (2004) JETP Lett 79(5):226
Bazhenov AV, Kveder VV, Ossipyan YuA, Nikolaev RK, Fursova TN, Shalynin AI (1998) JETP 86(5):1030
Bazhenov AV, Fursova TN, Antonov VE, Bashkin IO, Harkunov AI, Ponyatovsky EG (2004) Fullerenes Nanot Carbon Nanostruct 12(1):293
Bazhenov AV, Fursova TN, Bashkin IO, Antonov VE, Kondrat'eva IV, Krestinin AV, Shulga YuM (2006) Fullerenes nanotubes and carbon nanostructures 14(2&3):165
Bazhenov AV, Fursova TN, Bashkin IO, Moravskii AP, Shulga YuM (September 2008) Fullerenes Nanot Carbon Nanostruct 16(5&6):579
Bensasson RV, Hill TJ, Land EJ, Leach S, McGarvey DJ, Truscott TG, Ebenhoch J, Gerst M, Rüchardt C (1997) Chem Phys 215:111
Bini R, Ebenhoch J, Fanti M, Fowler PW, Leach S, Orlandi G, Ruchardt C, Sandall JPB, Zerbetto F (1998) Chem Phys 232:75
Bühl M, Thiel W, Schneider U (1995) J Am Soc 117:4623
Bulusheva LD, Okotrub AV, Antich AV, Lobach ASJ (2001) Mol Struct 562:119
Cataldo F (2003) Fullerenes Nanot Carbon Nanostruct 11(4):295
Chiang IW, Brinson BE, Huang AY, Willis PA, Bronikowski MJ, Margrave JL (2001a) J Phys Chem B 105:8297
Chiang IW, Brinson BE, Smalley RE, Margrave JL, Hauge RH (2001b) J Phys Chem B 105:1157
Darwish AD, Abdul-Sada AK, Langley GJ, Kroto HW, Taylor R, Walton DR (1995) J Chem Soc Perkin Trans 2:2359
Dewar MJS, Thiel WJ (1977) Am Chem Soc 99:4899
Dillon AC, Heben MJ (2001) Appl Phys A 72:133
Fuller MP, Griffiths PR (1978) Anal Chem 50:1906
Fullerenes, Ed. "Examen" (2005) Moscow, pp 245, 263, 285
Goldshleger NF, Moravskii AP (1997) Russ Chem Rev 66:323
Hall LE, McKenzie DR, Attalla MI, Vassallo AM, Davis RL, Dunlop JB, Cockayne DJH (1993) J Phys Chem 97:5741
Hamada N, Savada SI, Oshiyama A (1992) Phys Rev Lett 68:1579
Hirsch A (1994) Chemistry of fullerenes, Ch. 5. George Thieme Verlag, Stuttgart, p 117
Jin C, Hettich R, Compton R, Joyce D, Blencoe J, Burch TJ (1994) Phys Chem 98:4215
Krestinin AV, Raevskii AV, Kiselev NA, Zvereva GI, Zhigalina OM, Kolesova OI (2003a) Chem Phys Lett 381:529
Krestinin AV, Kiselev NA, Raevskii AV, Ryabenko AG, Zakharov DN, Zvereva GI (2003b) Eurasian Chem Tech J 5:7
Kwok S (2004) Nature 430:985
Lin MF, Shung KW-K (1994) Phys Rev B 50:17744
Loutfy RO, Lowe TP, Hutchison JL, Kiselev NA, Zakharov DN, Krinichnaya EP, Muradyan VE, Tarasov BP, Moravsky AP (1999) Abstracts of IV Workshop on Fullerenes and Atomic Clusters (IWFAC'99), St. Petersburg, 1999, pp. 117

Meletov KP, Kourouklis GA (2005) JETP 127:860
Meletov KP, Maksimov AA, Tartakovskii II, Bashkin IO, Shestakov VV, Krestinin AV, Shulga YuM, Andrikopoulos KS, Arvanitidis J, Christofilos D, Kourouklis GA (2007) Chem Phys Lett 433:335
Mintwire JM, Dunlap BI, White CT (1992) Phys Rev Lett 68:631
Pendelton YJ, Sandford SA, Allamandola LJ, Tielens AGGM, Seldgren LJ (1994) Astroph J 437:683
Pradhan BK, Sumanasekera GU, Adu CKW, Romero HE, Williams KA, Eklund PC (2002) Physica B (Amsterdam) 323:115
Russel RW, Soifer BT, Merrill KM (1977) Astroph J 213:66
Saito R, Fujita F, Dresselhaus G, Dresselhaus MS (1992) Appl Phys Lett 60:2204
Sharma SM, Karmakar S, Sikka SK, Teredesai PV, Sood AK, Govindaraj A, Rao CNR (2001) Phys Rev B 63:205417
Shigematsu K, Abe K, Mitani M, Tanaka K (1993) Chem Express 8:37
Shulga Yu M, Bashkin IO, Krestinin AV, Martynenko VM, Zvereva GI, Kondratieva IV, Ossipyan Yu A, Ponyatovsky EG (2004) JETP Lett 80:752
Stoldt CR, Maboudian R, Carraro C (2001) Astroph J 548:L225
Talyzin AV, Tsybin YuO, Purcell JM, Schaub TM, Shulga YuM, Noreus D, Sato T, Dzwilewski A, Sundqvist B, Marshall AG (2006a) J Phys Chem A 110(27):8528
Talyzin AV, Dzwilewski A, Sundqvist B, Tsybin YuO, Purcell JM, Marshall AG, Shulga YuM, McCammon C, Dubrovinsky L (2006b) Chem Phys 325(2):445
Taylor R, Walton DRM (1993) Nature 363:685
Thomsen C, Reich S (2000) Phys Rev Lett 85:5214
Venkateswaran UD (2004) Phys Stat Sol b 241:3345
Webster A (1991) Nature 352:412
Webster A (1993) Mon Not R Astron Soc 264:121
Willey RR (1976) Appl Spectrosc 30:593

Chapter 12
Topological Modeling of $C_{60}H_{36}$ Hydrides

Ottorino Ori[1], **Franco Cataldo**[1,2], **Susana Iglesias-Groth**[3], **and Ante Graovac**[4,5,6]

Abstract In recent literature, different $C_{60}H_{36}$ isomers have been proposed to interpret NMR experimental findings; this chapter ranks those fulleranes in terms of relative molecular stability using topological invariants computed on their chemical graphs. Our topological modeling exploits in fact Wiener index contributions from individual molecular sites and measures total graph topological efficiency. Moreover, it gives correct numbers of NMR resonance peaks and relative intensities. Comparisons with previous ab-initio predictions are presented together with some interesting original $C_{60}H_{18}$ and $C_{60}H_{36}$ isomers. Th-symmetric molecule appears a valid candidate for $C_{60}H_{36}$ fullerane.

12.1 Introduction

Current interests on $C_{60}H_{36}$ are boosted by possible practical applications in designing effective hydrogen storage molecular system (Withers et al. 1997; Peera et al. 2004; Schur et al. 2008) and for the astrochemical implications (Iglesias-Groth 2004, 2006; Cataldo 2003; Stoldt et al. 2001). In a relevant experimental paper Gakh et al. (2003) pointed out the existence of at least three $C_{60}H_{36}$ fullerane

[1]Actinium Chemical Research, Via Casilina 1626/A, 00133, Rome, Italy
e-mail: ottorino.ori@alice.it

[2]INAF, Osservatorio Astrofisico di Catania, Via S. Sofia 78, 95123, Catania, Italy
e-mail: franco.cataldo@fastwebnet.it

[3]Instituto de Astrofisica de Canarias, Via Lactea s/n, E-38200, La Laguna, Tenerife, 2 CSIC, Spain
e-mail: sigroth@iac.es

[4]IMC, University of Dubrovnik, Branitelja Dubrovnika 29, HR-20000, Dubrovnik, Croatia
e-mail: ante.graovac@irb.it

[5]The "R. Bošković" Institute, 180HR-10002, Zagreb, Croatia

[6]Faculty of Science, University of Split, Nikole Tesle 12, HR-21000, Split, Croatia

F. Cataldo and S. Iglesias-Groth (eds.), *Fulleranes: The Hydrogenated Fullerenes*, Carbon Materials: Chemistry and Physics 2, DOI 10.1007/978-1-4020-9887-1_12,

isomers in the mixture obtained by high temperature hydrogenation reactions of the physically stable "buckyball" C_{60} fullerene. NMR resonance spectra of these molecules allowed furthermore the authors to propose three different chemical structures with symmetries C_1, C_3, T.

Experimental abundances favor $C_1 > C_3 >> T$ stability trend and this fact somehow opposes theoretical ab-initio predicted $T \approx C_3 > C_1$ sequel (Nossal et al. 2001). Gakh and coworkers (2003) confirmed in their article theoretical relevance of the tetrahedral T-symmetric fullerene hydride already proposed by Bini and collaborators (1998) in their extended paper on Raman vibrational spectroscopy on $C_{60}H_{36}$ samples. The work of Bini et al. (1998) also establishes that, depending from the chemical synthesis adopted, it is possible to produce different isomers of $C_{60}H_{36}$. In particular, the hydrogen transfer hydrogenation yields isomers with D_{3d} and S_6 symmetry, the high-pressure hydrogenation favors almost exclusively the D_{3d} isomer. Taylor and coworkers (2006) were the first to synthesize $C_{60}H_{36}$ with nascent hydrogen proposing for it the T symmetry for the almost unique isomer formed with that synthesis. In fact, the T isomer was predicted to be by far the isomer with the lowest energy even than the more symmetric T_h by the fact that T isomer has four benzenoid rings in its structure (Fig. 12.1). No other isomer presents these characteristics (Book and Scuseria, 1994; Palit et al. 1998). The T_h symmetry (Fig. 12.2) was proposed only for the $C_{60}H_{36}$ isomer prepared by the Birch reaction with involves a conjugated and quite unusual addition of hydrogen (Palit et al. 1998). However, a recent revisitation of the C_{60} fullerene hydrogenation, with nascent hydrogen, and deuteration with nascent deuterium, shows that the most probable $C_{60}H_{36}$ isomer of appears to be the one with T_h symmetry, i.e. the most symmetrical one. This statement is based on the analysis of its electronic absorption spectrum in n-hexane which presents a maximum of absorption typical of molecules with isolated double bonds, excluding presence of benzenoid rings (Cataldo et al. 2009).

Figure 12.1 shows Schlegel diagram for tetrahedral isomer $C_{60}H_{36}$(T); for sake of simplicity, only 60 circles are represented, the black ones corresponding in fact to C–H bonds. More recently, this T isomer re-appears as the most stable of nine $C_{60}H_{36}$ structures listed by Karpushenkava and coworkers (2007) using density functional theory (DFT) to derive best candidates for $C_{60}H_{36}$ hydrides. These authors, on the basis of experimental ideal-gas enthalpy of formation for the hydrogen-transfer reaction, consider the resulting $C_{60}H_{36}$ as a mixture of two isomers, with symmetry T and C_1 respectively, also if this last molecule appears relatively less stable than other structures. It is worth to notice here that, based on their results, Bini and collaborators (1998) tend instead to exclude any physical relevance of the $C_{60}H_{36}$(T) isomer by introducing two different isomers, both having D_{3d} symmetry, to explain infrared and Raman spectroscopy data.

This brief panoramic introduction shows current level of interest in understanding the actual molecular structure of $C_{60}H_{36}$ fullerenic hydrides and this paper contributes to this search by determining the topological stability scale for several $C_{60}H_{36}$ isomers adding new chemical considerations. Topological methods apply to the study of different fulleranes configurations. The first move consists in producing molecular

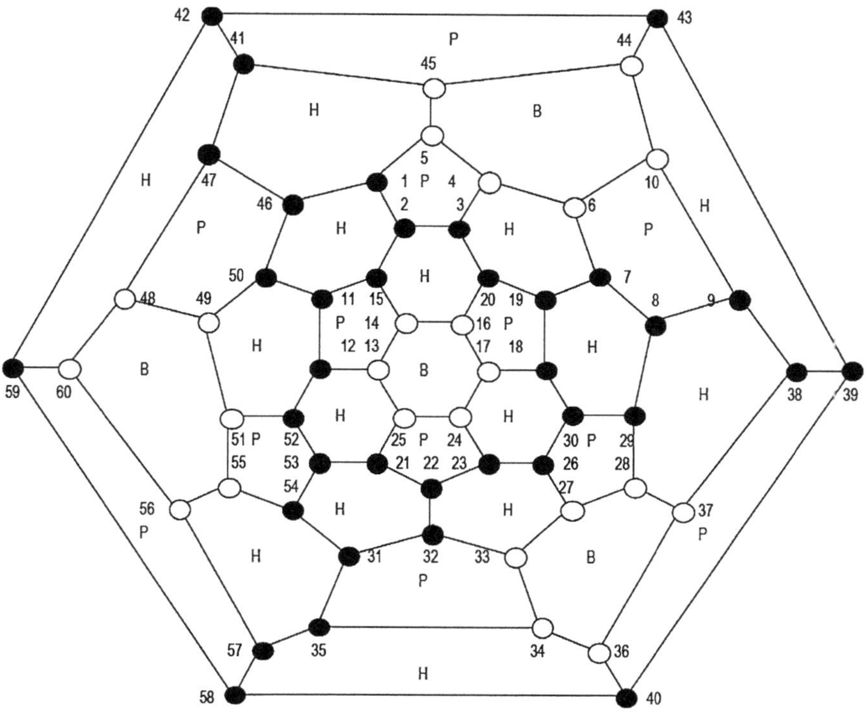

Fig. 12.1 $C_{60}H_{36}$(T) fullerane is represented as a Schlegel diagram; black circles refer to the 36 Carbon–Hydrogen bonds. Site labeling is arbitrary; graph has four benzenoid six-member rings B, 16 6-member rings H (one of these coincides with the graph external perimeter), 12 5-member rings P. This molecule includes C_{3V} $C_{60}H_{18}(\alpha)$ carbon–hydrogen bonds, Fig. 12.4. Hydrogen atoms are added par-wise to double bonds of underlying C60 fullerene

connectivity lists, by looking to a given $C_{60}H_{36}$ molecule as a 96-vertices *molecular graph,* having nodes with connectivity 4 (describing 36 carbon atoms 4-connected), 3 (24 carbon atoms 3-connected) or 1 (36 hydrogen atoms).

Figures 12.1 and 12.2 provide good examples of Schlegel diagrams for Fulleranes. Based on these graphs we will compute a set of graph invariant descriptors deriving physical–chemical information, see next sections. Schlegel diagrams with filled circles represent practical shorthand for $C_{60}H_{36}$ structures, being the 36 edges C–H simply represented by 36 black circles. These planar graphs are topologically fully represented by their *connectivity lists* made by 96 nodes originating from the 60 carbon atoms of $C_{60}(I_h)$ Buckminsterfullerene (including 12 pentagonal and 20 hexagonal rings) plus the 36 hydrogen atoms bounded to 36 distinct carbon sites. Connectivity lists of $C_{60}H_{36}$ isomers include 36 1-connected nodes (representing hydrogens) whereas 60 carbons atoms are grouped in two sets of 24 3-connected and 36 4-connected vertices respectively; $C_{60}H_{36}$ has total number of 126 graph edges (chemical bonds), 90 of them come from $C_{60}(I_h)$

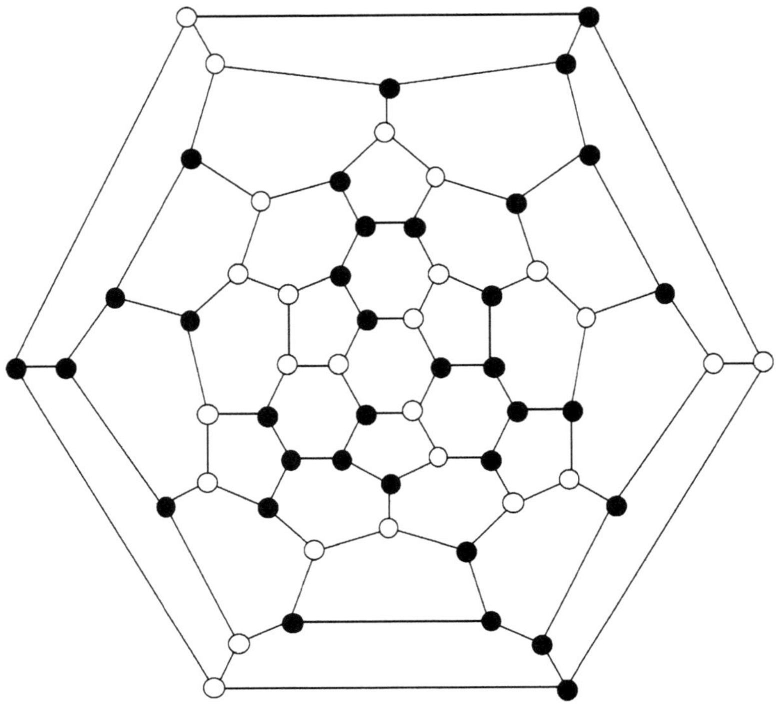

Fig. 12.2 $C_{60}H_{36}$(Th) fullerane Schlegel diagram; black circles refer to the 36 Carbon–Hydrogen bonds; this graph does not have benzenoid rings

original cage. Tables 12.1 and 12.2 give connectivity list for T and T_h symmetric $C_{60}H_{36}$ molecules. Vertices V_i are labeled as in Fig. 12.1 and sorted by connectivity values c_i (4 or 3 or 1); connected vertices V_j are shown. The 36 nodes with connectivity 4 correspond to carbon atoms bounded to one hydrogen atom. Site labeled 61 is for example the hydrogen atom bounded to carbon site number 1, see the corresponding black circle in Fig. 12.1. The 60 carbon atoms are usefully *numbered by quintets*, e.g. the first five individuate the first pentagon, vertices from 6 to 10 the second one and so on, until twelve pentagonal rings are populated, see Fig. 12.1.

From connectivity lists, we perform detailed calculations of Wiener index W(N), the graph-invariant positive integer number defined as the semi-sum of shortest distances d_{ij} between vertices V_i and V_j. The founding approximation of our topological modeling method consists in the usage of W(N) as molecular *topological long-range potential energy, able to measure molecular compactness among similar structures*. As a *topological potential*, W obeys to a minimum principle, i.e. stable structures evolve toward low values of W. This paper applies this W-minimum approach (that corresponds to search for the maximum structural topological compactness) to rank topological stability of $C_{60}H_{36}$ derivatives.

Table 12.1 $C_{60}H_{36}(T)$ fullerane connectivity list; its graph in Fig. 12.1 has a total of 96 vertices V_i with connectivity c_i equal to 4 in case of Carbon–Hydrogen bond (for example hydrogen atom vertex V_{61} bounds carbon atoms vertex V_1)

V_i	c_i	Connected vertices				V_i	c_i	Connected vertices			
1	4	2	5	46	61	31	4	32	35	54	79
2	4	1	3	15	62	32	4	22	31	33	80
3	4	2	4	20	63	33	3	27	32	34	–
4	3	3	5	6	–	34	3	33	35	36	–
5	3	1	4	45	–	35	4	31	34	57	81
6	3	4	7	10	–	36	3	34	37	40	–
7	4	6	8	19	64	37	3	28	36	38	–
8	4	7	9	29	65	38	4	9	37	39	82
9	4	8	10	38	66	39	4	38	40	43	83
10	3	6	9	44	–	40	4	36	39	58	84
11	4	12	15	50	67	41	4	42	45	47	85
12	4	11	13	52	68	42	4	41	43	59	86
13	3	12	14	25	–	43	4	39	42	44	87
14	3	13	15	16	–	44	3	10	43	45	–
15	4	2	11	14	69	45	3	5	41	44	–
16	3	14	17	20	–	46	4	1	47	50	88
17	3	16	18	24	–	47	4	41	46	48	89
18	4	17	19	30	70	48	3	47	49	60	–
19	4	18	20	7	71	49	3	48	50	51	–
20	4	3	16	19	72	50	4	11	46	49	90
21	4	22	25	53	73	51	3	49	52	55	–
22	4	21	23	32	74	52	4	12	51	53	91
23	4	22	24	26	75	53	4	21	52	54	92
24	3	17	23	25	–	54	4	31	53	55	93
25	3	13	21	24	–	55	3	51	54	56	–
26	4	23	27	30	76	56	3	55	57	60	–
27	3	26	28	33	–	57	4	35	56	58	94
28	3	27	29	37	–	58	4	40	57	59	95
29	4	8	28	30	77	59	4	42	58	60	96
30	4	18	26	29	78	60	3	48	56	59	–

We are collecting more and more indications about the validity of this approximated method especially on fullerenes; we saw in fact that fullerenes tend to minimize W(N) index minimization (Ori and D'Mello, 1992); for example C_{60} fullerene has 1812 non-isomorphic isomers and physically stable icosahedral "buckyball" $C_{60}(I_h)$ with isolated pentagons minimizes W value to $W_{min} = 8{,}340$ for $C_{60}(I_h)$. This minimum W property is valid for other fullerenes as C_{70} where D_{5h} molecule has $W_{min} = 12{,}375$. This topological modeling will surely incur in many limitations but has the indubitable power of quickly generating good candidates, guiding isomers selection among thousands of possibilities; this fast enumeration is not practicable by other more refined techniques. A recent publication on C_{28} isomers confirms the predictive power of Wiener index for to determine topologically independent sites of the Fullerene cages, predicting also a certain grade of carbon atoms chemical reactivity (Ori et al. 2009).

Another characteristic property of this topological descriptor appears somehow more intriguing: in fact, if ones decomposes each single contribution to W(N) coming

Table 12.2 $C_{60}H_{36}$(Th) Fullerene derivative; its graph in Fig. 12.1 has a total of 96 vertices V_i with connectivity c_i equal to 4 in case of Carbon–Hydrogen bond (for example, hydrogen atom vertex V_{61} bounds carbon atoms vertex V_1)

V_i	c_i	Connected vertices				V_i	c_i	Connected vertices			
1	4	2	5	46	61	31	3	32	35	54	–
2	4	1	3	15	62	32	3	22	31	33	–
3	4	2	4	20	63	33	4	27	32	34	79
4	3	3	5	6	–	34	4	33	35	36	80
5	3	1	4	45	–	35	4	31	34	57	81
6	4	4	7	10	64	36	4	34	37	40	82
7	3	6	8	19	–	37	4	28	36	38	83
8	3	7	9	29	–	38	3	9	37	39	–
9	4	8	10	38	65	39	3	38	40	43	–
10	4	6	9	44	66	40	4	36	39	58	84
11	4	12	15	50	67	41	3	42	45	47	–
12	3	11	13	52	–	42	3	41	43	59	–
13	3	12	14	25	–	43	4	39	42	44	85
14	4	13	15	16	68	44	4	10	43	45	86
15	4	2	11	14	69	45	4	5	41	44	87
16	3	14	17	20	–	46	3	1	47	50	–
17	4	16	18	24	70	47	4	41	46	48	88
18	4	17	19	30	71	48	4	47	49	60	89
19	4	18	20	7	72	49	4	48	50	51	90
20	3	3	16	19	–	50	3	11	46	49	–
21	4	22	25	53	73	51	3	49	52	55	–
22	4	21	23	32	74	52	4	12	51	53	91
23	3	22	24	26	–	53	4	21	52	54	92
24	3	17	23	25	–	54	4	31	53	55	93
25	4	13	21	24	75	55	3	51	54	56	–
26	4	23	27	30	76	56	4	55	57	60	94
27	3	26	28	33	–	57	3	35	56	58	–
28	3	27	29	37	–	58	3	40	57	59	–
29	4	8	28	30	77	59	4	42	58	60	95
30	4	18	26	29	78	60	4	48	56	59	96

from single atoms immediately gets indications about non-equivalent molecular sites that is exactly what NMR experiments determine. Next paragraph gives computational details of our topological method, whereas the remaining sections will be devoted to present topological modeling results compared with ab-initio theoretical calculations. Resonance lines and relative intensities will be also derived for different $C_{60}H_{36}$ isomers, giving evidences about the chemical relevance of $C_{60}H_{36}$(Th) molecule.

12.2 Topological Model

In our topological approach, the *most-compact structures are seen as best candidates to be the most chemically stable ones*. This statement implies a topological minimum principle on W: *chemically stable molecular structures tend to minimize their W index* according to the mechanism shown in Fig. 12.3. W behaves in practice as a physical potential in the topological space. This tendency to reach configurations with low values of W is heuristically paralleled by the behaviour of a set of harmonic oscillators that tends to minimize the sum of the squares of all distances. For a molecular graph G with N vertices Wiener index is the semi sum of minimum distances d_{ij} between all couples of vertices V_i and V_j and finds a computationally useful expression as:

$$W(N)= \tfrac{1}{2}\, \Sigma_i\, w_i \qquad i = 1,N \tag{12.1}$$

Where the contribution w_i coming from vertex v_i:

$$w_i = \Sigma_m\, m\, b_{im} \qquad m = 1,M \tag{12.2}$$

Wiener coefficients b_{im} are the number of vertices V_j representing the m-neighbors of graph site V_i (then they are at a distance d_{ij} = m from V_i) and M is the maximum distance present in the graph (M = max$\{d_{ij}\}$). We call vertex V_i the *center* of w_i and w_i itself the Wiener-weight (WW) of that vertex. On a pure Fullerene b_{i1} = 3 for each center; on a $C_{60}H_{36}$ molecule b_{i1} = 3, 4, 1 if we take in consideration non-hydrogenated carbons, hydrogenated carbons, hydrogen atoms respectively.

The resulting ordered strings of integers $\{b_{im}\}$:

$$\{b_{im}\}= \{b_{i1}\ b_{i2}\ \ldots\ldots b_{im}\ldots\ldots\ b_{iM-1}\ b_{iM}\} \qquad m = 1,M \tag{12.3}$$

of vertex V_i give a fast sorting of molecular independent sites, since one may easily recognize that equivalent atoms have the same string. The opposite statement may be not true as in larger Fullerenes where and accidental degeneracy may arise and is eliminated by introducing graph-coloring method (Ori and D'Mello, 1992, 1993). $C_{60}(I_h)$ molecular graph is made of equivalent atoms only and all WW (and WS) have the same value $\underline{w}$ = 278, W = 8,340. Previous Eq. 12.1 then becomes:

$$W(N)= \tfrac{1}{2}\, N\, \underline{w} \tag{12.4}$$

For a given molecule *w* indicates minimal WW value:

$$\underline{w} = \text{Min}\ \{w_i\} \tag{12.5}$$

and corresponding center V_i is then called the *minimal vertex of G*; obviously a graph may presents more than one minimal vertices and often they are equivalent sites, e.g. with the same $\{b_{im}\}$ set (12.3).

Topological minimum principle on W(N) in Fig. 12.3 implies *that minimal vertices are good candidates to be the most stable molecular sites*; in turn our model measures the structural overall topological efficiency of a given isomer with a simple non-integer index – *topological efficiency index* – based on *minimal vertex* contribution (12.4) to W(N):

$$\rho = 2W / N\, \underline{w} \qquad \textit{with}\ \rho \geq 1 \tag{12.6}$$

Topological efficiency index measures in practice how a given isomer grows around its minimal sites. As a reference, buckyball $C_{60}(I_h)$ has W = 8340, $\underline{w}$ = 278,

WIENER INDEX MINIMIZATION PRICIPLE:

W MINIMIZATION → MOLECULAR COMPACTNESS SELECTION

W MINIMIZATION → CHEMICAL STABLE ISOMERS

Fig. 12.3 Minimum Wiener-index principle; Wiener Index is able to select compact isomers, chemically stable

with M = 9 and $\{b_{im}\}$ = {3 6 8 10 10 10 8 3 1}. This molecule has L = 90 bonds and it generates a single peak in the ^{13}C-NMR spectrum with multiplicity $l = 60$. Symmetry considerations produce $\rho = 1$.

In this article we use *topological potential W(N) in* ranking relative stability of $C_{60}H_{36}$ isomers, selecting the *most-compact structures*. Our model is also capable, by sorting (12.3) strings, to predicting the shape of their NMR spectra. Moreover we will see that W(N) and ρ allow to individuate, among thousands of $C_{60}H_{36}$ isomers, new molecules with interesting features.

12.3 Topological Modeling of $C_{60}H_{18}$

We start our topological modeling from studying $C_{60}H_{18}$ molecules. Figure 12.4 represents the chemical graph of two distinct $C_{60}H_{18}$ isomers, (α) and (β), with C_{3V}-symmetry; Schlegel diagram presents 60 fullerenic nodes, being 18 Carbon–Hydrogen bonds represented by shading the involved carbon atoms. C_{3V} $C_{60}H_{18}$(α) molecule corresponds to the structure previously introduced in literature (Briggs et al. 2005) and shows a characteristic inner crown (α) of Hydrogen atoms placed around the central hexagons; C_{3V} $C_{60}H_{18}$(β) is originally proposed here as possible alternative and topologically it keeps the original C_{3V} symmetry just moving 12 Hydrogen atoms on the more external crown (β). Both molecules adopt a T-scheme for hydrogen addition (see Fig. 12.6), with 3 pairs of common Hydrogen atoms represented by black circles in Fig. 12.4. By comparing Fig. 12.4 with Figs. 12.1 and 12.7 it appears evident that C_{3V} $C_{60}H_{18}$(α) 18 carbon–hydrogen bonds are included in several $C_{60}H_{36}$ molecules.

Topologically, molecular symmetry-equivalent vertices are quickly evaluated by computing for each graph site V_i the number of its m-neighbors b_{im}, e.g. the number of vertices V_j having distance $d_{ij} = m$ from V_i. Ordered strings $\{b_{im}\}$ allow an easy sorting of equivalent atoms since different strings correspond to symmetry-inequivalent atoms, predicting the shape of NMR spectra. We have many ways to take in to account that some carbon nodes in the molecular graph bond one hydrogen atom. The most simple way – adopted in this paper – consists in just adding a fourth edge producing a graph with 18 4-connected carbon sites (corresponding to C–H bonds), 42 3-connected carbon

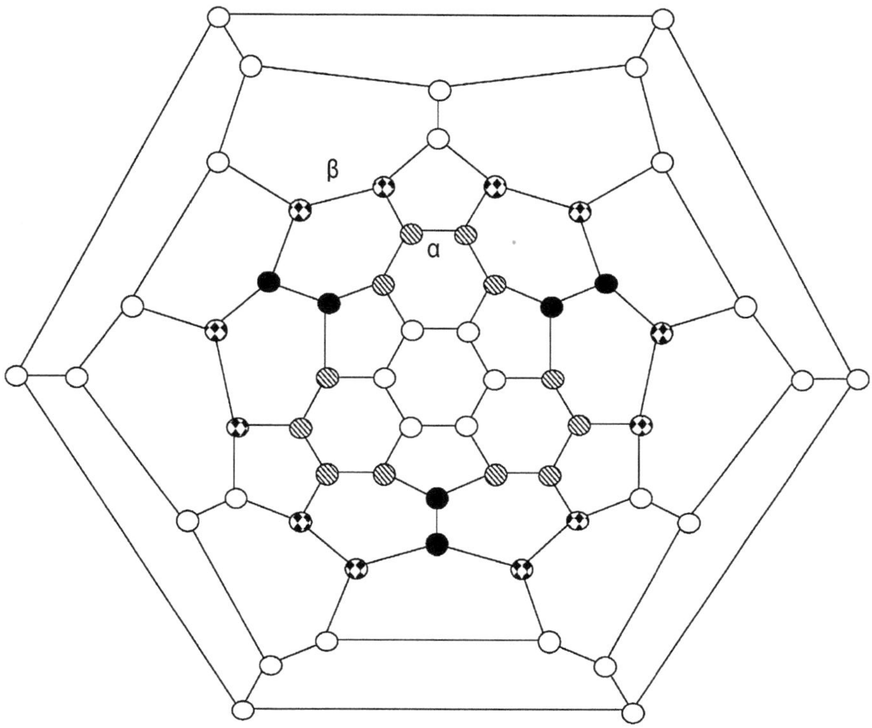

Fig. 12.4 Two distinct C_{3v} $C_{60}H_{18}$ hydrides are represented with Schlegel diagrams; black circles refer to the six Carbon–Hydrogen common bonds; molecules are distinguished by the position of the remaining 12 Carbon–Hydrogen bonds: the first one places the Hydrogen atoms along the shaded inner crown (α), whereas the in second isomer Hydrogen atoms follow the outer crown (β)

sites and 18 1-connected hydrogen sites. This assumption produces an immediate sorting of hydrogen atoms bonded to symmetry equivalent 4-connected carbon. Table 12.3 fully confirms this mechanism; for example one may verify that in $C_{60}H_{18}(\alpha)$ hydrogen atoms connected to independent carbon sites C,E (black nodes in Fig. 12.5) and B,D are grouped in four corresponding distinct sets C_H, E_H, B_H, D_H.

Table 12.3 gives the full topological characterization of C_{3v} $C_{60}H_{18}(\alpha)$ fullerene derivative. As represented in Fig. 12.5, it has 12 independent sets S of Carbon atoms, four of them B,D,C,E chemically bonded Hydrogen atoms, also grouped in 4 independent sets. Atoms with same $\{b_{im}\}$ belong in fact to the same topologically class S with multiplicity l_i; this molecular graph is then made of 78 vertices grouped in 16 sets S with W = 15,333, M = 10, total bonds in the graph are B = 108, e.g. the 90 bonds of buckyball $C_{60}(I_h)$, plus 18 C–H bonds.

By summing b_{i1} entries one get B value twice; by summing all products $(l_i\, w_i)$ one has W twice. The six vertices A in the cental hexagon of Fig. 12.2 have minimum w_i, so $\underline{w} = 353$. Then, resonance theoretical spectrum has 12 ^{13}C-NMR experimental lines (eight of them A,B,D,F,G,I,K,L twice intense of the remaining 4 C,E,H,J) and 4

Table 12.3 Topological characterization of C_{3v} $C_{60}H_{18}$ Fullerene derivatives. Sites with same $\{b_{im}\}$ belong to the same topologically class S with multiplicity l_i. In $C_{60}H_{18}(\alpha)$ the six hydrogen atoms in B_H set, for example, bound carbon atoms of the B set (note the differences in M values)

C_{3v} $C_{60}H_{18}(\alpha)$ W = 15,333; M = 10; B = 108; $\underline{w}$ = 353				C_{3v} $C_{60}H_{18}(\beta)$ W = 15,456; M = 11; B = 108; $\underline{w}$ = 374			
$S(l_i)$	$\{b_{im}\}$ M = 10	w_i	Site	$S(l_i)$	$\{b_{im}\}$ M = 11	w_i	Site
A(6)	3 7 12 16 15 12 8 3 1	353	C	A(6)	3 6 9 13 16 15 11 3 1	374	C
B(6)	4 8 11 12 14 14 10 3 1	358	C–H	B(6)	3 7 10 13 13 13 12 5 1	375	C
C(3)	4 9 10 12 12 14 12 3 1	361	C–H	C(3)	4 7 10 12 12 14 10 7 1	375	C–H
D(6)	4 8 10 13 13 13 11 4 1	363	C–H	D(6)	3 7 11 13 12 12 11 6 2	377	C
E(3)	4 7 10 12 12 12 14 5 1	375	C–H	E(3)	4 9 10 10 12 12 12 5 3	375	C–H
F(6)	3 7 11 12 13 12 11 6 2	378	C	F(6)	4 7 10 13 13 11 10 5 3 1	378	C–H
G(6)	3 7 10 13 12 13 10 7 2	381	C	G(6)	4 8 10 12 11 13 10 6 2 1	377	C–H
H(3)	3 6 10 14 12 12 10 7 2 1	387	C	H(3)	3 8 10 12 14 12 8 5 4 1	381	C
I(6)	3 6 9 12 14 12 11 6 3 1	394	C	I(6)	3 7 11 12 11 12 12 7 2	383	C
J(3)	3 6 8 12 14 12 12 5 4 1	399	C	J(3)	3 6 10 12 14 14 8 7 2 1	387	C
K(6)	3 6 8 11 13 15 10 7 3 1	401	C	K(6)	3 6 9 14 13 12 12 6 2	385	C
L(6)	3 6 8 10 11 14 15 8 2	405	C	L(6)	3 6 8 11 15 16 13 4 1	386	C
B_H(6)	1 3 8 11 12 14 14 10 3 1	434	H	C_H (3)	1 3 7 10 12 12 14 10 7 1	451	H
C_H(3)	1 3 9 10 12 12 14 12 3 1	437	H	E_H (3)	1 3 9 10 10 12 12 12 5 3	451	H
D_H(6)	1 3 8 10 13 13 13 11 4 1	439	H	F_H (6)	1 3 7 10 13 13 11 10 5 3 1	454	H
E_H(3)	1 3 7 10 12 12 12 14 5 1	451	H	G_H (6)	1 3 8 10 12 11 13 10 6 2 1	453	H

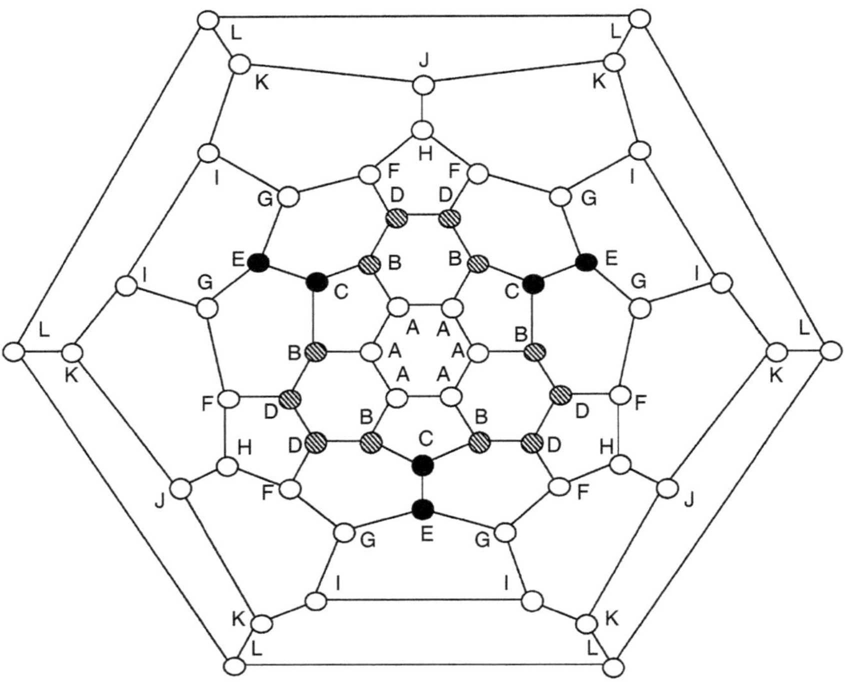

Fig. 12.5 Twelve topological sets A,B, …, L made of symmetry-equivalent atoms of C_{3v} $C_{60}H_{18}(\alpha)$ hydride; eight sets A,B,D,F,G,I,K,L have six nodes each, whereas C,E,H,J include three nodes; ^{13}C-NMR theoretical spectrum in fact presents 12 lines with relative intensities 8(6):4(3). Sets B(6);C(3);D(6);E(3) correspond to C–H bonds originating ^{1}H-NMR theoretical spectrum with four lines with relative intensities 2(6):2(3). C_{3v} $C_{60}H_{18}(\beta)$ hydride has similar resonance patterns due to the fact that 12 hydrogen bonds switch from sets B and D to F and G (see Fig. 12.4)

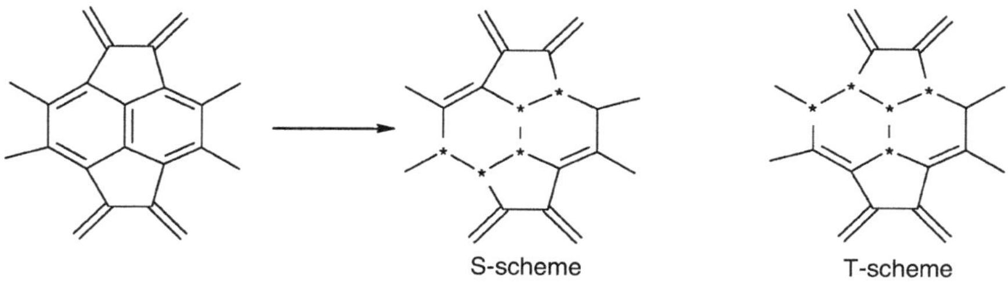

Fig. 12.6 Hydrogen S and T addiction schemes on the fullerene cage

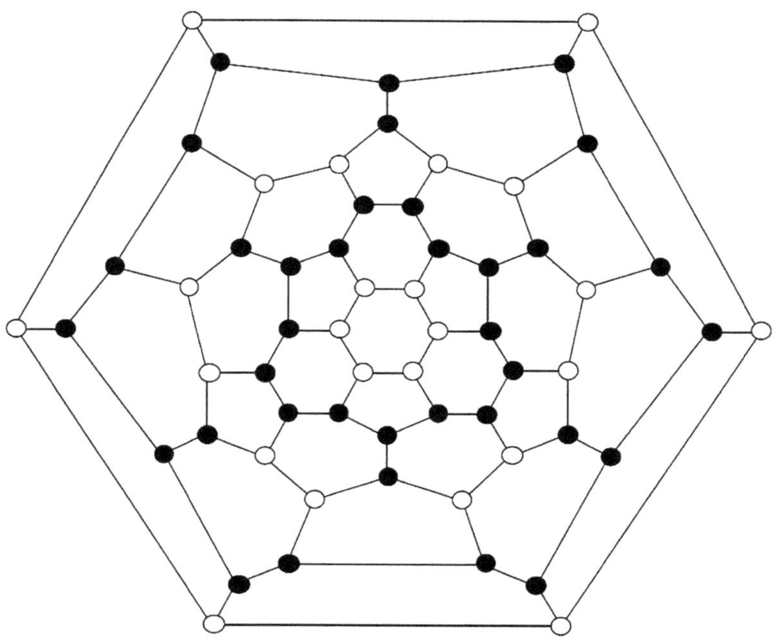

Fig. 12.7 $C_{60}H_{36}(D_{3d})$(CK) hydride Schlegel diagram; black circles refer to the 36 Carbon–Hydrogen bonds; graph has two benzenoid rings; it contains C_{3v} $C_{60}H_{18}(\alpha)$ hydrogen atoms sub-set (see Fig. 12.4)

^{1}H-NMR peaks (two of them B,D twice intense of the remaining 2 C,E). C_{3v} $C_{60}H_{18}(\alpha)$ has $\rho(\alpha) = 1{,}11375$.

C_{3v} $C_{60}H_{18}(\beta)$ newly proposed structure results in a slightly less compact structure with a larger graph diameter M = 11 and W = 15,456 and this fact may topologically favors C_{3v} $C_{60}H_{18}(\alpha)$ formation. What in turn enhances C_{3v} $C_{60}H_{18}(\beta)$ relative topological stability is the lower value of its efficiency index $\rho(\beta) = 105{,}965$. Our topological structural studies, in Table 12.4, show that both C_{3v} isomers present

Table 12.4 Topological theoretical resonance spectra of both C_{3v} symmetric $C_{60}H_{18}$ molecules

Isomer	^{13}C NMR number of peaks [relative intensity]	^{1}H NMR number of peaks [relative intensity]	W	ρ
C_{3v} $C_{60}H_{18}(\alpha)$	Total 12 [8(6): 4(3)]	Total 4 [2(6): 2(3)]	15333	1,11375
C_{3v} $C_{60}H_{18}(\beta)$	Total 12 [8(6): 4(3)]	Total 4 [2(6): 2(3)]	15456	1,05965

the same numbers of lines with equal relative intensities. C_{3v} $C_{60}H_{18}(\beta)$ has in fact resonance theoretical spectrum with 12 ^{13}C-NMR experimental lines (atoms subsets A,B,D,F,G,I,K,L give also in this case peaks twice intense than C,E,H,J) and 4 ^{1}H-NMR peaks (F,G twice intense of 2 C,E). These theoretical resonance data are consistent with C_{3v} symmetry as shown in Fig. 12.5 and Table 12.3. In isomer (β) graph diameter M jumps, noticeably, from 10 to 11 as one may expect being hydrogen atoms spread on a larger portion of the fullerene surface. M = 11 is in fact the value that, structurally, one should expect by adding hydrogen bonds on a fullerene cage with starting value M = 9. All $C_{60}H_{36}$ molecules studied in this article show M = 11 diameter, as we will verify in the following section.

We may theoretically and tentatively conclude that both isomers with C_{3v} symmetry may participate in the mixtures experimentally produced.

12.4 Topological Modeling of $C_{60}H_{36}$

One of the most interesting candidates among $C_{60}H_{36}$ isomers is tetrahedral molecule $C_{60}H_{36}(T)$ whose molecular graphs, Fig. 12.1, has the connectivity list given in Table 12.1. Table 12.5 gives topological characterization of $C_{60}H_{36}(T)$ molecular graph with W = 24738; M = 11; total number of bonds being B = 126. Twelve vertices C2, C8, C11, C19, C22, C30, C31, C39, C42, C46, C53, C58 present minimum $\underline{w}$ = 477. Sites with same string $\{b_{im}\}$ belong to the same topologically class with multiplicity *l*. Therefore our model quickly predicts NMR resonance spectrum with five equally intense ^{13}C-NMR experimental lines K_C, A_C, H_C, X_C, Y_C and three equally intense ^{1}H-NMR peaks K_H, A_H, H_H. These results agree with experimentally and theoretical data reported in (Gakh et al. 2003) (labels K,A,H have been introduced by that article) and Balasubramanian (2004) respectively. As mentioned in the previous paragraph, one should generally expect that graph diameter M topologically comes from the buckyball one augmented by 2: our calculations confirm M = 11 for all $C_{60}H_{36}$ derivatives studied so far. Hydrogen atoms bonded to carbon atoms belonging to a given class (say A_C in Table 12.5) constitute a class of equivalent atoms, say A_H. The twelve hexagonal vertices C1, C7, C15, C18, C21, C29, C32, C40, C43, C50, C54, C59 do form the class A_C whereas the bonded hydrogen H61, H64, H69, H70, H73, H77, H80, H84, H87, H90, H93, H96 constitute the set A_H of equivalent atoms. Table 12.5 summarizes the Wiener-based resonance peaks distribution in perfect agreement with Balasubramanian's (2004) theoretical analysis

Table 12.5 $C_{60}H_{36}$(T) fullerane topological characterization. W = 24738; M = 11; B = 126; by summing vertex connectivity b_{i1} we get B value twice. Sites with same $\{b_{im}\}$ belong to a topologically equivalent class with multiplicity *l*. NMR theoretical spectrum then has five equally intense ^{13}C-NMR experimental lines and three equally intense ^{1}H-NMR peaks as experimentally reported (graph shows distribution of equivalent sites)

Vertex	$\{b_{im}\} \rightarrow M = 11$	*l*	Peak	$w_l/2$
C2 C8 C11 C19 C22 C30 C31 C39 C42 C46 C53 C58	4 9 11 14 14 17 16 8 2	12	K_C	238,5
C1 C7 C15 C18 C21 C29 C32 C40 C43 C50 C54 C59	4 8 12 13 15 16 17 8 2	12	A_C	240
C3 C9 C12 C20 C23 C26 C35 C38 C41 C47 C52 C57	4 8 11 14 17 17 13 7 3 1	12	H_C	240
C5 C6 C14 C17 C25 C28 C33 C36 C44 C49 C55 C60	3 7 12 17 16 16 11 9 3 1	12	X_C	241,5
C4 C10 C13 C16 C24 C27 C34 C37 C45 C48 C51 C56	3 7 12 16 18 14 13 7 4 1	12	Y_C	242
H62 H65 H67 H71 H74 H78 H79 H83 H86 H88 H92 H95	1 3 9 11 14 14 17 16 8 2	12	K_H	285,5
H61 H64 H69 H70 H73 H77 H80 H84 H87 H90 H93 H96	1 3 8 12 13 15 16 17 8 2	12	A_H	287
H63 H66 H68 H72 H75 H76 H81 H82 H85 H89 H91 H94	1 3 8 11 14 17 17 13 7 3 1	12	H_H	287

based on a totally different computational approach. From graph invariants reported in Table 12.5 and Eq. 2.6 one may directly calculate efficiency index $\rho = 1{,}08045$ of $C_{60}H_{36}$(T) graph. One of the most outstanding chemical features of this tetrahedral fullerane is that in this isomer all hydrogen atoms are added par-wise to double bonds of underlying C_{60} fullerene. This structural property is valid also for both $C_{60}H_{18}$ isomers presented in the previous paragraph.

Figure 12.1 shows the positions of carbon atoms grouped in the five equivalence classes as in Table 12.5. It gives also a pictorial view of the proposed rule by Balasubramanian (2004) for hydrogen distribution that assigns three hydrogen atoms to each one of the 12 Fullerene pentagons, bringing the total number of Hydrogen atoms to 36. Hydrogen population on $C_{60}H_{36}$ isomers is supposed to show at least one fully hydrogenated hexagon. T isomer has four isolated benzenoid rings and four fully hydrogenated hexagons (including the external one). Balasubramanian's *"3 hydrogen per pentagon"*-rule practically reverberates the two standard chemical addition ways that hydrogen atoms pairs usually adopt in bonding fullerene localized double bounds (see Fig. 12.6).

Let's focus now on $C_{60}H_{36}$(Th) molecule, Fig. 12.2, firstly introduced by Bini and coworkers (1998). Table 12.6 resumes graph invariants computed on this chemical graph that has a higher symmetry in comparison to the tetrahedral one of Fig. 12.1. Present topological model predicts $C_{60}H_{36}$(Th) theoretical resonance spectrum made by only 3 ^{13}C-NMR experimental lines P_C, Q_C, R_C with relative intensities (12:24:24) and 2 ^{1}H-NMR peaks P_H, Q_H with relative intensities (12:24), see Table 12.6. Equation 2.6 produces for $C_{60}H_{36}$(Th) the value $\rho = 1{,}07845$ that reflects a quite effective topological structure in respect to the tetrahedral molecule

Table 12.6 $C_{60}H_{36}$(Th) fullerane topological characterization. W = 24744; M = 11; B = 126; by summing vertex connectivity b_{i1} we get B value twice. Sites with same $\{b_{im}\}$ belong to a topologically equivalent class with multiplicity *l*. NMR theoretical spectrum then has three ^{13}C-NMR experimental lines and two ^{1}H-NMR peaks as experimentally reported (graph shows distribution of equivalent sites)

Vertex	$\{b_{im}\} \to M = 11$	*l*	Peak	$w_i/2$
C2 C10 C15 C18 C21 C30 C34 C36 C44 C48 C53 C60	4 9 10 14 18 16 14 5 4 1	12	P_C	239
C1 C3 C6 C9 C11 C14 C17 C19 C22 C25 C26 C29 C33 C35 C37 C40 C43 C45 C47 C49 C52 C54 C56 C59	4 7 12 16 14 17 14 8 2 1	24	Q_C	240
C4 C5 C7 C8 C12 C13 C16 C20 C23 C24 C27 C28 C31 C32 C38 C39 C41 C42 C46 C50 C51 C55 C57 C58	3 8 12 14 17 15 14 9 3	24	R_C	241,5
H62 H66 H69 H71 H73 H78 H80 H82 H86 H89 H92 H96	1 3 9 10 14 18 16 14 5 4 1	12	P_H	286
H61 H63 H64 H65 H67 H68 H70 H72 H74 H75 H76 H77 H79 H81 H83 H84 H85 H87 H88 H90 H91 H93 H94 H95	1 3 8 12 13 15 16 17 8 2	24	Q_H	287

Table 12.7 Topological modeling results on various $C_{60}H_{36}$ molecules; isomer have different values of *w*, W and ρ: the most compact molecule is the T one; both T and Th molecules exhibit low ρ values having effective topological organization around the most stable sites. D_{3d} isomers are the best structures in terms of ρ

Isomer	W	Stability	*w*/2	Graph minimal vertices (see Fig. 12.1)	ρ
$C_{60}H_{36}$(T)	24,738	Max↑ Min	238,5	C2 C8 C11 C19 C22 C30 C31 C39 C42 C46 C53 C58	1,08045
$C_{60}H_{36}$(Th)	24,744		239,0	C2 C10 C15 C18 C21 C30 C34 C36 C44 C48 C53 C60	1,07845
$C_{60}H_{36}(C_3)$	24,744		238,0	C11 C19 C22	1,08298
$C_{60}H_{36}(C_1)$	24,747		237,0	C11	1,08768
$C_{60}H_{36}(D_{3d})$(CK)	24,774		239,5	C2 C3 C9 C10 C26 C30 C34 C35 C47 C48 C52 C53	1.07751
$C_{60}H_{36}(D_{3d})$	24,774		239,5	C1 C4 C6 C8 C27 C29 C31 C33 C46 C49 C51 C54	1.07751

$C_{60}H_{36}$(T); this last molecule assumes vice-versa a slightly most stable value for W. Both isomers appear topologically competitive and further experimental considerations, as NMR results, may help in selecting one of them.

Similarly, we applied our topological modeling to the remaining $C_{60}H_{36}$ isomers C_1, C_3 presented by Balasubramanian (2004) and to the pair of molecules D_{3d} symmetry proposed in (Bini et al. 1998), called $C_{60}H_{36}(D_{3d})$ and $C_{60}H_{36}(D_{3d})$(CK) respectively; Table 12.7 reports overall results. We get the confirmation that the Wiener index seen as a pure topological potential W immediately presents $C_{60}H_{36}$(T) isomer as the most stable representative for this fullerane in respect to C_1,C_3. We derive a theoretical stability sequence T > Th,C_3 > C_1 > D_{3d}, D_{3d}(CK) based on Wiener index minimization that matches similar findings on T, C_1, C_3 isomers coming from ab-initio theoretical predictions (Balasubramanian 2004). It is worth to remember here that the adopted *topological minimum principle on W(N)* is, computationally, very easy to implement than any other theoretical tools and it allows a quickly evaluation of a *realistic* stability trend among similar chemical structures. Moreover, it may be used to find out proper candidates for $C_{60}H_{36}$ isomers, as next paragraph will show.

In order to provide further quantitative indications about isomer stability, Table 12.7 reports molecular descriptor ρ (12.6) that is able to discriminates those isomers compactly growing around their minimal sites. C_{60} *fullerene case with* $\rho = \underline{1}$ *tells that smallest values of* ρ *usually correspond topologically efficient structures.* Our computation on ρ ranks theoretical probability of formation of $C_{60}H_{36}$ isomers in a different ways leading to the stability sequence D_{3d}, D_{3d}(CK) > Th > T > C_3 > C_1. It is very interesting to notice that by searching for isomers able to minimize the topological efficiency index ρ, with the constraint $\rho \geq 1$, it is possible to predict a certain grade of chemical relevance for D_{3d} symmetry molecules already proposed in Bini et al. (1998) on the basis of different theoretical considerations. Figures 12.7 and 12.8 give Schlegel representations of these D_{3d}, D_{3d}(CK) isomers and Tables 12.8 and 12.9 present their full topological characterizations. Also for these two molecules, our computations shows that for a given class of equivalent carbon atoms that generate a ^{13}C-NMR resonance line (say peak 1_C), bonded hydrogen atoms give rise

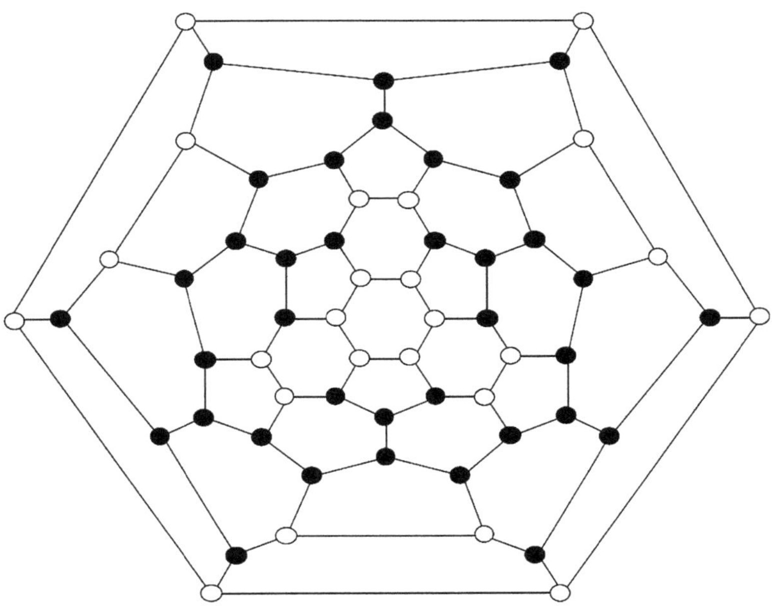

Fig. 12.8 $C_{60}H_{36}(D_{3d})$ hydride Schlegel diagram; black circles refer to the 36 Carbon–Hydrogen bonds; graph has two benzenoid rings; it contains C_{3V} $C_{60}H_{18}(\beta)$ hydrogen sub-set (see Fig. 12.4)

Table 12.8 $C_{60}H_{36}(D_{3d})$(CK) fullerane topological characterization. W = 24774; M = 11; B = 126; by summing vertex connectivity b_{i1} we get B value twice. Sites with same $\{b_{im}\}$ belong to a topologically equivalent class with multiplicity *l*. NMR theoretical spectrum then has six ^{13}C-NMR experimental lines and three corresponding ^{1}H-NMR peaks

Vertex	$\{b_{im}\} \rightarrow M = 11$	*l*	Peak	$w_i/2$
C2 C3 C9 C10 C26 C30 C34 C35 C47 C48 C52 C53	4 8 11 15 17 15 14 7 3 1	12	1_C	239,5
C13 C14 C16 C17 C24 C25 C39 C40 C42 C43 C58 C59	3 7 12 16 16 16 15 8 2	12	2_C	240
C1 C4 C6 C8 C27 C29 C31 C33 C46 C49 C51 C54	3 8 13 15 15 15 13 10 3	12	3_C	240,5
C12 C15 C18 C20 C21 C23 C36 C38 C41 C44 C57 C60	4 8 11 13 17 19 12 7 3 1	12	4_C	240,5
C11 C19 C22 C37 C45 C56	4 9 10 14 16 16 16 5 4 1	6	5_C	241
C5 C7 C28 C32 C50 C55	4 7 12 16 14 14 16 9 2 1	6	6_C	242
H61 H62 H65 H66 H76 H78 H80 H81 H88 H89 H91 H92	1 3 8 11 15 17 15 14 7 3 1	12	1_H	286,5
H68 H69 H70 H72 H73 H75 H82 H84 H85 H86 H95 H96	1 3 8 11 13 17 19 12 7 3 1	12	4_H	287,5
H67 H71 H74 H83 H87 H94	1 3 9 10 14 16 16 16 5 4 1	6	5_H	288
H63 H64 H77 H79 H90 H93	1 3 7 12 16 14 14 16 9 2 1	6	6_H	289

Table 12.9 $C_{60}H_{36}(D_{3d})$ fullerane topological characterization. W = 24774; M = 11; B = 126; by summing vertex connectivity b_{i1} we get B value twice. Sites with same $\{b_{im}\}$ belong to a topologically equivalent class with multiplicity *l*. NMR theoretical spectrum then has six ^{13}C-NMR experimental lines and three corresponding ^{1}H-NMR peaks

Vertex	$\{b_{im}\} \rightarrow M = 11$	*l*	Peak	$w_i/2$
C1 C4 C6 C8 C27 C29 C31 C33 C46 C49 C51 C54	4 8 12 16 15 14 13 9 3 1	12	1_C	239,5
C13 C14 C16 C17 C24 C25 C39 C40 C42 C43 C58 C59	3 7 11 15 18 17 16 6 2	12	2_C	240
C5 C7 C28 C32 C50 C55	4 9 12 14 14 16 14 7 4 1	6	3_C	240
C2 C3 C9 C10 C26 C30 C34 C35 C47 C48 C52 C53	3 8 13 15 15 14 15 9 3	12	4_C	240,5
C12 C15 C18 C20 C21 C23 C36 C38 C41 C44 C57 C60	4 7 11 15 16 18 13 8 2 1	12	5_C	241
C11 C19 C22 C37 C45 C56	4 9 10 12 18 18 12 7 4 1	6	6_C	242
H61 H62 H65 H66 H76 H78 H80 H81 H88 H89 H91 H92	1 3 8 12 16 15 14 13 9 3 1	12	1_H	286,5
H63 H64 H77 H79 H90 H93	1 3 9 12 14 14 16 14 7 4 1	6	3_H	287
H68 H69 H70 H72 H73 H75 H82 H84 H85 H86 H95 H96	1 3 7 11 15 16 18 13 8 2 1	12	5_H	288
H67 H71 H74 H83 H87 H94	1 3 9 10 12 18 18 12 7 4 1	6	6_H	289

to a corresponding ^{1}H-NMR line, say peak 1_H and similar, see Tables 12.8 and 12.9. Our topological computations shows that the theoretical resonance spectra of $C_{60}H_{36}$ *D_{3d} and D_{3d}(CK)* isomers consist of 6 ^{13}C-NMR experimental lines 1_C–6_C with relative intensities (12:12:12:12:6:6) and 4 ^{1}H-NMR peaks 1_H, 3_H, 5_H, 6_H with relative intensities (12:12:6:6). Coordination strings $\{b_{im}\}$ differences in Tables 12.8 and 12.9

Table 12.10 $C_{60}H_{36}$ fullerane topological characterization; for each isomer topological invariants W, ρ and NMR theoretical spectrum are given, being *l* multiplicity of resonance lines. Highly symmetric Th molecule, in bold, has low W, ρ values

Sorted by W

Isomer	^{13}C NMR lines (*l*)	^{1}H NMR lines (*l*)	W	ρ	[Ref]
$C_{60}H_{36}(C^*_1)$ (2)	60 (1)	36 (1)	24,734	1,09879	This work
$C_{60}H_{36}(C^*_1)$	60 (1)	36 (1)	24,736	1,09879	This work
$C_{60}H_{36}(T)$	5 (12)	3 (12)	24,738	1,08045	[10,15,20]
$C_{60}H_{36}(C^*_3)$	50 (1), 5 (2)	28 (1), 4 (2)	24,742	1,08290	This work
$C_{60}H_{36}(T_h)$	**3 (24:24:12)**	**2 (24:12)**	**24,744**	**1,07845**	**[10,15]**
$C_{60}H_{36}(C_3)$	20 (3)	12 (3)	24,744	1,08298	[20]
$C_{60}H_{36}(C_3)$ (64)	58 (1), 1 (2)	34 (1),1 (2)	24,744	1,08526	[15]
$C_{60}H_{36}(C_3)$ (3)	20 (3)	12 (3)	24,744	1,08755	[15]
$C_{60}H_{36}(C_1)$	58 (1), 1 (2)	34 (1),1 (2)	24,745	1,08990	[15]
$C_{60}H_{36}(C_1)$	60 (1)	36 (1)	24,747	1,08768	[20]
$C_{60}H_{36}(C_3)$ (3)	20 (3)	12 (3)	24,750	1,09707	[15]
$C_{60}H_{36}(S_6)$ (88)	10 (6)	6 (6)	24,756	1,07897	[15]
$C_{60}H_{36}(D_{3d})$	6 (12:12:12:12:6:6)	4 (12:12:6:6)	24,774	1,07751	[10,15]
$C_{60}H_{36}(D_{3d})$ (CK)	6 (12:12:12:12:6:6)	4 (12:12:6:6)	24,774	1,07751	[10,15]
$C_{60}H_{36}(S_6)$ (91)	10 (6)	6 (6)	24,780	1,07777	[10,15]
$C_{60}H_{36}(D_{3d})$	6 (12:12:12:12:6:6)	4 (12:12:6:6)	24774	1,07751	[10,15]
$C_{60}H_{36}(D_{3d})$(CK)	6 (12:12:12:12:6:6)	4 (12:12:6:6)	24774	1,07751	[10,15]
$C_{60}H_{36}(S_6)$(91)	10 (6)	6 (6)	24780	1,07777	[10,15]
$C_{60}H_{36}(T_h)$	**3 (24:24:12)**	**2 (24:12)**	**24744**	**1,07845**	**[10,15]**
$C_{60}H_{36}(S_6)$(88)	10 (6)	6 (6)	24756	1,07897	[15]
$C_{60}H_{36}(T)$	5 (12)	3 (12)	24738	1,08045	[10,15,20]
$C_{60}H_{36}(C^*_3)$	50 (1), 5 (2)	28 (1), 4 (2)	24742	1,08290	This work
$C_{60}H_{36}(C_3)$	20 (3)	12 (3)	24744	1,08298	[20]
$C_{60}H_{36}(C_3)$ (64)	58 (1), 1 (2)	34 (1),1 (2)	24744	1,08526	[15]
$C_{60}H_{36}(C_3)$ (3)	20 (3)	12 (3)	24744	1,08755	[15]
$C_{60}H_{36}(C_1)$	60 (1)	36 (1)	24747	1,08768	[20]
$C_{60}H_{36}(C_1)$	58 (1), 1 (2)	34 (1),1 (2)	24745	1,08990	[15]
$C_{60}H_{36}(C_3)$ (3)	20 (3)	12 (3)	24750	1,09707	[15]
$C_{60}H_{36}(C^*_1)$ (2)	60 (1)	36 (1)	24734	1,09879	This work
$C_{60}H_{36}(C^*_1)$	60 (1)	36 (1)	24736	1,09879	This work

Sorted by ρ

clearly show that these isomers are two structurally distinct structures also if they have many topological markers in common as W and ρ values. This fact evidences how our approximated proposed topological method is able to discriminate very similar molecular structures if all computational details are taken in to account. Resonance spectra of C_1 and C_3 isomers proposed by Karpushenkava et al. (2007) and Balasubramanian (2004) are also easily computed, predicting a totally unsymmetrical situation for C_1 that is in fact formed by only inequivalent molecular sites and an organization by triplets for C_3 that will then show 20 equally intense ^{13}C-NMR lines and 12 equally intense ^{1}H-NMR experimental peaks. See Table 12.10 for the complete list of theoretical predictions of NMR resonance spectra, based on $\{b_{im}\}$ string sorting method adopted in this article; all results agree with literature articles (Gakh et al. 2003; Briggs et al. 2005; Balasubramanian 2004).

Table 12.7 also points out the relevant topological features of the structure with Th symmetry that shows good scores in terms of both topological markers W and ρ, pointing out that this isomer is a perfect $C_{60}H_{36}$ fullerane theoretical candidate. The electronic absorption spectrum of $C_{60}H_{36}$ synthesized with "nascent hydrogen" (see Chapters 7 and 8 of this book) shows only an electronic transition at 217 nm with $\varepsilon_{217} = 17140$ L cm^{-1} mol^{-1} fully consistent with the presence of isolated double bonds as expected for the Th isomer. Instead, for the other isomers possessing benzenoid rings, an additional "benzenoid" electronic transition is expected at 260–280 nm with $\varepsilon \approx 2{,}000$ L cm^{-1} mol^{-1} but has not been observed at all.

12.5 $C_{60}H_{36}$ Isomers with Extremal W or ρ Values

After above complete topological characterizations of C_1, C_3, T, Th, D_{3d} $C_{60}H_{36}$ isomers, non-exhaustive random generations have been performed searching for other isomers able to minimize topological descriptors W or ρ. In such a way, the topological modeling generated a set of new theoretical molecules with interesting features. We report here some computationally interesting new structures.

12.5.1 $C_{60}H_{36}(C_1^*)$

This is a variant, Fig. 12.9, of C_1 molecule given in Balasubramanian (2004), with just one hydrogen atom that changes position from V_{40} to V_{13}. It has a lower Wiener index value W = 24,736 if compared to C_1 (W = 24,747) and T (W = 24,738) molecules in Table 12.7. This C_1^* isomer has one pentagon with four carbon–hydrogen bonds, one pentagon with two and three benzenoid rings. The number of resonance lines is still 60 for ^{13}C-NMR and 36 for ^{1}H-NMR being all molecular atoms symmetrically distinct. According to the Wiener index minimization principle, this topologically compact isomer (and other similar ones) may concur in explaining the experimentally detected relative abundance of C_1 – like $C_{60}H_{36}$ structures (Balasubramanian 2004). The molecule has two distinct minimal sites C_{11} (as for C1) and C_{15} with $\underline{w} = 469$, whereas C_1 molecule, less compact, has $\underline{w} = 474$. About molecular topology, the relatively high value of topological marker ρ = 1,09879 points out that this isomer grows around its minimal vertices C11 and C15 with a worst topological efficiency of the other molecules proposed in Table 12.7.

12.5.2 $C_{60}H_{36}(C_1^*)(2)$

This is the fullerane with minimum W. It is a variant of the C_1 molecule reported by Karpushenkava et al. (2007) with one carbon–hydrogen bonds that changes position again from graph node V_{40} to V_{13}, see Fig. 12.10. This isomer exhibits

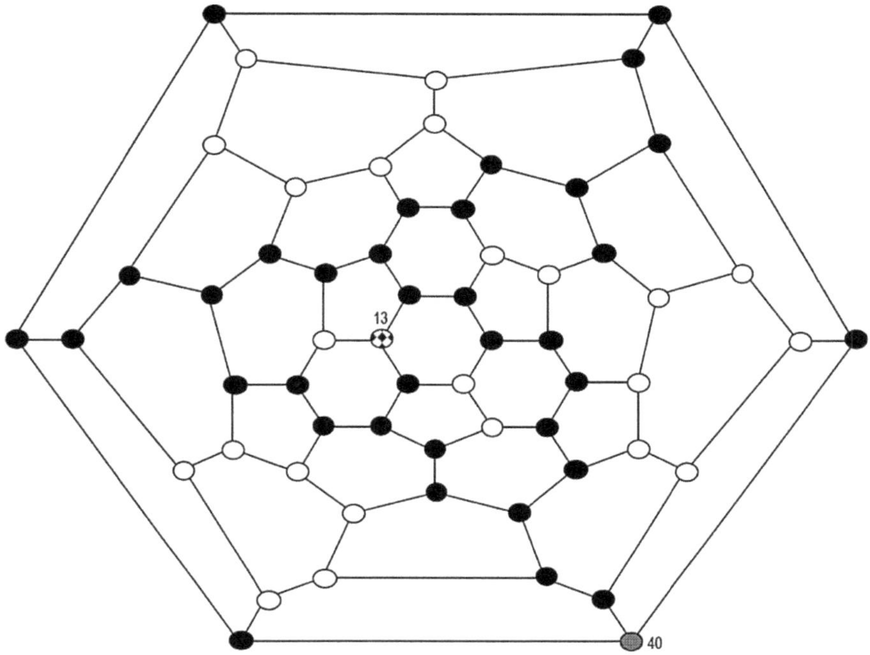

Fig. 12.9 $C_{60}H_{36}(C_1^*)$ hydride Schlegel diagram; black circles refer to the Carbon–Hydrogen bonds including site labeled 13; graph has three benzenoid rings. By moving carbon–hydrogen bond from site V_{13} to V_{40} one obtains C_1 graph. This structure has one pentagon with four carbon–hydrogen bonds and one pentagon with two

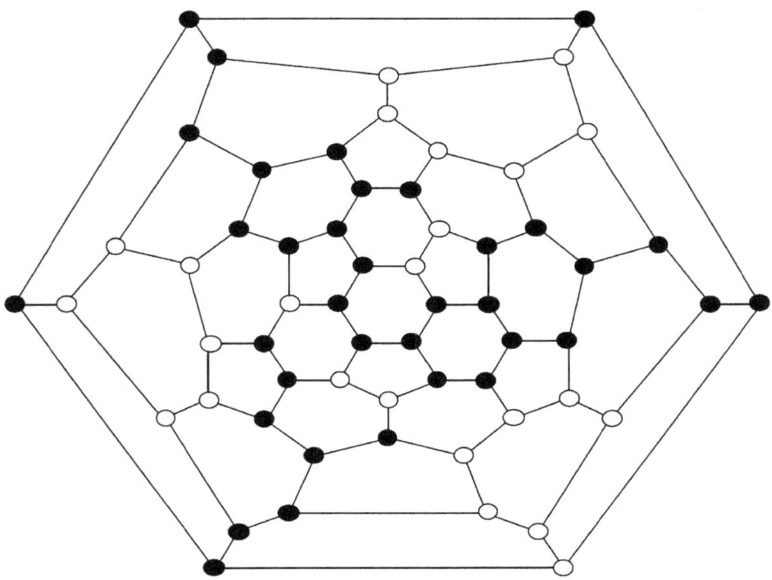

Fig. 12.10 $C_{60}H_{36}(C_1^*)(2)$ hydride Schlegel diagram; black circles refer to the 36 Carbon–Hydrogen bonds; graph has three benzenoid rings. This structure has one pentagon with four carbon–hydrogen bonds and one pentagon with two

the minimum value for the Wiener index sieved so far in our calculations, with W_{min} = 24734, still with a low topological efficiency ρ = 1.09404 and low symmetry resonance patterns with 60 for ^{13}C-NMR and 36 for ^{1}H-NMR lines. This variant has one minimal site C19 with $\underline{w}$ = 471 but globally it grows in a more compact way referred to above $C_{60}H_{36}(C_1^*)$ isomer and this is in fact testified by lower ρ and W values. This molecule presents, as $C_{60}H_{36}(C_1^*)$ isomer, an unusual mechanism of hydrogen-addition so both have only a theoretical relevance since this molecule is not present in sample synthesized so far.

12.6 $C_{60}H_{36}(C_3^*)$

Also this isomer may be seen as a variant of C_3 structure given by Balasubramanian (2004), generated by moving two carbon–hydrogen bonds, see Fig. 12.11. Structure has still a quite low symmetry with a theoretical ^{13}C-NMR spectrum made by five double intense and 50 single lines, for a total of 55 peaks, whereas the ^{1}H-NMR pattern consists four double intense and 28 single lines. W = 24,742 and ρ = 1.08290.

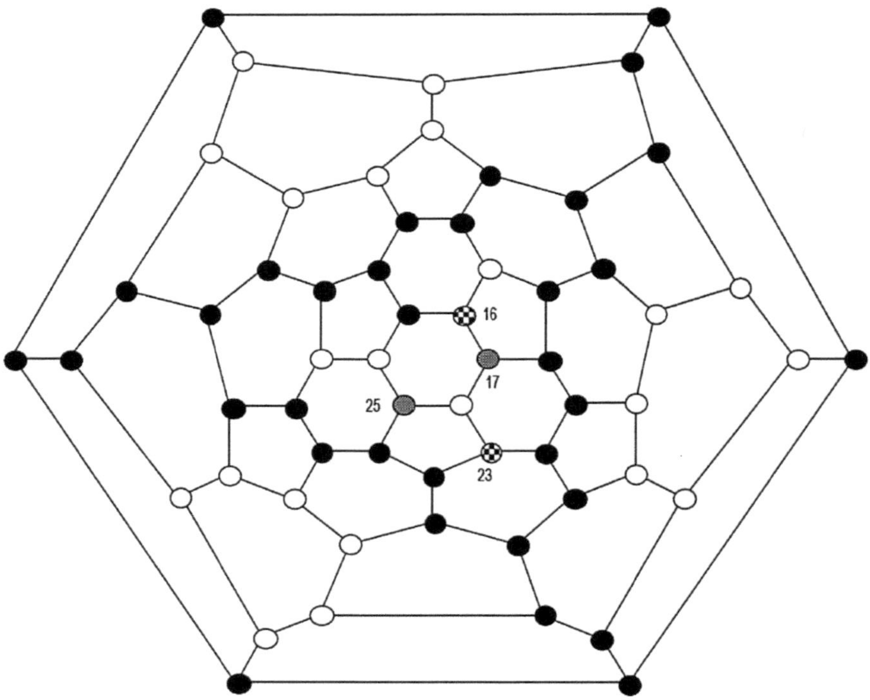

Fig. 12.11 $C_{60}H_{36}(C_3^*)$ hydride Schlegel diagram; black circles refer to the Carbon–Hydrogen bonds including sites labeled 16 and 23; graph has three benzenoid rings. By moving carbon–hydrogen bond from site V_{16} to V_{17} and from V_{23} to V_{25} one obtains C_3 graph

$C_{60}H_{36}(C_3^*)$ has 17 pairs of hydrogen atoms disposed along fullerenic double bonds, same situation one verifies for Th isomer in Fig. 12.2, providing to this molecule a certain theoretical importance.

Working on minimization of the topological efficiency index ρ, the random isomer generation produces a pair of different molecules with three hydrogen atoms on each pentagon and with several pairs of pentagons bridged by a fully hydrogenated C–C double bond, being this kind of configurations chemically favored. The Wiener index value for both isomers is W = 24,774 and their graphs shows on the contrary a relevant structural topological efficiency ρ = 1,07751 being this the lowest value found so far; this fact evidences the theoretical possibility to produced this two fulleranes. Complete topological analyses reveal that these isomers correspond to both D_{3d} structures already proposed by Bini et al. (1998) (see Figs. 12.7 and 12.8).

We end our topological study on $C_{60}H_{36}$ molecules pointing out that in all cases the total contribution to W coming from the 60 carbon atoms only is a constant W_{Carbon} = 14,424 reflecting the complete equivalence of all carbon sites on the underlying $C_{60}(I_h)$ graph. In case of $C_{60}H_{36}$ that constant is W_{Carbon} = 11,382.

12.7 Conclusions

Table 12.10 resumes topological calculations performed on $C_{60}H_{36}$ isomers reported in literature and on structures newly presented in this article. Among these molecules, several low symmetry structures with symmetry C_1, T, Th, C_3 appear also very compact, with low Wiener index values, and for the effect of *topological minimum principle on Wiener-index* (Fig. 12.3) they have a good possibility to be chemically stable isomers. Similarly, by optimizing topological efficiency invariant ρ our computations select in particular D_{3d}, S_6, Th, T structures. Topological modeling attributes then low values of W, ρ to T, Th structures, predicted as chemically stable. Experimental confirmation of these theoretical data derives from the recent successful experimental evidences (see Chapter 7 and 8 of this book) pointing out the relevance of $C_{60}H_{36}$(Th) isomer in explaining the structure of this fullerane.

Acknowledgements We wish to acknowledge ASI, the Italian Space Agency for the partial support of the present work under the contract n.I/015/07/0 (Studi di Esporazione del Sistema Solare).

References

Balasubramanian K (2004) Chem Phys Lett 400:78–85

Bini R, Ebenhoch J, Fanti M, Fowler PW, Leach S, Orlandi G, Ruechardt Ch, Sandall JPB, Zerbetto F (1998) Chem Phys 232:75–94

Book LD, Scuseria GE (1994) J Phys Chem 98:4283–4286

Briggs JB, Montgomery M, Silva LL, Miller GP (2005) Org Lett 7:5553–5555

Cataldo F (2003) Fullerenes Nanot Carbon Nanostruct 11:295–316
Cataldo F, Iglesias-Groth S, Manchado A (2009) Month Not Roy Astronom Soc 400:291–298
Gakh AA, Romanovich AY, Bax A (2003) J Am Chem Soc 125:7902–7906
Iglesias-Groth S (2004) Astrophys J 608:L37–L40
Iglesias-Groth S (2006) Mon Not RoyAstronom Soc 368:1925–1930
Karpushenkava LS, Kabo GJ, Diky VV (2007) Fullerenes Nanot Carbon Nanostruct 15:227–247
Nossal J, Saini RK, Sadana AK, Bettinger HF, Alemany LB, Scuseria GE, Billups WE, Saunders M, Khong A, Weisemann R (2001) J Am Chem Soc 123:8482–8495
Ori O, D'Mello M (1993) Appl Phys A Solid Surf 56:35–39
Ori O, D'Mello M (1992) Chem Phys Lett 197:49–54
Ori O, Cataldo F, Graovac A (2009) Fullerenes Nanot Carbon Nanostruct 17:308–323
Palit DK, Mohan H, Mittal JP (1998) J Phys Chem A 102:4456–4461
Peera AA, Alemany LB, Billups WE (2004) Appl Phys A78:995–1000
Schur DV, Zaginaichenko S, Veziroglu TN (2008) Int J Hydrogen Energy 33:3330–3345
Stoldt CR, Maboudian R, Carraro C (2001) Astrophys J 548:L225–L228
Taylor R (2006) Compt Rend Chimie 9:982–1000
Withers J, Loufty R, Lowe T (1997) Fullerene Sci Technol 5:1–4

Index

MIX
Papier aus verantwortungsvollen Quellen
Paper from responsible sources
FSC® C105338

If you have any concerns about our products,
you can contact us on
ProductSafety@springernature.com

In case Publisher is established outside the EU,
the EU authorized representative is:
Springer Nature Customer Service Center GmbH
Europaplatz 3, 69115 Heidelberg, Germany

Printed by Libri Plureos GmbH
in Hamburg, Germany